AF755054

Chemistry of the Textiles Industry

CLOTHWORKERS LIBRARY
UNIVERSITY OF LEEDS

Chemistry of the Textiles Industry

Edited by

C.M. CARR
Department of Textiles
UMIST

BLACKIE ACADEMIC & PROFESSIONAL
An Imprint of Chapman & Hall
London · Glasgow · Weinheim · New York · Tokyo · Melbourne · Madras

SHORT LOAN
REFERENCE COLLECTION

CHEMISTRY

CLASS MARK
SRT 26631

Published by
Blackie Academic & Professional, an imprint of Chapman & Hall,
Wester Cleddens Road, Bishopbriggs, Glasgow G64 2NZ

Chapman & Hall, 2–6 Boundary Row, London SE1 8HN, UK

Blackie Academic & Professional, Wester Cleddens Road, Bishopbriggs, Glasgow G64 2NZ, UK

Chapman & Hall GmbH, Pappelallee 3, 69469 Weinheim, Germany

Chapman & Hall USA, 115 Fifth Avenue, Fourth Floor, New York NY 10003, USA

Chapman & Hall Japan, ITP-Japan, Kyowa Building, 3F, 2-2-1 Hirakawacho, Chiyoda-ku, Tokyo 102, Japan

DA Book (Aust.) Pty Ltd, 648 Whitehorse Road, Mitcham 3132, Victoria, Australia

Chapman & Hall India, R. Seshadri, 32 Second Main Road, CIT East, Madras 600 035, India

First edition 1995

© 1995 Chapman & Hall

Typeset in 10/12pt Times by AFS Image Setters Ltd, Glasgow
Printed in Great Britain by The University Press, Cambridge

ISBN 0 7514 0054 8

Apart from any fair dealing for the purposes of research or private study, or criticism or review, as permitted under the UK Copyright Designs and Patents Act, 1988, this publication may not be reproduced, stored, or transmitted, in any form or by any means, without the prior permission in writing of the publishers, or in the case of reprographic reproduction only in accordance with the terms of the licences issued by the Copyright Licensing Agency in the UK, or in accordance with the terms of licences issued by the appropriate Reproduction Rights Organization outside the UK. Enquiries concerning reproduction outside the terms stated here should be sent to the publishers at the Glasgow address printed on this page.

The publisher makes no representation, express or implied, with regard to the accuracy of the information contained in this book and cannot accept any legal responsibility or liability for any errors or omissions that may be made.

A catalogue record for this book is available from the British Library

Library of Congress Catalog Card Number: 94-73587

∞ Printed on acid-free text paper, manufactured in accordance with ANSI/NISO Z39.48-1992 (Permanence of Paper)

Preface

The manufacture and processing of textiles is a complex and essential industry requiring many diverse skills to ensure profitability. New products are continually being developed, and reflect the energy and innovation of those working in the field. This book focuses on the technological aspects of the chemical processing of textiles, and on the modifications necessary for specific work environments. Coverage ranges from fibre structure and its relationship to tensile properties, textile aesthetics, comfort physiology, and end-use performance, through to the effect of domestic processing by the consumer on the textile product. The industry is constantly under environmental pressure, and the book examines the nature of environmental control and the development of alternative technology to produce less environmental impact.

In order to provide a balanced view of the current situation, authors have been drawn from academia, research institutes and industry to produce a text that will be useful to both industrial readers and university students. In conclusion I would like to thank the authors for their dedication and their contributions.

CMC

Contributors

Dr D.P. Bishop School of Design and Manufacture, Department of Textiles and Fashion, De Montfort University, The Gateway, Leicester LE1 9BH

Dr K.M. Byrne International Wool Secretariat, Ilkley Development Centre, Valley Drive, Ilkley, West Yorkshire LS29 8PB

Dr J.M. Cardamone Hides, Lipids and Wool Research Unit, United States Department of Agriculture, Eastern Regional Research Center, 600 E. Mermaid Lane, Philadelphia, PA 19118, USA

Dr C.M. Carr Department of Textiles, UMIST, PO Box 88, Manchester M60 1QD

Dr D.L. Connell Precision Processes Textiles, Dylan Laboratories, Ambergate, Belper, Derbyshire DE56 2EY

Dr J. Ellis Precision Processes Textiles, Dylan Laboratories, Ambergate, Belper, Derbyshire DE56 2EY

Mr D.W. Heywood Sacola Associates, Cheadle Hulme, Cheshire SK8 6HW

Dr W.N. Marmer Hides, Lipids and Wool Research Unit, United States Department of Agriculture, Eastern Regional Research Center, 600 E. Mermaid Lane, Philadelphia, PA 19118, USA

Mr D.P. Oulton Department of Textiles, UMIST, PO Box 88, Manchester M60 1QD

I.D. Rattee Northwold, Woodacre Crescent, Bardsey, West Yorkshire LS17 9DQ (formerly Professor of Colour Chemistry and Dyeing, University of Leeds)

Dr R.A. Scott SCRDE, Flagstaff Road, Colchester, Essex CO2 7SS

Dr M.A. Wilding Department of Textiles, UMIST, PO Box 88, Manchester M60 1QD

Contents

1 Introduction: The structure of fibres

M.A. WILDING

1.1 Introduction

The specific chemistry of a given type of fibre is undeniably crucial in determining many features associated with its processing, performance and end-uses; but it may also be asserted that its physical structure—i.e. the organisation and geometrical arrangement of its component parts—is at least equal in importance. This accounts for the largely physical approach taken in the first chapter of a work devoted to the *Chemistry of the Textiles Industry*. The fact is that the two aspects cannot, and ought not to, be separated.

Hearle (1963) quotes the definition of a fibre according to The Textile Institute (1960) thus: "units of matter characterised by fineness, flexibility and a high ratio of length to thickness". As he pointed out, this is an incomplete definition as far as textile fibres are concerned. It is deficient in several respects, but notably in that it omits any reference to what constitutes the actual make-up of a fibre, viz. its structure.

The reason for this undoubtedly lies in part in the huge diversity of features encountered in fibres of differing types and origins. It is possible to make some general statements, however, regarding the nature of textile fibres. Certain key features are now beyond dispute, and can be summarised simply. In the main, textile fibres are composed of:

Partially ordered, partially oriented assemblies of linear polymer molecules.

This definition serves well for the vast majority of textile fibres, but it prompts many questions. For example: "What exactly do 'partially ordered' and 'partially oriented' mean?"; "How are these concepts realised in practice?"; "Do the same features apply equally across a wide range of fibre types?"; and, importantly, "What bearing do such features have on the performance characteristics of fibres?".

Since the inception of fibre structure research near the beginning of this century, such questions have challenged the minds of an impressive array of investigators, but the period between 1930 and the late 1960s, in particular, seems to have been characterised more by disagreement than by unanimity. At this point ideas regarding fibre structure were emerging almost as rapidly as the new fibres themselves were appearing on the market, and a great deal of confusion and controversy reigned. Many, often conflicting, models were produced, each representing an attempt to explain certain observed phenomena.

To a large degree, such graphical representations should be regarded as 'artists' impressions', since they were entirely qualitative, and not generally verified against any measurable parameters. Hearle (1991) warns against the dangers of "... confusing our aids to understanding improperly with reality". Although he made this comment in the specific context of nylon and polyester fibres, it does have more general applicability. This is not to detract from the value of pictorial representations, however: there is now a realisation that whereas no single view of fibre structure can of itself tell the whole story, a grain of truth almost certainly lies in all of them. Moreover, in recent years there has been a more concerted effort to find ways of quantifying the features described by the early models.

An area of increasing commercial importance is the exploitation of high-performance fibres. However, these materials differ almost in kind from the commodity fibres, and although brief consideration is given to them in section 1.2.3.2, no attempt is made to cover them in detail.

1.2 Classification of textile fibres

Textile fibres have been, and continue to be, derived from an enormous range of materials. Figure 1.1, showing just a few examples, can scarcely do justice to this diversity; nor is it possible here to do more than give an indication of the main chemical types and of their sources.

As was stated earlier, the overwhelming majority of fibres are polymeric in nature. In broad terms, they may be classified thus:

Natural
- Animal (protein) fibres
- Vegetable (cellulosic) fibres

Man-made
- Regenerated cellulose fibres
- Chemical derivatives of cellulose

Synthetic polymer
- 'First-generation' (up to 1970s): commodity domestic and industrial fibres
- 'Second-generation' (1970s on): high-performance and specialised fibres

Other
- Metal fibres
- Ceramic fibres

1.2.1 *Natural fibres*

1.2.1.1 *Animal fibres.* The most important textile fibres of animal origin are those based on the α-keratins, which are the proteins found in skin, horn,

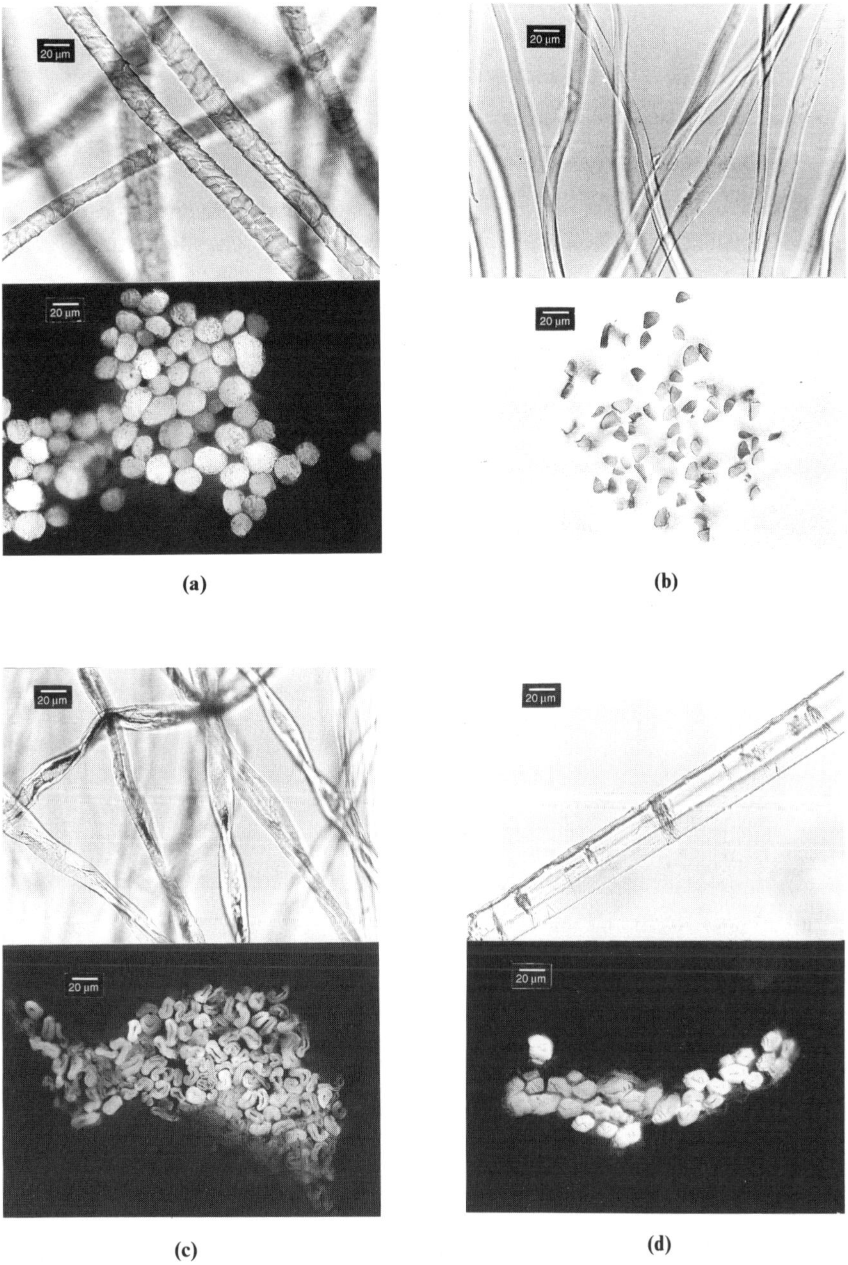

Figure 1.1 Optical micrographs of various fibres: (a) wool; (b) silk; (c) cotton; (d) flax.

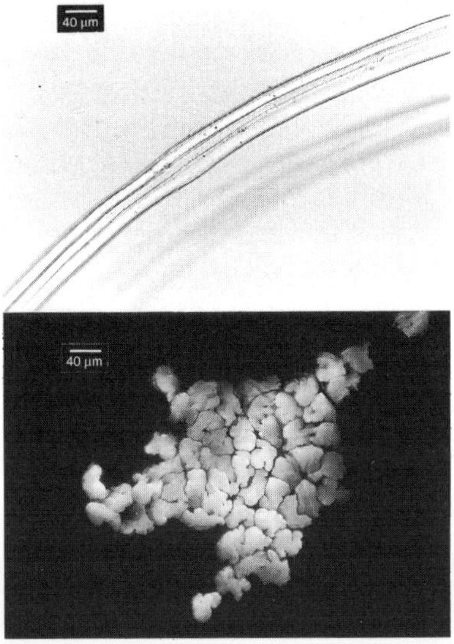

(e)

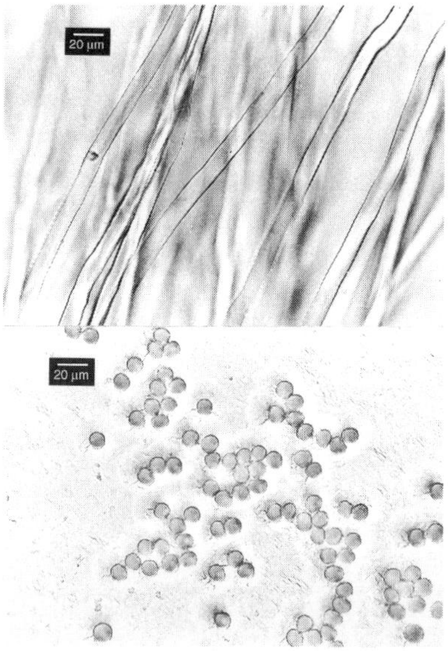

(f)

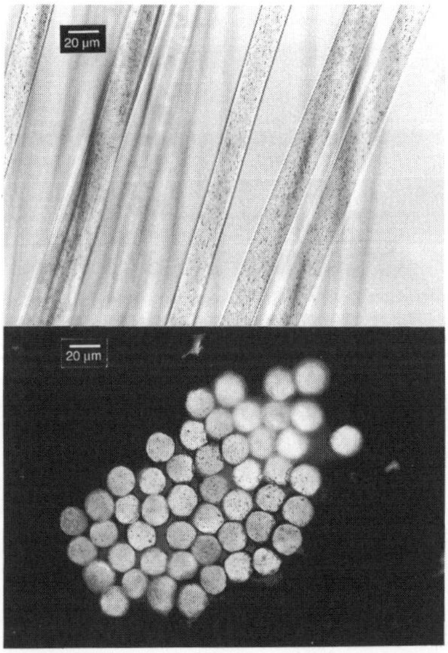

(g)

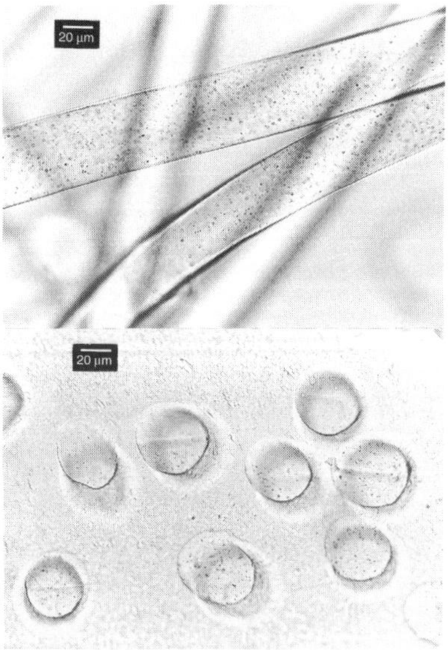

(h)

Figure 1.1 (Continued) (e) viscose carpet fibre; (f) Tencel solution-spun cellulosic; (g) nylon 6.6; (h) polyacrylonitrile ('acrylic').

nails, hair and other tissue, and which form the main chemical constituents of wool.

The wool keratin molecule consists of a highly complex sequence of amino acids. Typically the protein is mainly composed of eighteen amino acids with cystine, lysine, arginine, glutamic acid and aspartic acid important in the physicochemical nature of the fibre (Lewis, 1992). Moreover, this complexity is not confined to the individual molecules: the chemical composition varies over the length of the fibre itself. Near the root, for example, the cystine content is higher than it is near the tip.

The physical structure of wool is equally complex, being organised on several hierarchical levels, as shown in highly idealised form in Figure 1.2. The essential macroscopic components of the fibre are the cortex, which constitutes over 90% of the fibre mass, and the cuticle. In many varieties of wool the cortex has a bilateral structure in which approximately half of the cross-section is made up of *ortho*-cells, whilst the opposite side consists of *para*-cells.This unusual feature gives rise to several distinctive properties, including differential dye-affinity and differential swelling. The cortical cells are themselves composed of complex arrangements of smaller structural units including 'microfibrils' and 'macrofibrils'. The microfibrils are thought to contain bundles of left-handed 'coiled-coil ropes' which in turn are formed from the intertwining of the right-handed keratin helices.

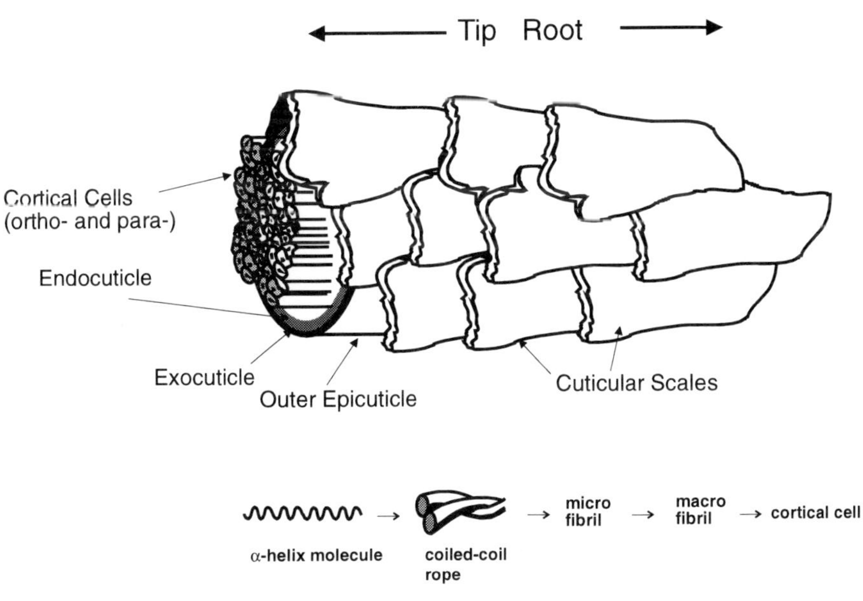

Figure 1.2 Highly simplified view of a wool fibre, showing cortical cells and cuticular scales.

The main feature of the cuticle is the overlapping scale cells, which point from root to tip. These impart the property of differential friction which, among other things, is responsible for the felting propensity of wool. The cuticle scales can easily be observed using light microscopy (see Figure 1.1a).

Keratin molecules are highly polar, rendering the fibre hygroscopic; wool has in fact a higher moisture regain than almost any other fibre. The same feature also accounts for the ease with which wool may be dyed.

The essentially helical configuration of the keratin chains and the microfibrils gives rise to very distinctive tensile properties. Since the helices may be extended with relative ease, the fibre is inherently compliant, extensible, and recoverable up to unusually large strains.

The other animal fibre of major commercial importance is silk, which is derived from the cocoon of the silk moth, *Bombyx mori*. Chemically, silk is a close relation of wool and hair, being a fibroin which shares many of the amino acids present in wool molecules. In contrast to wool and hair, however, the fibroin chains are believed to be nearly fully extended, highly crystalline, and almost perfectly aligned in the fibre direction, all of which contribute to the fibre's considerable stiffness and strength.

Bamford and Elliot (1963) give a detailed survey of the chemistry and structure of protein-based fibres, while Lewis (1992) presents an excellent collection of work relating to the chemistry of wool dyeing.

1.2.1.2 *Vegetable fibres.* The natural vegetable fibres are those based on cellulose, a substance found in the stems, leaves and seed-hairs of a wide range of plants, and formed in nature by photosynthesis under the assistance of a specific enzyme.

(*a*) *The cellulose molecule.* The cellulose molecule comprises a succession of glucose rings, as shown in highly idealised form in Figure 1.3. Several key features of this molecule give rise to important structural and mechanical characteristics in the fibre itself.

(i) It is ribbon-like, with an approximately rectangular cross-section.
(ii) It is relatively flexible, through rotation about the bridging oxygen links between the glucose rings.
(iii) It features a high concentration of hydroxyl groups along its edges.
(iv) It has directionality.

The flexibility of the cellulose molecule is much smaller than that of, say, polyethylene or rubber, but very much greater than those of high performance aromatic polyamides like Kevlar.

The ribbon-like form means that whereas bending out of plane (with respect to the glucose rings) is easy, in-plane bending is practically impossible except if accompanied by twisting.

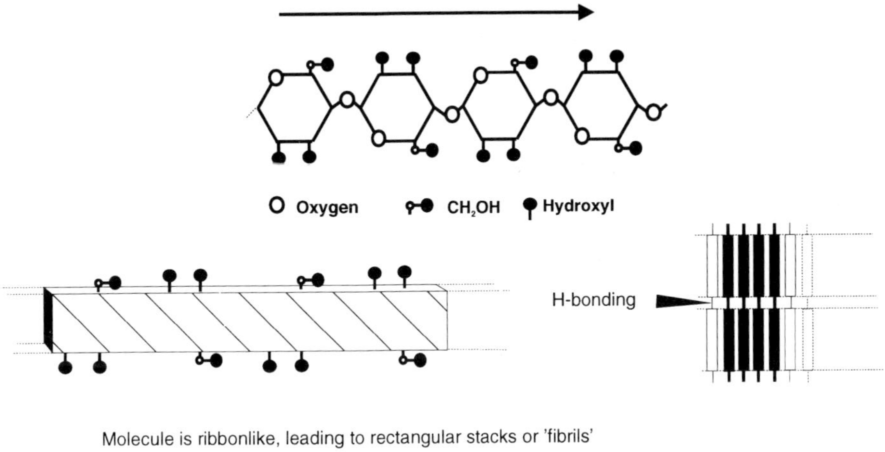

Figure 1.3 Key features of the cellulose molecule.

The adjacent edges of neighbouring chains can interact strongly through hydrogen bonding. This, together with the highly regular geometry of the molecule, provides a strong tendency towards the formation of crystals, and helps to stabilise the structure, once formed.

The molecules of cellulose in fact tend to group together to form microfibrils, which also have a ribbon-like form, and a more or less rectangular cross-section. These in turn can form even larger structural building blocks with, again, similar geometry.

Substances like cellulose, which contain hydrogen bonding, generally display effects at temperatures close to or above 100°C, as these bonds begin to dissociate. (The fact that water boils at this temperature is a manifestation of the same phenomenon.) Furthermore, any moisture present will have a dramatic plasticising effect, since water molecules can break down the hydrogen bonds. This has considerable implications, for example in laundering and other after-care treatments such as ironing.

The directionality of the molecules gives rise to the possibility of having at least two different crystal forms ('Form I' and 'Form II' respectively) in which they are arranged either parallel or antiparallel to one another.

(*b*) *The cotton fibre.* Of the many fibres of plant origin, cotton is without doubt that of greatest commercial importance, although flax, jute and hemp are also very significant.

Figure 1.1c shows the typical appearance of dry cotton fibres. The principal features are a 'kidney bean' cross-section, and the appearance of twists and kinks along the length (the so-called 'convolutions').

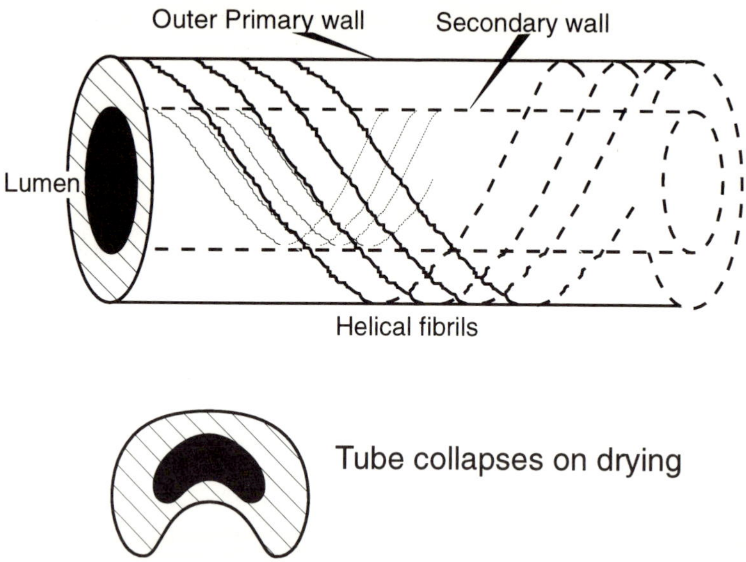

Figure 1.4 Highly simplified view of a cotton fibre, showing helical arrangement of the fibrils.

The structure of cotton and other vegetable fibres is described in detail by Bailey *et al.* (1963). During the first few days of growth, cotton fibres emerge as hollow cylinders 2–3 cm long, with a primary and secondary wall (shown greatly simplified in Figure 1.4). As the fibre matures, fibrils are laid down on the inner wall until the lumen — the hole down the centre — becomes quite small. When the boll opens, the fibres are cut off from the plant's transpiration system and dry out, whereupon the cylinder collapses. Hearle once likened this process to collapsing a roll of sticky tape.

Examination of the fibre whilst it is still growing, i.e. in moist conditions, reveals that the fibrils are actually laid down in a helical form — an interesting fact in the light of the comment made earlier regarding in-plane bending of the molecule.

The arrangement of the fibrils overall could be thought to resemble a twisted filament yarn with a hole down the centre, but it is further revealed that the helix angle remains constant throughout the thickness of the fibre. For an ideal twisted bundle of filaments the period would be constant, but the angle would vary from zero at the centre to a maximum at the surface. This means that the fibrils in cotton cannot be packed together perfectly; there must be occasional discontinuities of alignment. The value of the helix angle depends on the variety to some extent, but is normally about 21°.

It is also known that the sense of the helix reverses periodically along the length of the fibre. This is thought to provide torsional balance within the structure. The macroscopic convolutions tend to occur at points along the fibre corresponding to where the reversals take place.

The development of the fibre structure is determined by the genetics of the plant. In particular, the polymerisation, or conversion from glucose to cellulose, actually takes place on the surface of an enzyme molecule which has a well-defined small number (*ca.* 30) of active sites. Thus a few polymer molecules grow together directly into the crystalline microfibril. Because they all grow in the same direction, they produce the parallel, rather than the anti-parallel, crystal form (Form I).

Brief consideration is given in section 1.4.2.1 to the influence of the cellulose and cotton structures on the mechanical properties of cotton fibres.

1.2.2 *Man-made fibres: regenerated cellulose and its derivatives*

To quote from Sharples (1963): "The amount of published scientific work on the chemistry of cellulose and its derivatives is so great that any attempt to cover the entire field in a short chapter would result in little more than a list". How much more pertinent this statement is some 30 years on. It is clearly impossible here to do more than present some fairly basic information.

The identifying feature of a regenerated fibre is that it constitutes, in effect, a physical rearrangement of the molecules originally present in some naturally occurring source, such as wood pulp. For this reason, the term 'man-made' is used to distinguish such fibres from the true synthetics, in which the polymer molecules are artificially produced.

By far the most important and well-known example of a regenerated fibre is viscose rayon. This material, originally produced in 1891, was in part a result of attempts to create artificial silk—an endeavour which, even today, is regarded by many as the 'holy grail' of fibre technology.

Viscose may be formed by dissolving raw cellulose in carbon disulphide and sodium hydroxide to produce a highly viscous liquid. Continuous cellulose filaments are then precipitated by extrusion of this fluid into sulphuric acid followed by zinc sulphate. The commercialisation of viscose rayon was due, in the main, to Courtaulds Ltd, and it is still one of their most successful products. Present-day commercial processes are broadly based on the original procedure, although they generally incorporate special features, providing a certain degree of control over the structure and physical properties of the resultant fibres.

The dissolution process brings about a physical disruption of the original structure, and imparts mobility to the molecules. This enables them subsequently to reorganise into a new configuration, and to adopt what is in fact the energetically-preferred Form II crystal unit cell (antiparallel chains). Curiously, the crystals of native cellulose represent a higher free-energy state; it is the physical constraints within the plant itself which dictate that this should be so.

The physical appearance of viscose fibres is typically as in Figure 1.1e. A distinguishing feature is the irregular, crenellated cross-section. The fibres also display a marked skin-core difference, which can be observed microscopically, and which gives rise to differential surface/bulk properties such as dye affinity. The crenellations have a profound influence on the optical properties, and in particular, on the lustre of fabrics made from viscose.

Nowadays many viscose-type products are available, ranging from fairly conventional low-modulus, medium-strength commodity fibres, to tough, high strength, high modulus fibres such as Tenasco, which is used extensively in the tyre industry.

Although there are now numerous other variants, probably the most commercially important chemical modifications of cellulose are:

Cuprammonium rayon
Hydrolysed cellulose acetate ('acetate')
Cellulose triacetate (Tricel)
Cellulose secondary acetate (marketed in the USA as Celanese).

By and large, the appearance of these fibre types is similar to that of viscose, with the exception of cuprammonium rayon, which has a circular cross-section, and under the microscope appears homogeneous (classified as 'all-core').

An adverse side-effect of the conventional regeneration process is molecular degradation. This can be severe, resulting in average molecular weights as low as 10% of those in the source material, with an accompanying deterioration in strength and other properties. The recent development, again by Courtaulds, of the solution-spun cellulose Tencel has to a large extent eradicated this drawback. This, coupled with its superior dye-affinity and the fact that its manufacture uses non-toxic, fully-recyclable solvents, has given rise to substantial commercial interest in Tencel.

Wooding (1963) gives a very useful account of rayon and acetate fibres, as does Sharples (1963).

1.2.3 *Synthetic polymer fibres*

1.2.3.1 *First-generation fibres.* Many commodity synthetic fibres are based on polymers composed of chemical combinations of the same four groupings:

$-CH_2-$	$-\underset{\displaystyle O}{\overset{\|\|}{C}}-O-$	$-\underset{\displaystyle O}{\overset{\|\|}{C}}-\underset{\displaystyle H}{\overset{\|}{N}}-$	$-C_6H_4-$
Methylene	Ester	Amide	Benzyl

Each of these imparts certain characteristics to the polymer molecules, which render them particularly suitable for textiles use.

The methylene and ester linkages allow substantial freedom of rotation. There are usually only moderate van der Waals forces laterally, although a certain amount of polarisation is associated with the carbonyl group, and this may provide somewhat stronger attraction.

The amide group is reasonably flexible, but because it forms strong hydrogen bonding, it also encourages crystallisation and cross-linking.

The aromatic ring is quite stiff, and this provides a rigid character. Some electronic interactions probably occur between neighbouring rings, which also favours crystallisation.

The history of the true synthetic polymer fibres arguably began with the research of Carothers in the late 1930s, which culminated in the development of the polyamides. Carothers had been studying the reactions of difunctional reagents such as glycols, $HO(CH_2)_mOH$, with aliphatic dicarboxylic acids, $HOOC(CH_2)_nCOOH$, amino acids, $NH_2(CHR)_nCOOH$, and others. Such reactions are known as condensation reactions, and polymers formed by this route are known as 'condensation polymers'. The early work prompted much further research, and another major group of condensation polymers—the polyesters—grew as an indirect result. In the formation of polyesters, the exact chemical steps may vary, dependent on the precise manufacturing process used, but in essence the condensation reaction takes place between a diglycol and terephthalic acid.

The polyamides and polyesters are the principal examples of condensation polymer fibres. Each in fact represents a large homologous group of materials, although the majority of their members are of only minor commercial significance. In the polyamide group the chief products are poly(caprolactam) (nylon 6) and poly(hexamethylene adipamide) (nylon 6.6), while poly(ethylene terephthalate) (also known as PET, 2GT, or simply polyester), is by far the most important polyester.

Nylon 6

$$nNH_2(CH_2)_5COOH \rightarrow [NH(CH_2)_5CO]_n + nH_2O$$

6-Aminohexanoic acid

Nylon 6.6

$$nNH_2(CH_2)_6NH_2 + nHOOC(CH_2)_4COOH \rightarrow [NH(CH_2)_6NHCO(CH_2)_4CO]_n + nH_2O$$

Hexamethylene diamine   Adipic acid

PET

$$nHOOC(C_6H_4)COOH + nHO(CH_2)_2OH \rightarrow [OC(C_6H_4)COO(CH_2)_2O]_n + nH_2O$$

Terephthalic acid   Ethylene diglycol

It may be noted that two distinct types of polyamide exist, typified by nylon 6 and nylon 6.6, respectively. In the former type, the monomer has a directionality, whereas the monomer of the latter is symmetrical. This difference leads to some specific differences between the crystalline structures and some physical properties of the two types.

The condensation polymers form a relatively small class of polymers (most others being so-called 'addition' polymers), yet they dominate the synthetic textile fibres market. The polyamides and polyesters are also the most commercially important 'melt-spun' synthetics. This term arises from the fact that the filaments are produced by extrusion of molten polymer through a spinaret. The spun filaments are generally stretched and subjected to a variety of downstream processes. This gives rise to fibre products with a very wide range of physical characteristics. More will be said in section 1.4.2.2 concerning the structure–property relationships in melt-spun synthetics.

Notwithstanding their commercial dominance, it would clearly be wrong to imply that nylon and polyester are the only synthetic fibre types, and there are many others of commercial significance. As stated earlier, these are in the main addition polymers. Examples include the following.

Polyolefins. The most important polyolefins are polyethylene (PE) and polypropylene (PP).

$$\text{—CH}_2\text{—CH}_2\text{—CH}_2\text{—CH}_2\text{—CH}_2\text{—}$$

PE

$$\begin{array}{cccccccc} \text{—CH}_2\text{—} & \text{CH} & \text{—CH}_2\text{—} & \text{CH} & \text{—CH}_2\text{—} & \text{CH} & \text{—CH}_2\text{—} & \text{CH—} \\ & | & & | & & | & & | \\ & \text{CH}_3 & & \text{CH}_3 & & \text{CH}_3 & & \text{CH}_3 \end{array}$$

PP

Both materials consist of flexible chain molecules and a short repeat unit, by virtue of which they are highly crystalline. However, their melting and softening temperatures are too low to enable them to withstand the rigours of laundering, etc. They also suffer quite markedly from adverse time effects, and thus find limited applications as commodity textile fibres. However, polypropylene, in particular, has high strength and fracture toughness. These attributes, coupled with its very low density (less than $0.9\,\text{g}\,\text{cm}^{-3}$), make it a valuable fibre for ropes and many industrial products.

By suitable techniques, the polyolefins may be manufactured with special properties — ultra-high stiffness and strength for example — which render them important as high performance fibres in certain industrial and military textile applications. In this context, they should really be regarded as second-generation fibres.

Polyacrylonitrile (PAN)

$$\begin{array}{cccccccc} \text{—CH}_2\text{—} & \text{CH} & \text{—CH}_2\text{—} & \text{CH} & \text{—CH}_2\text{—} & \text{CH} & \text{—CH}_2\text{—} & \text{CH—} \\ & | & & | & & | & & | \\ & \text{CN} & & \text{CN} & & \text{CN} & & \text{CN} \end{array}$$

The 'acrylic' fibres, consisting of the homopolymer shown above together with a number of copolymers (modacrylics), and marketed under a variety of trade names — Courtelle, Orlon, Acrilan, Dynel and others, form a significant commercial group. As pointed out by Furness (1963), one of their chief assets is their ability to form bulked yarns.

Acrylic fibre is an important wool substitute in many applications, such as carpets and upholstery. However, it tends to be rather harsh, and with a regain of virtually zero, it is not generally regarded as a 'comfortable' fibre. Consequently it is now generally blended with other fibre types for apparel end uses.

An important and growing application of PAN is in the production of high-performance carbon fibres.

Vinyl fibres

Poly(vinyl chloride) (PVC)

$$-CF_2-CF_2-CF_2-CF_2-$$

Poly(tetrafluoroethylene) (PTFE)

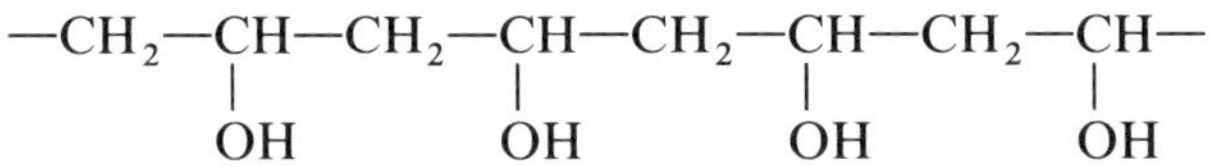

Poly(vinyl alcohol) (PVA)

PVC, PTFE and PVA are probably the best-known examples of this group. PVC was one of the earliest of the addition polymer fibres, but suffers from low melting point and high plasticity. However, it is valuable for its fire-retarding properties.

PTFE, marketed by DuPont as Teflon™, has remarkably high resistance to heat and chemical attack, coupled with an exceptionally low coefficient of friction. It is a valuable industrial fibre, but its high price generally prohibits its use where alternatives exist.

PVA is inherently water-soluble, but may be insolubilised by after-treatments, making it suitable for some textiles applications.

Furness (1963) gives a detailed account of the structure and properties of addition polymer fibres, while Goodman (1963) presents a valuable account of first-generation synthetic fibres in general.

1.2.3.2 *Second-generation fibres.* The second-generation fibres are those, developed from the 1970s onwards, whose properties enable them to be used where there are special or highly demanding requirements. The term 'high performance' is often applied to fibres in this category, several of which have now replaced conventional engineering materials such as steel in a variety of applications, especially in fibre-reinforced composites, and where weight is an important consideration. High performance generally implies high stiffness and/or strength, but other features of importance include chemical and biological stability, fire retardancy (and/or resistance to high temperatures), and high energy absorption. The latter is particularly crucial in situations requiring protection from impact, such as vehicle body panels.

Space does not permit a detailed examination of the high performance fibres here, but it is worth noting that the polymers of which they are composed fall broadly into three classes.

(*a*) *Liquid-crystalline ('rigid-rod') polymers.* These are typified by the aramid fibre Kevlar™. The term 'rigid-rod' arises from the fact that the monomer itself is inherently stiff. The macromolecular chains are, in consequence, rod-like. In solution they aggregate in much the same way that logs in a river tend to cluster together, to form 'liquid crystals'. Fibres produced from rigid-rod polymers are generally wet- or solution-spun.

Kevlar is DuPont's trade name for the fibre based on the aromatic polyamide:

$$—CO—(C_6H_4)—CO—NH—(C_6H_4)—NH—$$

Its properties include:

- Melting point in excess of 600°C
- Low combustibility
- Low relative density (*ca.* 1.44 compared with 2.55 for glass and 7.9 for steel)
- High specific strength (greater than 5 times that of mild steel)
- High specific modulus (greater than twice that of mild steel)
- Moderate ease of processing.

A good account of the performance characteristics of Kevlar is given by Wilfgong and Zimmerman (1977). Other fibres falling within the same class include the aromatic polyesters, and PEEK, poly(ether-ether ketone).

(*b*) *Flexible chain polymers.* This group of fibres is made up of the polyolefins (polyethylene and polypropylene), together with various copolymers. In these materials the molecular chains themselves are inherently flexible, which enables extremely high levels of chain extension and orientation to be achieved through stretching, solid-phase extrusion, and other methods. Polyolefin fibres can be produced with specific stiffness and strength close to or above that of the aramids, and with very low bulk density (less than $1\,g\,cm^{-3}$ for polyethylene). They are almost chemically inert, making them very suitable for biomedical applications, and as reinforcement in concrete in some situations. However, they are limited by their low softening and melting temperatures, and by their susceptibility to creep and stress-relaxation.

(*c*) *Carbon fibre.* Carbon fibre warrants a class of its own, in that it is neither a liquid-crystalline polymer, nor a flexible chain polymer. It is in fact a modified fibre in the sense that it is generally produced from a precursor such as poly(acrlyonitrile), pitch, or other source.

The essential process by which carbon fibre is formed is that of carbonisation and 'graphitisation', although there are generally several other stages involved. Carbon fibre has a composition similar to that of graphite, and is extremely stiff and virtually incombustible. High performance carbon fibre is very brittle, which makes textile processing difficult. However, there are lower-modulus grades available, which are easier to process into staple yarns, and which retain many of carbon's other desirable properties.

The interested reader is referred, for a fuller account of high performance fibres, to Lewin and Preston (1985, 1989) and to Mukhopadhyay (1992). The latter, in particular, also gives a valuable presentation of current thinking on fibre structure–property relationships in a broader sense.

1.2.4 *Other types of fibre*

Recently, there has been a considerable level of interest in the development of fibres composed of non-polymeric materials, notably glasses, ceramics, and metals. However, a discussion of these types is beyond the scope of this chapter.

1.3 The problem of fibre structure

1.3.1 *Approaches to describing fibre fine structure*

1.3.1.1 *The nature of the problem.* As was stated initially, textile fibres consist, in the main, of partly crystalline, partly ordered linear polymers. This definition encompasses a vast range of fibre types. At one end of the spectrum are the 'general-purpose' materials, including cotton, wool, man-made fibres, and many of the melt–spun synthetic fibres with average performance characteristics, while at the other lie the high-performance fibres, such as the aramids and carbon fibre. The range of structural types is therefore equally vast. However, it is useful to focus on some common general features.

The concept of a linear polymer presents little difficulty. Most textile fibres comprise polymers with a backbone of carbon atoms. Covalent bonding along the chains gives potentially high strength and stiffness. There is generally no significant branching or cross-linking, but some form of interaction generally exists between molecular segments. Although these forces are very much weaker than those along the chains, they are nevertheless important in a number of respects. For example, they play a central role in crystallisation processes, and provide much of the stability of the ultimate structure. They may vary in strength from weak van der Waals forces, as in polyolefins, to relatively strong hydrogen bonds, as in cellulosics.

What is meant by 'partially oriented' is rather less straightforward, but can be considered without too much difficulty. It refers to the extent to which

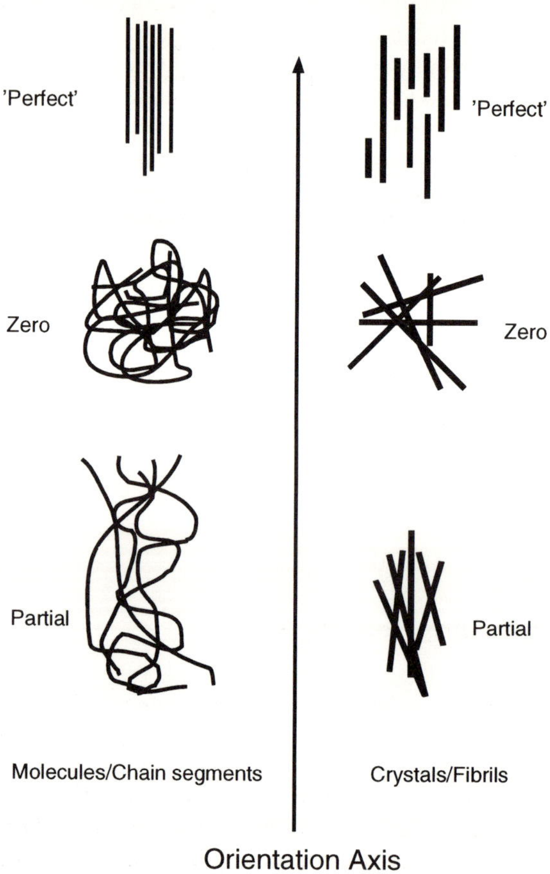

Figure 1.5 Various ways of visualising partial orientation.

molecular segments, whole molecules, or crystalline blocks are aligned with respect to the fibre axis. Perfect orientation implies that all molecular chain segments lie perfectly parallel to the fibre axis, providing considerable stiffness and strength in this direction. Figure 1.5 illustrates several ways of envisaging partial orientation.

The most challenging problem, however, is what is meant by 'partial crystallinity'. To begin with, the concept of a 'polymer crystal' warrants consideration. Crystals of low molecular weight materials are composed of a precisely regular three-dimensional lattice of points, each defining the position of a constituent particle (Figure 1.6). Except for its overall size, the crystal is completely specified by six unit cell parameters. It may be difficult to imagine how the enormously long chain molecules of a polymer could be

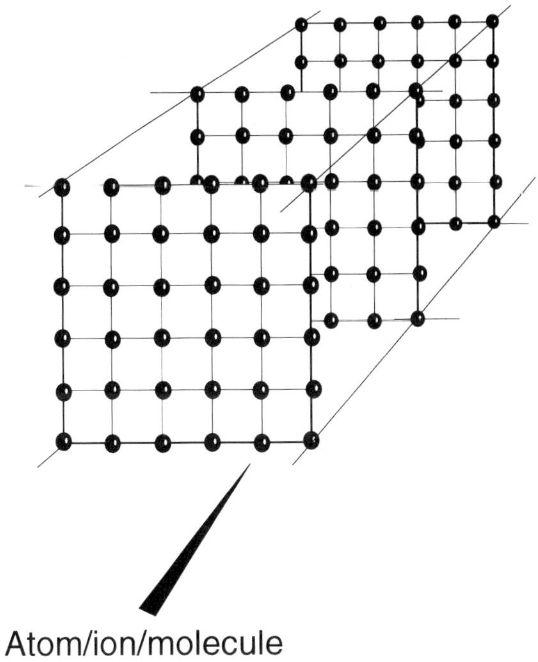

Figure 1.6 Representation of a simple crystal lattice.

organised into anything like such a lattice. For one thing, the molecules within any given sample vary in length one from another. However, if it is borne in mind that polymers (and certainly homopolymers, which account for the vast majority of textile fibres) comprise a repeating sequence of a relatively simple chemical unit — the monomer — then a way forward emerges. In most cases the molecular backbone provides some degree of rotational freedom at sufficiently high temperatures. Generally, the chain conformation of lowest free energy is also one of high geometrical regularity (e.g. the all-*trans* 'planar zigzag' form of polyethylene, based on the tetrahedral angle). This provides the regularity required for a crystal lattice. The geometrical repeat distance along a chain may not always coincide with the chemical repeat, but will generally be only a small integral multiple of it. It is this distance which usually defines one of the unit cell parameters. The remaining two lengths are easier to deal with, being simply related to the lateral packing of the molecules.

Figure 1.7 illustrates how the chains pack in a crystal of polyethylene. For clarity, the hydrogen atoms have been omitted. The unit cell is orthorhombic, with dimensions $a = 0.714$ nm; $b = 0.494$ nm; c (chain axis) $= 0.255$ nm. Interestingly, the dimension parallel to the molecular axis is actually the shortest.

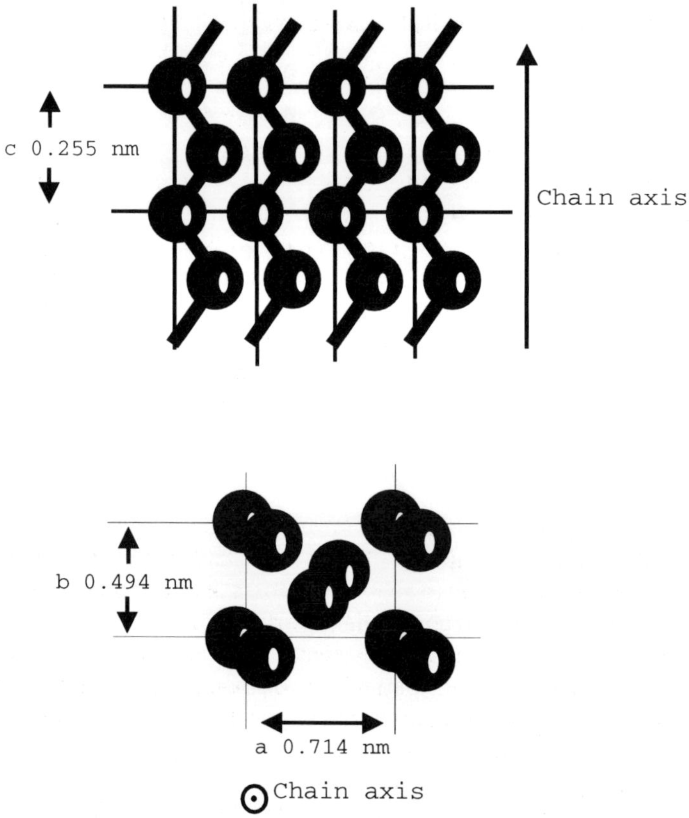

Figure 1.7 Packing of chain segments within the unit cell of a polyethylene crystal.

The unit cells of most textile fibres are now accurately characterised. As with conventional crystals, a variety of unit cell types exists, including monoclinic, triclinic and hexagonal.

However, it should be realised that polymer crystals represent a low state of order compared with those of conventional solids. They are almost always rather poorly developed, and certainly very much smaller — so much so that surface energy and surface irregularities generally produce a substantial depression of the melting point. This phenomenon can be exploited to assess the average size of polymer crystals.

Finally, and perhaps most significantly, the question of partial crystallinity presents a challenge simply because the number of ways in which the concept could be realised is virtually limitless.

1.3.1.2 *Model representations of fibre fine structure.* Attempts to describe fibre structure have traditionally focused on pictorial representations designed to account for one or more experimentally observed phenomena — the so-called 'model' approach. It is clear from the above discussion that no single view can hope to apply equally across such a diversity of materials, but this is not a criticism. There is no doubt that our current understanding of fibre structure would be considerably less advanced were it not for the vision of the early painters of these (with hindsight) over-simple pictures. The various models tended to divide into three broad types, which may be termed 'two-phase', 'continuous structure', and 'intermediate'. The boundaries between these are largely artificial, however. Over the years opinion has differed widely as to the extent to which order is localised within fibre structure. A wealth of ideas developed, ranging more or less uniformly from structures containing no definite ordered phase at one extreme, to ones with no clear disordered phase at the other — and with just about everything in between.

(*a*) *Two-phase models.* In the 1920s and 1930s much attention centred on understanding the structure of the natural and man-made fibres. The melt-spun synthetics were yet to appear, but much of our present understanding of their microstructure is owed to these early studies.

The prevailing ideas at that time stemmed from a theory due to Nägeli (1928), who proposed that starch and certain vegetable materials comprised extremely small crystalline particles which he called 'micelles'. These particles were bound together with other substances such as gums, giving an overall structure analogous to that of bricks in mortar.

X-ray diffraction techniques applied to textile fibres indicated significant structural regularity, but the presence of incoherent scattering implied this must co-exist with a substantial level of disorder. The initial conclusion was that textile fibres might also conform to a micellar structure. Measurements suggested the dimensions of the particles in various cellulosic fibres to be around 5 nm × 40–60 nm, which appeared to agree well with current estimates of the average length of a cellulose molecule.

It was generally assumed that fibres consisted of an amalgamation of hard, discrete and perfectly crystalline micelles embedded in a matrix of non-crystalline (amorphous) polymer, as illustrated in Figure 1.8.

A major characteristic of this structural view was the existence of two distinct phases, i.e. the crystals (or micelles) and the non-crystalline polymer. The term 'two-phase' applies to any structural model in which there is such a distinction.

(*i*) *The fringed micelle concept.* The arguments in favour of the micellar theory of fibre structure were persuasive, but there were a number of factors which did not accord well. One of the salient features of most commodity

Figure 1.8 Simplistic 'bricks-in-mortar' two-phase representation of fibre fine structure (similar to that due to Nägeli, 1928).

fibres is their ability to withstand quite large relative extensions (*ca.* 50%) without significant loss of physical form or elasticity. This range of deformation is quite unusual; the majority of non-polymeric solids fracture at a considerably lower strain — generally a few percent at most. It is difficult to envisage how a simple micellar structure could retain continuity at large deformations. It is likely that delamination would occur between the micelles and the matrix (to which they were assumed not to be intimately connected), leading to voids, cracks and other gross damage at comparatively small extensions.

A further, and more damning, criticism arose once better estimates of molecular length became available. It is now known, for example, that cellulose molecules are actually 400–1000 nm long, which is ten to twenty times longer than the early data suggested.

If the interpretation of the X-ray data is accepted as essentially correct, one is forced to the conclusion that the molecules must be ten or twenty times longer than the crystals themselves.

Abitz and co-workers (Abitz *et al.*, 1930; Gerngross *et al.*, 1930), who were studying gelatin and collagen at the time, suggested that individual molecules might pass alternately through several micelles, and through several intervening regions, to produce a structure similar to that illustrated in Figure 1.9. This view became known as the 'fringed micelle' concept, and was soon adopted by many other researchers as a plausible description of the structure of fibres in general.

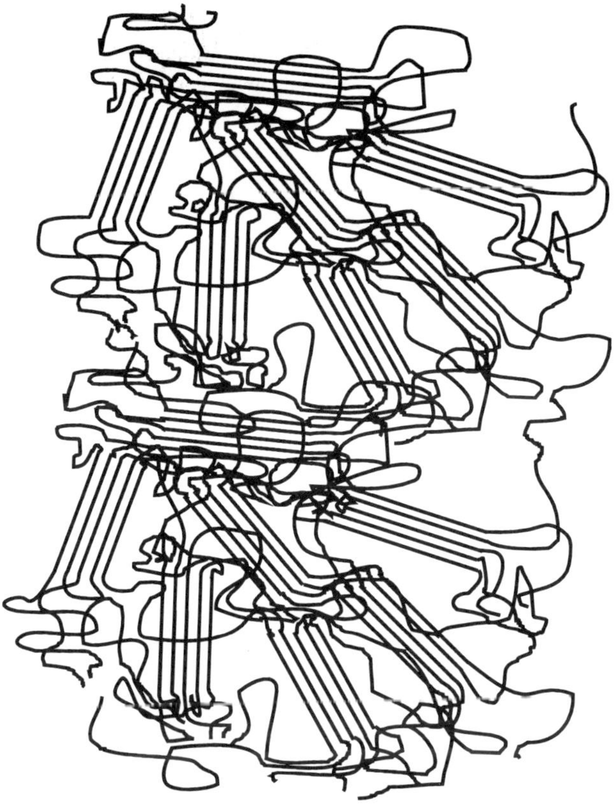

Figure 1.9 The fringed micelle model: an individual chain molecule may pass alternately through ordered and disordered regions.

A major advantage of this model was that it provided the structural continuity which was absent from those proposed earlier. Several alternative two-phase models later emerged, together with a good deal of controversy about which most closely described true fibre structure. The precise details of these vary, but most share the basic characteristic of two continuously linked, but geometrically seperate, crystalline and non-crystalline regions. Among the most notable contributors to research in this area were Kratky and Mark (1937), Frey-Wyssling (1938), Hess and Kiessig (1944), and Hess *et al.* (1957).

(*ii*) *The fringed fibril concept.* Hearle (1958) proposed a two-phase structure which differed from previous models in an important respect: all previous fringed micelle variants showed chain molecules emerging exclusively from the ends of the crystallites, as depicted in Figure 1.10. Hearle (1958)

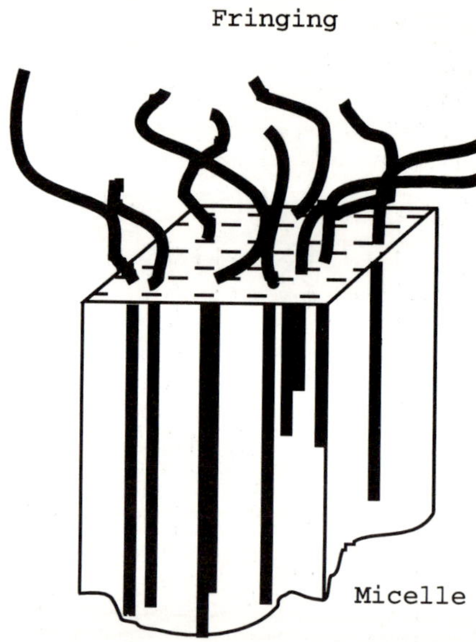

Figure 1.10 Commonly held view of the interface between ordered and disordered structural phases: chains emerge exclusively from the ends of crystallites.

pointed to the lack of rationale for this assumption, and proposed that chains could be expected to enter or leave a crystal at essentially any position along its length. His view of the structure of a fibre incorporated continuous 'fringed fibrils', giving rise to an arrangement similar to that shown in simplified form in Figure 1.11. This model had the advantage that it could be applied to fibres which various techniques revealed to possess long, needle-like features ('fibrils') as distinct from approximately cuboidal micelles.

(*iii*) *Chain folding.* More recent views of fibre structure included the notion of chain molecules folding back and forth on themselves. The work of Till (1957), Fischer (1958a,b), and others in the late 1950s, demonstrated that under suitable conditions single polymer crystals could be grown from dilute solution. It is now clear that the phenomenon applies quite generally over a wide range of crystallisable polymers. These single crystals appear as extremely thin, almost flat lozenges or 'lamellae'. Although they are sufficiently large to be observed easily using scanning electron microscopy, estimates suggest their thickness is confined to the range 10–20 nm. Electron diffraction experiments confirm, also, that their constituent molecules are invariably arranged with the main chain axes approximately parallel to the thickness. As there was no evidence for any substantial amorphous material associated with the single crystals, Keller (1957) proposed that the chains must fold, as illustrated in Figure 1.12. Chain folding is now in fact widely acknowledged

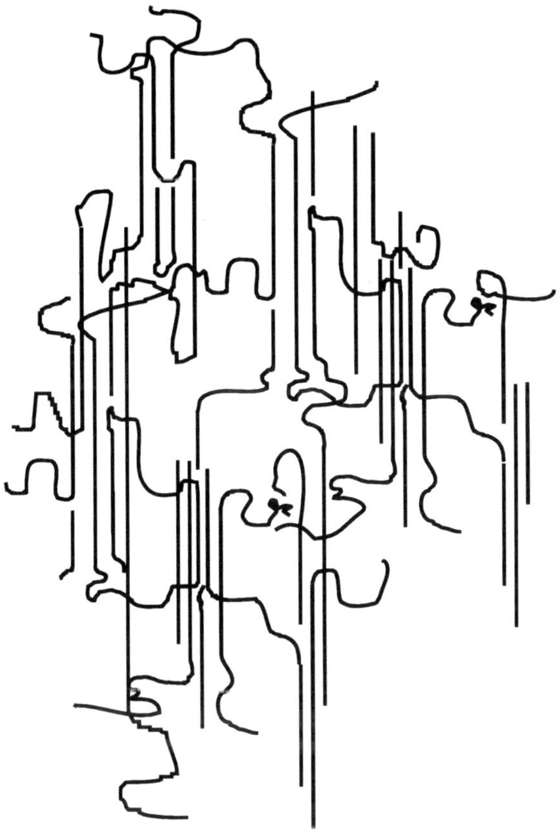

Figure 1.11 Fringed fibril representation (after Hearle, 1958), in which chains may enter or leave crystallites at any surface location.

to occur in many bulk polymers, such as thermoplastics. The inclusion of the phenomenon in fibre structure models was therefore a logical progression; there are, for example, modified fringed micelle and fringed fibril forms which incorporate chain folding.

Keller (1963) presents an excellent account of contemporary ideas regarding crystallinity in high polymers.

(*b*) *Continuous structure models.* Not all researchers accepted the notion of separate ordered and disordered phases. Staudinger (1932) believed that the molecules of high polymers would organise themselves into a single large, imperfect crystal as opposed to separate small, and essentially perfect, ones. His model of fibre structure was essentially as depicted in Figure 1.13. Other proponents of the continuous structure theory included Neale (1933), Carothers, and Hill (Carothers and Hill, 1932).

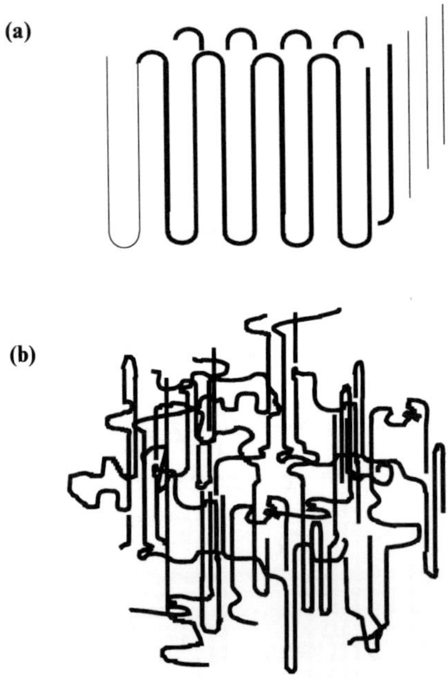

Figure 1.12 Schematic representations of chain folding: (a) folding in a polymer single crystal; (b) fringed fibril model incorporating some chain folding.

Although these models would not permit a separate disordered phase, it was acknowledged that some form of structural disorder nevertheless had to exist — the evidence from X-ray diffraction and other sources appeared to put this beyond any doubt. In Staudinger's representation, a key feature is the imperfection of the crystal itself. He maintained that such features as chain ends within the lattice could account for the experimental data.

(*i*) *Paracrystals.* Another continuous structure representation stems from the ideas of Hosemann (1950a,b, 1962), who put forward the existence of 'paracrystals'. These were essentially similar to the Staudinger crystal, but the disorder was supposed to arise from a somewhat different source. In a

Figure 1.13 Arrangement of molecules in a large, but imperfect, fibre crystal as envisaged by Staudinger (1932).

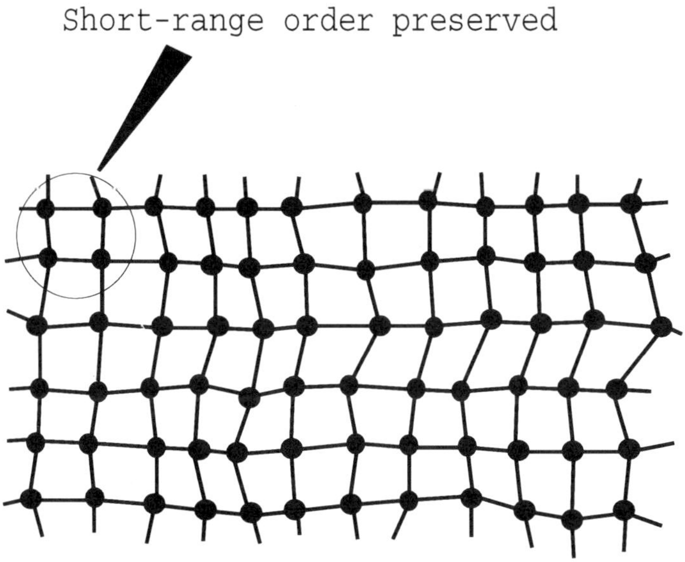

Figure 1.14 Representation of Hosemann's paracrystal in two dimensions: short-range order is preserved, but long-range order lost.

perfect crystal the lattice spacing is precise and exactly constant throughout the crystal. However, suppose there is a small, random variation in the lattice spacing, as illustrated in Figure 1.14 for the case of a hypothetical simple two-dimensional lattice. If the error is sufficiently small, the overall effect will be a loss of long-range coherence, but retention of the short-range regularity. The notion of order and disorder co-existing is preserved, but without the need to invoke two distinct phases.

(*ii*) *Amorphous-with-correlation.* An alternative point of view, expressed by Kargin (1958) and Mikhailov (1958), was that for the long (possibly entangled) molecules of a high polymer to crystallise in the conventional sense would require a degree of molecular freedom beyond what is likely to exist. This led to models, diametrically opposed to those of Staudinger, Hosemann, etc., in which the structure was thought of as essentially amorphous, the order arising from mutual aligning of molecular segments without their necessarily being in crystallographic register. While it is true to say that this view was not widely adopted by other researchers, there are sound arguments for incorporating the principle, at least in certain fibre types.

(*c*) *Intermediate views.* The two-phase and continuous models represent extreme views, but most current ideas tend to fall somewhere between the two, and to include more textural features than perhaps were previously recognised.

(*i*) *Defects.* Fibre crystals (whether of the fringed micelle type or not) are now thought to contain defects analogous to those observed in other materials such as metals — vacancies, impurities, dislocations, and so on. Indeed, given the physical form of polymer molecules, it would be surprising if such abnormalities were not present. Among other things, defects are thought to be responsible for plastic deformation processes in many crystalline fibres, since they impart mobility to the lattice.

(*ii*) *Assembly of crystalline fibrils.* Structural disorder need not arise from the presence of a true amorphous material, nor even from imperfections within crystals. A collection of otherwise perfect crystals would appear disordered if they themselves were imperfectly arranged. Figure 1.15 illustrates this point for a collection of rods of varying dimensions. Discontinuities, and possibly voids, at the crystal boundaries, distributed over the bulk of the material, can give rise to many of the features associated with disorder, such as the diffuse scattering and broadening observed in X-ray diffraction experiments on fibres.

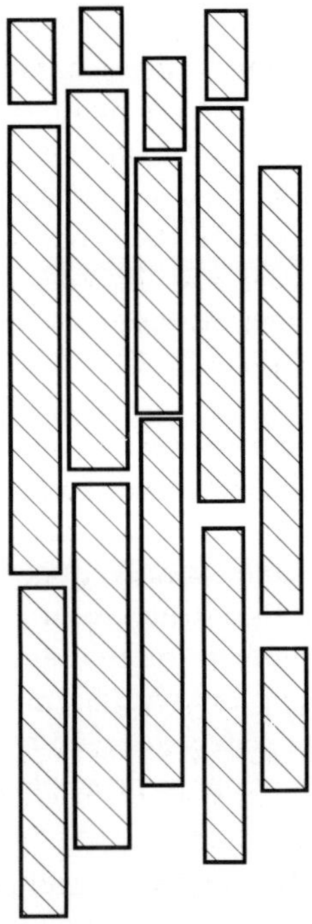

Figure 1.15 Structural disorder associated with the imperfect packing of various-sized rods.

1.3.1.3 *Structure parameters.* A major drawback to the model representation of fibre structure is its inherent subjectivity. Reference was made, in the introduction to this chapter, to 'artists' impressions', and it is certainly the case that such representations are generally capable of yielding at best qualitative information.

In applying a knowledge of structure to the elucidation of mechanical and other fibre properties a more concrete description is needed. For example, reliable prediction of the tensile behaviour of a fibre possessing a specific structural architecture requires hard numerical data rather than vague pictorial notions.

To deal with this criticism, Hearle and co-workers (Morton and Hearle, 1975; Hearle, 1977) devised a scheme whereby fibre structure would be described in terms, not of general geometrical arrangements, but of certain quantitative parameters.

In a simple crystalline solid the constituent particles are regularly arranged throughout the extent of the material, and only a very small number of unit cell parameters are required to specify its structure completely. Alternatively, if the material has a random configuration, such as in a simple liquid or a gas, it is possible to apply the statistics of large assemblies as was done, for example, in Maxwell's Kinetic Theory of Gases. Again, a small number of parameters is able to provide a complete description.

Conducting the equivalent exercise for a textile fibre is by no means a trivial matter. The problem is more difficult to address precisely because of their two-phase nature (partly ordered, partly disordered), coupled with the complexities associated with long-chain polymer molecules. At first sight it might appear necessary to locate all the various structural components, but to specify exactly the co-ordinates of the constituent atoms within the structure would require the determination of a vast number of parameters (of the order of three times Avogadro's Number), which would clearly be impossible.

However, Hearle (1977) asserted that as few as ten general parameters might suffice, of which six could be expected to exert the greatest control over a fibre's ultimate physical behaviour. The most important parameters having been identified, the task then becomes one of setting up suitable numerical scales upon which to measure them.

1. Degree of order. This is perhaps the most important parameter. Often referred to (inaccurately) as 'crystallinity', it expresses the fraction of material considered to exist in an ordered state within the fibre structure. In the case of a solid polymer, the perfect crystal represents the most efficient chain-packing arrangement possible, and the highest density. On the other hand, the imperfect packing of the amorphous chains leads to a lower density. Density is therefore a good indicator of the degree of order, and is typically used as a basis for the scale upon which degree of order is measured.

II. Degree of localisation of order. The two-phase and continuous structure models discussed earlier represent extreme views; in the former there is a clearly defined ordered phase, whereas in the latter there is no such concept, and the 'order' is deemed to permeate the entire structure in a rather diffuse way. One could regard these two extremes as defining two fixed points on a scale of 'localisation of order'. Experimentally, this parameter is more difficult to deal with than the degree of order, although it can be estimated from crystal size measurements using, say, X-ray diffraction techniques.

III. Length:width ratio of ordered units. The third important parameter is the average aspect ratio of the ordered units, specifically with respect to the fibre direction. For example, the units could be cuboidal micelles with similar width and length. Alternatively, they might be long fibrils, or thin, flat lamellae. The measurement scale for this parameter could run from zero for infinitely thin lamellae, to infinity for infinitely long fibrils. Micelles would have a value around unity. For the sake of symmetry, however, it is preferable to adopt a logarithmic scale, when the value for cubic crystals would be zero.

This third parameter may also be estimated from various diffraction techniques.

Hearle demonstrated the value of representing these first three parameters on a three-dimensional plot. The various fibre types are found to be located within fairly well-defined domains, as illustrated in Figure 1.16.

IV. Degree of orientation of ordered units. Any of the structures defined by the first three parameters can be envisaged with varying degrees of orientation of the ordered units. Orientation has a profound influence on most fibre properties, but notably on mechanical properties such as modulus, strength and extensibility.

Orientation can be estimated by using a variety of analytical techniques, including wide-angle X-ray diffraction (WAXD) and vibrational spectroscopy.

V. Size of ordered units. The size of the ordered units determines the overall scale of the structure, and is effectively a measure of its fineness or coarseness. It is thus analogous to the 'magnification' of the structure, and is quite distinct from the length:width ratio. Any of the structures discussed so far could be imagined to exist with a wide range of sizes.

VI. Molecular extent. The sixth parameter is analogous to fibre extent in slivers or yarns, and may apply to either ordered or disordered chains, or to some average over all the molecules within the structure (Figure 1.17).

The measurement scale for this parameter would range from near zero for a tightly coiled molecule, to unity for a fully-extended one.

The distinction between extent and orientation must be preserved. Figure 1.18 illustrates that it is possible to combine high orientation and low extent on the one hand, or low orientation and high extent on the other.

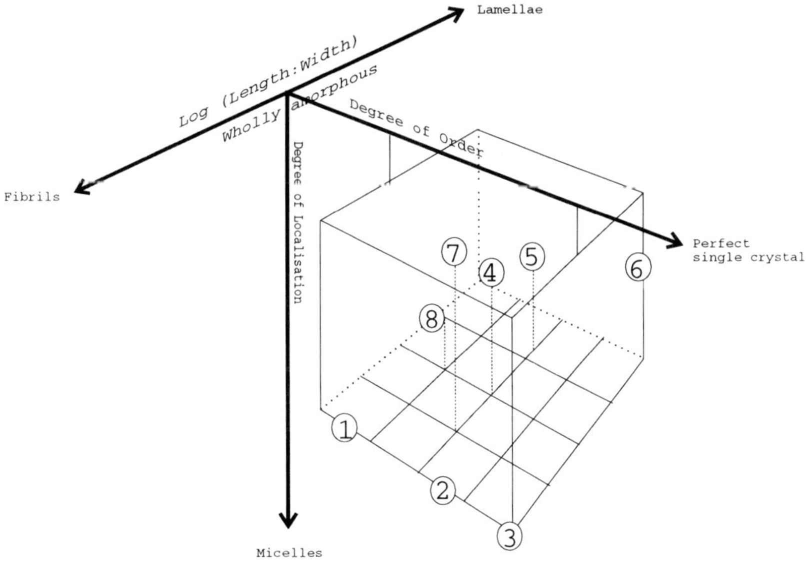

Figure 1.16 Representation of fibre structure parameters on a three-dimensional plot (after Morton and Hearle, 1975). Approximate location: 1, wool and hair; 2, silk; 3, cotton; 4, polyesters; 5, polyamides; 6, linear polyethylene; 7, acrylic; 8, viscose.

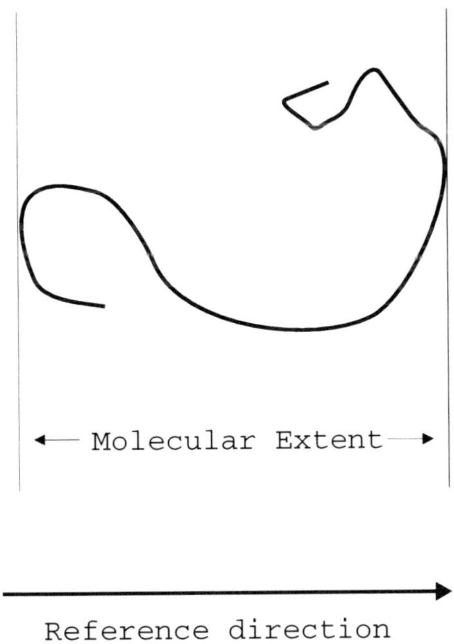

Figure 1.17 Definition of molecular extent.

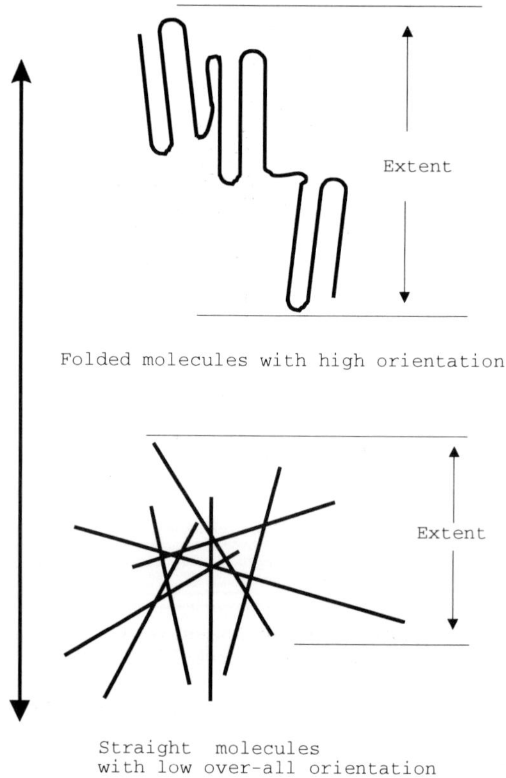

Figure 1.18 Two contrasting combinations of molecular orientation and molecular extent.

As with orientation, molecular extent will have a dramatic influence on properties such as strength and modulus.

VII. Shape of ordered units. Most simple models say little about the precise shape of the ordered units, either laterally, or when viewed in cross-section. The micelles (or other entities) tend to be assigned a vague cuboidal shape. However, certain aspects of fibre behaviour will no doubt be sensitive to whether the units have, for example, cylindrical, rectangular, or even lozenge-shaped profiles. The latter case in particular may be appropriate in melt-spun fibres, where there is evidence that the normal to the chain-fold surface may be regularly inclined to the fibre axis (see, for instance, Hearle and Greer, 1971). Such features may lead to the possibility of interlamellar shear within the structure of a fibre loaded in tension (Figure 1.19).

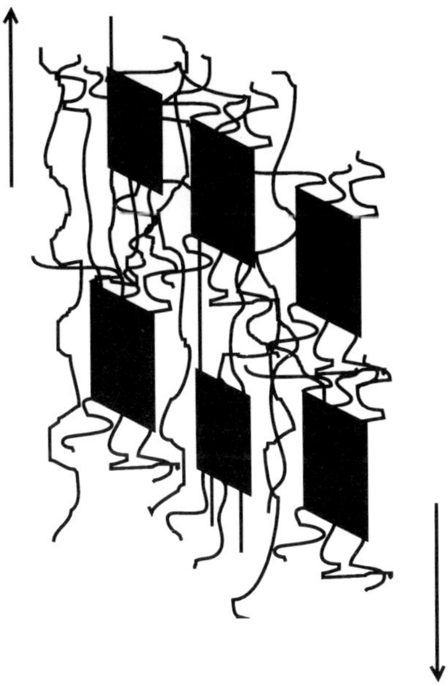

Figure 1.19 Two-phase structure model incorporating lozenge-shaped ordered units.

VIII. Relative orientation of disordered and ordered material. A well-developed fibre would feature equally high levels of amorphous and crystalline chain orientation. However, there will be cases where the ordered units are themselves well oriented, but the amorphous chain segments are essentially isotropic (Figure 1.20). Such a combination can arise when an initially highly-drawn melt-spun fibre, say, is heat treated at a temperature above its glass transition, enabling the amorphous regions to relax, and shrinkage to take place.

IX. Packing pattern of ordered units. The way in which the ordered units themselves aggregate will play a role in the overall fibre structure. As with shape, the packing both in cross-section and in profile needs to be taken into account. Whereas a random lateral arrangement such as that shown in Figure 1.21a might yield a uniformly distributed coarse-grain structure, more regular stacking could produce a 'pseudofibrillar' effect (Figure 1.21b). These two structures could be expected to have differing influences on, for example, the mechanics of fracture in tension. A further level of detail is the extent of staggering of the stacks (Figure 1.21c).

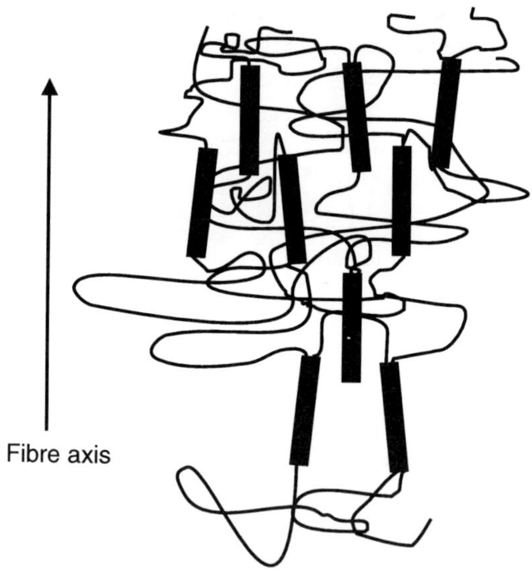

Figure 1.20 Combination of oriented ordered and unoriented disordered phases.

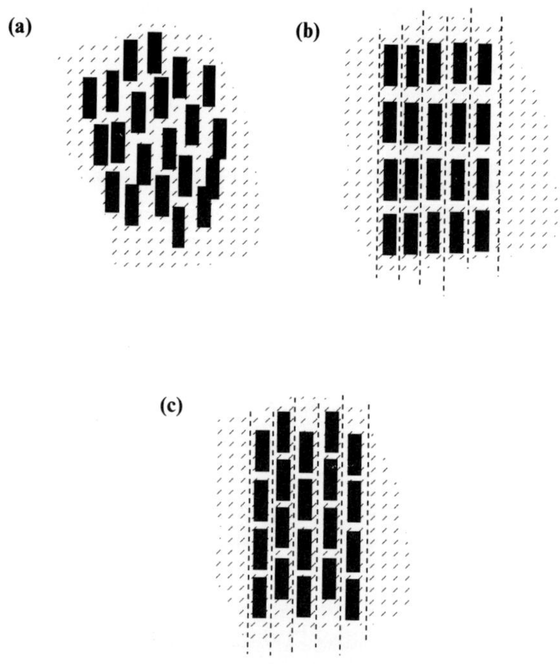

Figure 1.21 Various ways of visualising the lateral packing of ordered structural units: (a) random stacking; (b) regular stacking; (c) staggered stacking.

X. Degree of linking between ordered units. The fringed micelle and other two-phase views of fibre fine structure emphasise the continuity and cohesion provided by having individual molecules pass alternately through ordered and disordered regions. However, not every chain emerging from one crystalline surface will enter another; some will simply terminate in the intervening space between crystallites; others may re-enter the same crystal, forming loose loops; yet others may terminate within crystallites. These chain segments will contribute nothing towards the overall load-bearing capacity of the structure. Those chains that are effective are known as 'tie molecules'. Some of these may be stretched tightly between crystallites, and it is these that will bear the initial stress when the structure is loaded.

Figure 1.22 illustrates different types of chain segment which could emerge from the crystal surfaces. Hearle *et al.* (1988) attempted to develop quantitative structure models for melt-spun synthetic fibres which incorporate such features.

1.3.1.4 *Final comment on microstructure.* Views of fibre fine structure nowadays generally assume that for any given fibre type, and for any particular processing history, a combination of features will be present, which may include any or all of those discussed, and many for which space does

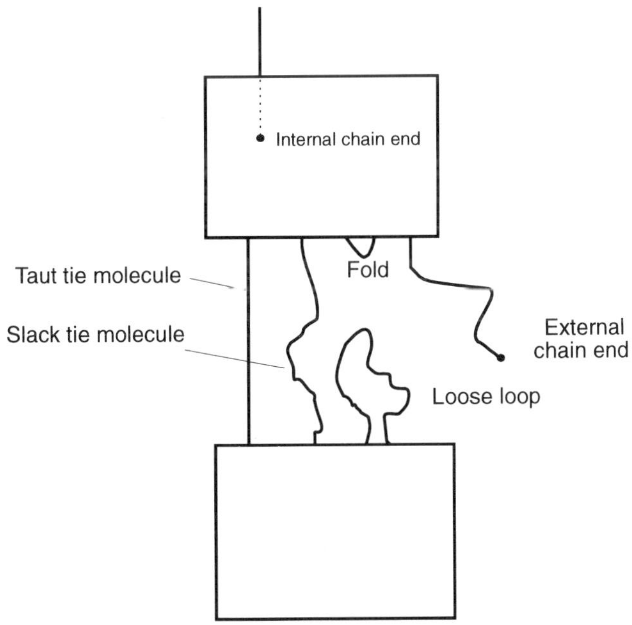

Figure 1.22 Illustration of different types of chain segments associated with the crystal–amorphous boundary.

not permit description. Hearle and Peters (1963) give an exhaustive and detailed account of the development of ideas relating to fibre fine structure.

1.3.2 *Comments on fibre macrostructure*

Over and above the submicroscopic fine structure detail of fibres lie larger-scale features which often form the distinguishing characteristics of a particular fibre type. These may be dictated by genetics in animal and vegetable fibres (for example the cuticular scales in wool, and the lumen in cotton), or they may arise from the crystallisation kinetics and other factors associated with the manufacturing process.

At a more macroscopic level still, there will often be physical features associated, for example, with the damage caused during processing and in end-use. It is inappropriate here to consider such aspects in detail. Rather, brief consideration will be given to two noteworthy phenomena. However, the interested reader is referred to Hearle and Peters (1963) for further discussion.

1.3.2.1 *Spherulites.* The 'spherulite' is perhaps the most frequently observed structural arrangement on a large scale in melt-crystallised polymers. However, although the existence of similar objects in metals and certain other substances had been recognised earlier, it was not until 1945 that their presence in high polymers was reported by Bunn and Alcock (1945).

Spherulites are, ideally (Figure 1.23a), spherical crystalline aggregates, whose diameter may range from the submicroscopic up to several millimetres. When viewed using polarised light they almost always display a 'Maltese cross' extinction pattern, indicating them to be birefringent entities in which the refractive index along a radius differs from that tangentially.

During crystallisation from the polymer melt, spherulites grow radially until either they impinge upon one another or they are stopped by other factors, such as quenching or the presence of non-crystallisable amorphous material. Under a polarising microscope they usually appear as either discrete circular features or as domains with line boundaries (Figures 1.23b and 1.23c).

Although spherulites are generally associated with isotropic bulk polymers, they are sometimes present in undrawn melt-spun fibres such as polyamides. When they do occur their dimensions can be comparable with the filament diameter. In this case they present a serious problem in downstream processing because they tend to act as regions of stress concentration, causing threadline breakage.

1.3.2.2 *Kink bands.* Many fibre types exhibit a phenomenon known as 'kink-banding', characterised by the appearance of linear criss-cross striations. These may be observed in either optical or scanning electron microscopy. Figure 1.24 shows an example of the effect in a polyester fibre.

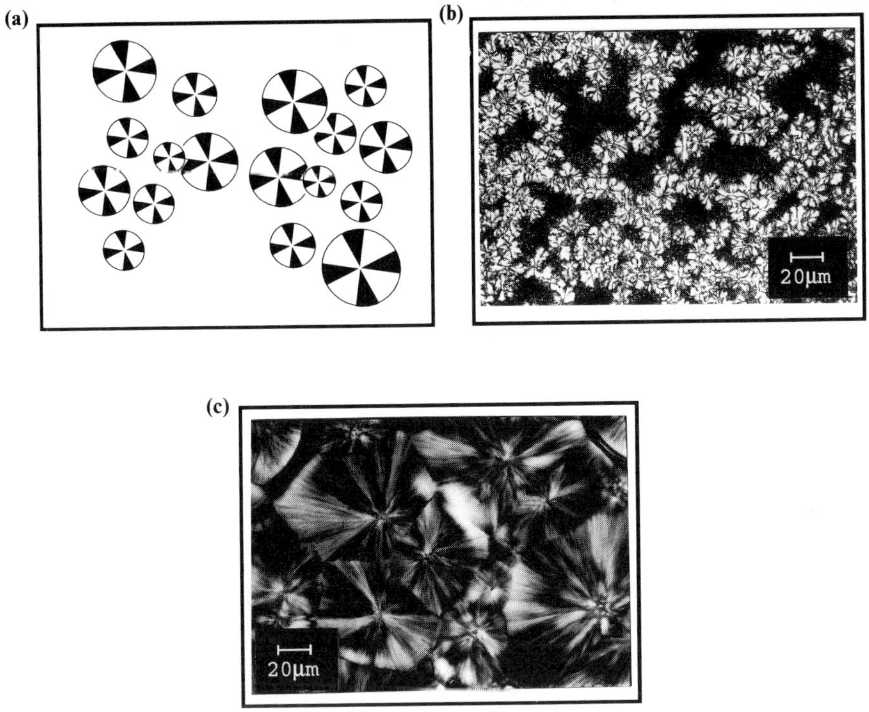

Figure 1.23 Spherulites in fibrous polymers: (a) idealised; (b) as observed in remelted, recrystallised polyester fibres; (c) as observed in remelted, recrystallised polypropylene film.

Kink bands are not inherent structural features in themselves, but arise from structural damage accompanying compressive loading. This is often encountered where bending occurs. In bending, the outer surface is subjected to extensional strain, but the inside surface is placed under compression. This compression causes buckling of the internal structure. Precisely what within the structure undergoes this buckling is not always clear, but is probably dependent on the fibre type. In the case of stiff fibres such as aramids it could be the straight-chain fibrils themselves which kink, whereas in melt-spun synthetics it is more likely to be shearing of intercrystalline amorphous material.

Contrary to what might be expected, textile fibres most commonly fail in use through compression rather than extension. This is because fibres generally yield more easily in compression than in tension. The fibres themselves are rarely stretched close to the point of tensile rupture, but repeated flexing within a fabric structure can induce compressional damage, leading eventually to failure.

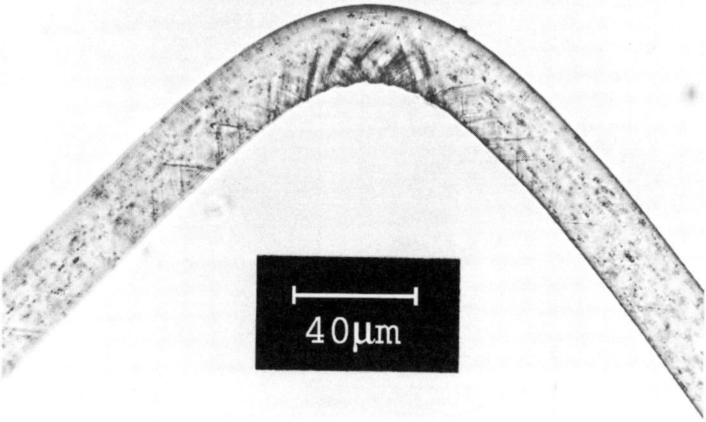

Figure 1.24 Optical micrograph showing kink-bands in a bent polyester fibre.

1.4 Relationship of structure to tensile properties

The salient technological feature of a fibre is its response to mechanical loading and/or deformation along its length, i.e. its tensile behaviour. Indeed, the tensile stress–strain curve is probably the single most important characteristic of a textile fibre.

Textile fibres display many phenomena which put them in a class apart from other materials. The stress–strain graph for a fibre is rarely a straight line as would be expected from a Hookean solid. Although it is often ascribed a 'typical' shape (Figure 1.25), in practice there are wide variations, both from one fibre type to another, and from one fibre to another within the same type. Except at quite small levels of extension, most fibres are inelastic, and so will fail to return completely to their original dimensions after stretching. The tensile behaviour is also highly temperature- and humidity-dependent in many cases.

It is appropriate to ask what influence the fine structure has on the ultimate physical behaviour of a fibre, and in particular, on its tensile properties. What structural features, for example, determine the characteristic shape of a fibre's stress–strain curve? It should be realised that the influence of any particular feature will be highly complex, and it is not possible to do more than present a few general ideas here.

1.4.1 *General influence of structural features*

If the fine structure of a fibre may be described in terms of a two-phase assembly of ordered and disordered material, then both fractions must

CLOTHWORKERS LIBRARY
UNIVERSITY OF LEEDS

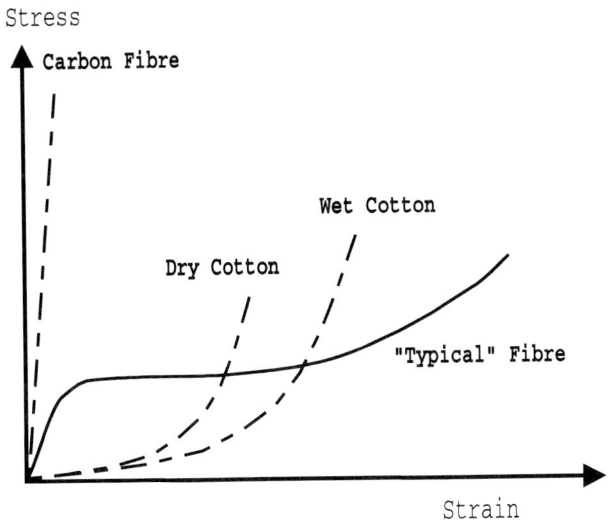

Figure 1.25 The 'idealised' fibre stress–strain curve, compared with those of cotton and carbon fibre (schematic).

contribute to some extent to the overall mechanical behaviour. At the very least, it ought to be possible to predict the initial modulus with some certainty.

Various researchers have attempted to develop suitable mechanical models to describe fibre tensile behaviour, and to relate these to actual structural features. Since the chain segments in a fibre crystal are more efficiently packed, and more fully extended, than they are in the amorphous regions, the crystalline modulus is generally significantly greater than the amorphous modulus. Bearing this in mind, one approach, due to Takayanagi (1964), is to devise mechanical models comprising crystalline volumes in combination with homogeneous amorphous blocks, as illustrated in Figure 1.26. The simplest model of this type assumes only two material properties—the crystalline and amorphous moduli. Depending on the relative abundance of the two phases, and how they are coupled mechanically, it is possible to predict a variety of results for the tensile modulus of the composite structure.

For fibres whose structure is continuous rather than two-phase, it may still be possible to arrive at a mechanical description. For example, if the structure is essentially crystalline, one might start by considering a single perfect crystal with its main chain axis lying exactly parallel to the fibre axis, and then make adjustments to allow for structural imperfections, misalignments and known macroscopic features (e.g. cell boundaries in animal and vegetable fibres).

1.4.1.1 *Time, temperature and anisotropy.* There are many complicating factors in relating fibre mechanics to structure, one of which is time

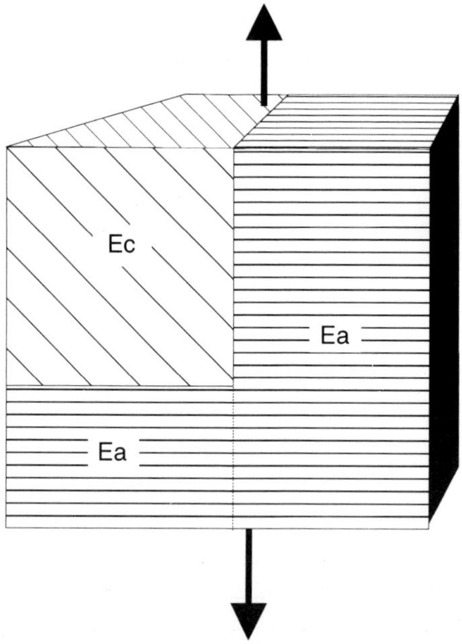

Figure 1.26 Simple Takayanagi model for the calculation of fibre modulus. Ea, amorphous modulus; Ec, crystalline modulus.

dependence. Most fibres are viscoelastic to a greater or lesser degree, which means their mechanical properties display a combination of elastic and plastic features. The phenomena of creep and stress-relaxation in fibres are manifestations of their viscoelastic nature.

Under normal conditions of use, time effects may be insignificant in many fibre types, but in others creep and stress-relaxation cause serious problems. It is important to find mechanical and structural models for such features, in order to understand the underlying causes better, and to offer ways of reducing such effects. An exhaustive study was made by Wilding and Ward (Wilding and Ward 1978, 1981, 1984; Ward and Wilding, 1984), who proposed a model for time-dependent phenomena in high-performance linear polyethylene, based on flow processes occurring in two parallel molecular phases (Figure 1.27). They showed that parameters such as molecular weight and the degree of cross-linking play an important role in suppressing creep and stress-relaxation.

Temperature has a profound effect on the tensile behaviour of most fibres. Where a two-phase structure is thought to exist, it is customary to regard the amorphous phase as a tangled network of disordered chains. At low temperatures this will display so-called 'energy-dependent' elasticity, and will be essentially glassy. However, at relatively high temperatures, where molecular mobility is sufficient, the amorphous phase may take on some of

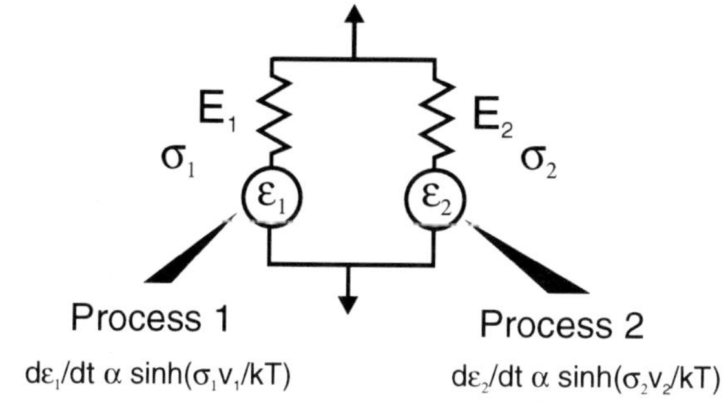

Figure 1.27 Mechanical model due to Wilding and Ward (1981, 1984) for time-dependence in high performance linear polyethylene fibres. Total strain $\varepsilon = \sigma_1/E_1 + \varepsilon_1 = \sigma_2/E_2 + \varepsilon_2$. E = spring stiffness, ε = strain, σ = stress, v = activation volume.

the characteristics of elastomers. At intermediate temperatures, it will display some features of both glasses and rubbers.

The ordered phase is likely to be much less sensitive to temperature. However, since plastic flow processes are generally highly temperature-dependent, the overall influence of temperature on the mechanical behaviour is far from obvious.

The influence of time and temperature on fibres having structures other than two-phase may be just as important, but the approach to analysing them may have to be quite different. For example, if there is no true amorphous phase, then rubber-like effects presumably cannot occur. However, such features as interfibrillar shear may take place, and these will be affected by the rate of deformation, temperature and, very often, whether or not moisture is present.

Additional factors such as the degree of anisotropy complicate the problem even further. The overriding principal, however, irrespective of the precise structural details, is that the higher the crystallinity and the more perfect the orientation, the higher will be the stiffness and strength.

Other parameters, such as yield stress and extensibility, tend to be influenced in the way one might intuitively imagine (e.g. yield strength increased, extensibility decreased).

1.4.2 *Examples*

This discussion of fibre structure has been of necessity brief, but it would be seriously incomplete if it did not pay some attention to structural features present in real fibres.

Since space is limited, attention will be restricted to two commodity fibres whose structural architectures are widely different, but which are commercially of equal importance, viz. cotton and melt-spun synthetics (typified by polyamides and polyester). The reader is once again referred to the excellent work by Hearle and Peters (1963), and also to that by Woods (1955) for a wider discussion.

1.4.2.1 *Cotton.* The tensile curve of cotton (Figure 1.25) is very different from those of most other fibres in that it is smoothly concave to the point of rupture. Moreover, unlike the majority of other textile fibres, cotton is stronger when wet than when dry. These observations, if no other, require some explanation.

Since, in cotton, all the cellulose molecules produced grow directly into crystalline fibrils, true amorphous cellulose is virtually non-existent within the structure. For this reason, cotton fine structure is not generally regarded as two-phase. Nonetheless, X-ray and other data do show the existence of significant disorder. It is thought that this arises from factors such as the imperfect registration of neighbouring fibrils, and possibly also to crystal defects, together with the presence of small amounts of non-cellulosic substances.

In attempting to describe the tensile curve of cotton, since the fibre is essentially crystalline one could choose the tensile behaviour of a single cellulose crystal as a starting point. Using this assumption would imply a steep straight line, very different from the actual tensile curve. However, the inclination of the fibrils to the fibre axis will bring about a reduction in modulus. The simplest analysis yields:

$$E = E_{\mathrm{f}} \times \cos^4(\alpha)$$

where E is the fibre modulus, E_{f} that of the fibril, and α is the helix angle.

The helix angle is therefore predicted to have a dramatic influence. Nevertheless, even allowing for this effect, the tensile graph would still be linear, and of a steeper slope than is encountered in practice.

A further factor is that because the helix angle reverses periodically, there is no real twist in the structure, only 'false' twist. This being the case, it is possible that when the structure is extended, the fibrils can rotate with comparative ease. This leads to a further reduction in the modulus. If there is moisture present, the process is made easier still, so that wet cotton would be expected to be less stiff than dry, as is indeed the case.

However, these effects alone cannot account for the characteristic concave shape of the true tensile curve. The explanation almost certainly lies in the convolutions. When the fibre is extended, one of the first processes to occur is the removal of the convolutions. This is similar to the removal of crimp in a fabric; it is easy initially, but becomes more difficult as the fibre is straightened.

The other curious property of cotton is its high wet strength. The explanation for this is thought to be that when the fibre is dry, extension generates high shear stresses between the helical layers of fibrils, promoting tearing of the structure, in which fibrils break in succession. In wet conditions, the reorganisation of the fibrils is assisted by the lubricating effect of the water molecules. This favours more uniform load-sharing, and the strength tends towards that of a bundle of fibrils, rather than that of a single fibril.

1.4.2.2 *Melt-spun synthetics.* As was stated in section 1.2.3, the melt-spun synthetic fibres are dominated by the polyamides and polyesters. In broad terms, the properties of these two classes of material are similar, although there are some important differences. The typical tensile curves are similar to those shown in Figure 1.28.

(*a*) *Likely fibre structure.* The fibres are generally formed by melt extrusion followed by hot drawing. Because the molecules are relatively flexible, they reorganise under the influence of the applied shear stresses to produce a moderately high level of orientation. However, the considerable chain entanglements present in the melt will hinder crystallisation. As a result, melt-spun fibres are rarely more than about 50% crystalline. The remaining material is assumed to exist as an entangled amorphous network. This is close to the situation required for a two-phase system. The currently-accepted structural view is a combination of fringing and folding, as described by Hearle and Greer (1971). In their model the chain axes of the molecules within a lamella lie approximately parallel to the fibre axis, but owing to staggering of the folds, the normal to the lamellar surface is inclined to the fibre axis by approximately 42°. Similar models have been put forward by Peterlin (1971) and by Prevorsek *et al.* (1973).

Spherulite growth may be observed in undrawn melt-spun fibres, and has also been reported to exist in undrawn sections of drawn films and fibres (Keller, 1963).

There is at least one crucial distinction between the synthetic (including man-made) and natural fibres; in the former the constituent polymer molecules already existed prior to the fibre-formation process, which as a result becomes one of physical rearrangement; in the latter, the molecules develop simultaneously with the structure. The important consequence of this is that the fibre structure is effectively fixed by nature, whereas in the synthetics, it can be modified almost limitlessly in processing by control of such variables as temperature, draw ratio, and wind-up speed.

(*b*) *Structure–tensile behaviour of melt-spun synthetics.* Because a wide range of structural variations is possible for synthetic fibres, there is a correspondingly wide variation in tensile parameters. However, suffice it to say that, by and large, these materials can be considered as composites consisting of hard crystalline blocks embedded in an amorphous matrix.

The anticipated behaviour will depend upon the temperature range in which the fibres are operating. For example, at temperatures well below room temperature, the amorphous network might be expected to be glassy and brittle, so that the fibre itself will be brittle. At normal temperatures in, say,

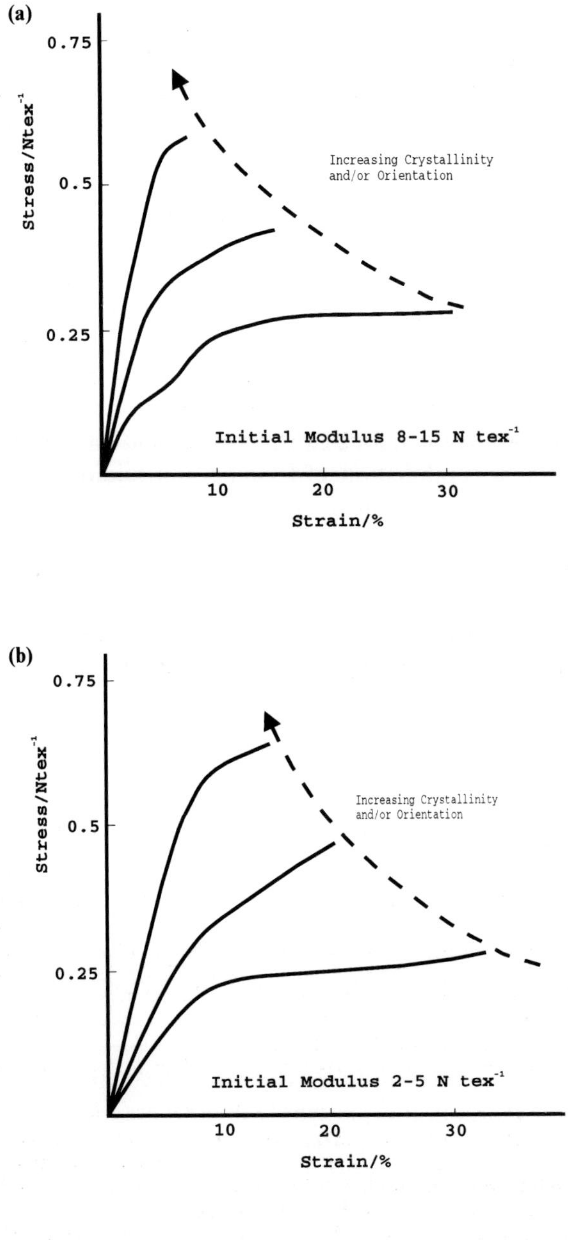

Figure 1.28 Typical stress–strain curves of polyester and nylon fibres, showing general effects of increasing crystallisation and orientation.

nylon, the amorphous molecules will have a degree of mobility, but there will still be hydrogen bonds. In this regime the matrix could be likened to a toughened rubber. The stress–strain curve would show some ductility. At temperatures above about 100°C, the hydrogen bonds will be dissociated, and the material will become significantly softened. (This effect is important, incidentally, in the heat-setting of nylon, and in ironing, since when the temperature is subsequently lowered, new hydrogen bonds can form, thus stabilising the deformed structure.)

Similar phenomena could be expected to occur in polyesters, with benzene ring interactions and steric hindrance replacing the hydrogen bonding. In addition, the inherent stiffness of the benzene rings ought to be reflected in a high tensile modulus as compared with the polyamides. This is in fact the case.

What has been discussed has been largely qualitative. Ideally, one would like to be able to make quantitative predictions for the tensile curves over their entire range. This is a daunting task and, not surprisingly, there have been few reported attempts, although Hearle *et al.* (1987) had a measure of success in describing the tensile curve for nylon 6.6 in terms of this composite model.

Acknowledgement

Sincere thanks are due to Mr J.T. Jones, Ms L. Lu and Mr P. Noone for their invaluable help in the preparation of photographs.

References

Abitz, W., Gerngross, O. and Herrmann, K. (1930) X-ray examination of gellatin micelles. *Naturwiss.* **18**, 754–755.

Bailey, T.L.W. jun., Tripp, V.W., and Moore, A.T. (1963) Cotton and other vegetable fibres. In *Fibre Structure* (Eds J.W.S. Hearle and R.H. Peters), Butterworth & Co., London, pp. 422–454.

Bamford, C.H. and Elliott, A. (1963) Synthetic polypeptides and fibrous proteins. In *Fibre Structure* (Eds J.W.S. Hearle and R.H. Peters), Butterworth & Co., London, pp. 46–110.

Bunn, C.W. and Alcock, T.C. (1945) Texture of polythene. *Trans. Faraday Soc.*, **41**, 317–325.

Carothers, W.H. and Hill, J.W. (1932) Polymerization and ring formation. XI. Use of molecular evaporation as a means for propagating chemical reactions. *J. Amer. Chem. Soc.*, **54**, 1557–1559.

Fischer, E.W. (1958a) *Disc. Faraday Soc.*, **25**, 205.

Fischer, E.W. (1958b) Oriented crystallization of polyethylene on sodium chloride. *Kolloid-Z.*, **159**, 108–118.

Frey-Wyssling, A. (1938) *Submikroskopische Morphologie des Protoplasmas und seiner Derivate*, Gebrüder Borntraeger, Berlin; revised and translated as *Submicroscopic Morphology of Protoplasm* (1st English edn 1948, 2nd English edn 1953), Elsevier, Amsterdam.

Furness, V.I. (1963) Fibres from addition polymers. In *Fibre Structure* (Eds J.W.S. Hearle and R.H. Peters), Butterworth & Co., London, pp. 524–533.

Gerngross, O., Herrmann, K. and Abitz, W. (1930) Fine structure of gelatin micelles. *Biochem. Z.*, **228**, 409–425.

Goodman, I. (1963) Synthetic fibre-forming polymers and co-polymers. In *Fibre Structure* (Eds J.W.S. Hearle and R.H. Peters), Butterworth & Co., London, pp. 111–180.

Hearle, J.W.S. (1958) Fringed-fibril theory of structure in crystalline polymers. *J. Polymer Sci.*, **28**, 432–435.
Hearle, J.W.S. (1963) Structure, properties, and uses. In *Fibre Structure* (Eds J.W.S. Hearle and R.H. Peters), Butterworth & Co., London, p. 621.
Hearle, J.W.S. (1977) On structure and thermomechanical responses of fibres, and the concept of a dynamic crystalline gel as a separate thermodynamic state. *J. Appl. Polym. Sci.: Applied Polymer Symposia*, **31**, 137–161.
Hearle, J.W.S. (1991) Understanding and control of textile fiber structure. *J. Appl. Polym. Sci.: Applied Polymer Symposia*, **47**, 1–31.
Hearle, J.W.S. and Greer, R. (1971) Fibre structure. *Textile Progress*, **2**(4), 53.
Hearle, J.W.S. and Peters, R.H. (eds) (1963) *Fibre Structure*, Butterworth & Co., London.
Hearle, J.W.S., Prakash, R. and Wilding, M.A. (1987) Prediction of mechanical properties of nylon and polyester fibres as composites. *Polymer*, **28**, 441–448.
Hearle, J.W.S., Prakash, R., Wilding, M.A. and Davis, H.A. (1988) The structural mechanics of nylon and polyester fibres: Approaches to theoretical understanding. In *Integration of Fundamental Polymer Science and Technology—2*, (Eds P.J. Lemstra and L.A. Kleintjens), Elsevier, London, pp. 540–544.
Hess, K. and Keissig, H. (1944) Long-period interferences and micelle fiber—fine structure of synthetic polymers (polyamides and polyesters). *Z. physik. Chem.*, **A193**, 196–217.
Hess, K., Mahl, H. and Gütter, E. (1957) Electron microscopic representation of long periodic intervals in cellulose fibers and comparison with the periods of other kinds of fibers. *Kolloid-Z.*, **155**, 1–19.
Hosemann, R. (1950a) Röntgen interferences in materials with lattice perturbations obeying liquid statistics. *Z. Phys.*, **128**, 1–35.
Hosemann, R. (1950b) Ideal paracrystal and the coherent Röntgen radiation scattered by it. *Z. Phys.*, **128**, 465–492.
Hosemann, R. (1962) Crystallinity in high polymers, especially fibers. *Polymer*, **3**, 349–392.
Kargin, V.A. (1958) Structure and phase state of polymers. *J. Polymer Sci.*, **30**, 247–258.
Keller, A. (1957) Single crystals in polymers: Evidence of a folded chain configuration. *Phil. Mag.*, **2**, 1171–1175.
Keller, A. (1963) The Crystallinity of High Polymers. In *Fibre Structure* (Eds J.W.S. Hearle and R.H. Peters), Butterworth & Co., London, pp. 332–390.
Kratky, O. and Mark, H. (1937) Individual cellulose micelles. *Z. physik. Chem.*, **B36**, 129–139.
Lewin, M. and Preston, J. (eds) (1985, 1989) *High Technology Fibers, Parts A and B*, International Fiber Science and Technology Series, (ISSN 0570-4898), Marcel Dekker Inc., New York.
Lewis, D.M. (ed.) (1992) *Wool Dyeing*, The Society of Dyers and Colourists, Bradford.
Mikhailov, N.V. (1958) On the phase structure of cellulose. *J. Polymer Sci.*, **30**, 259–269.
Morton, W.E. and Hearle, J.W.S. (1975) *Physical Properties of Textile Fibres*, 2nd edn, Textile Institute and William Heinemann, London, pp. 32–38.
Mukhopadhyay, S.K. (ed.) (1992) *Advances in Fibre Science*, The Textile Institute, Manchester, and Hobbs Ltd., Southampton.
Nägeli, C. (1928) Micellartheorie. Original papers reprinted as *Ostwalds Klassiker*, No. 227 (Ed. A. Frey), Leipzig.
Neale, S.M. (1933) The modification of natural cotton cellulose by swelling and by degradation. *Trans. Faraday Soc.*, **29**, 228–238.
Peterlin, A. (1971) Molecular model of drawing polyethylene and polypropylene. *J. Mater. Sci.*, **6**, 490–508.
Prevorsek, D.C., Harget, P.J., Sharma, R.K. and Reimschuessel, A.C. (1973) Nylon 6 fibers: changes in structure between moderate and high draw ratios. *J. Macromol. Sci.*, **B-8**, 127–156.
Sharples, A. (1963) Cellulose and its derivatives. In *Fibre Structure* (Eds J.W.S. Hearle and R.H. Peters), Butterworth & Co., London, pp. 21–45.
Staudinger, H. (1932) *Die hochmolekularen Organischen Verbindungen*, Springer-Verlag, Berlin.
Takayanagi, M. (1964) Viscoelastic behaviour of crystalline polymers. *Kautschuk Gummi, Kunststoffe, Plastomere, Elastomere, Duromere*, **17**, 164–173.
The Textile Institute (1960) *Textile Terms and Definitions, 4th edn*, The Textile Institute, Manchester, p. 65.
Till, P.H. (1957) Growth of single crystals of polyethylene, *J. Polymer. Sci.*, **24**, 301–306.

Ward, I.M. and Wilding, M.A. (1984) Creep behaviour of ultra-high modulus polyethylene: Influence of draw ratio and polymer composition. *J. Polymer Sci., Phys. Ed.*, **22**, 561–575.

Wilding, M.A. and Ward, I.M. (1978) Tensile creep and recovery in ultra-high modulus linear polyethylenes. *Polymer*, **19**, 969–976.

Wilding, M.A. and Ward, I.M. (1981) Creep and recovery of ultra-high modulus polyethylene. *Polymer*, **22**, 870–876.

Wilding, M.A. and Ward, I.M. (1984) The creep and stress-relaxation in ultra-high modulus linear polyethylene. *J. Materials Sci.*, **19**(2), 629–636

Wilfgong, R.E. and Zimmerman, J. (1977) Strength and durability characteristics of Kevlar aramid fiber. In *Fiber Science* (Ed. M. Lewin), International Fiber Science and Technology Series (ISSN 0570-4898), Marcel Dekker Inc., New York, pp. 1–21.

Wooding, N.S. (1963) Rayon and acetate fibres. In *Fibre Structure* (Eds J.W.S. Hearle and R.H. Peters), Butterworth & Co., London, pp. 455–479.

Woods, H.J. (1955) *Physics of Fibres*, The Institute of Physics and John Wright & Sons, Bristol.

2 The whitening of textiles

J.M. CARDAMONE and W.N. MARMER

2.1 Introduction

Textiles are whitened by the bleaching process. This process is generally applied to grey goods to finish them as white or to fabrics to prepare them for dyeing to pastel shades. Bleaching is a refurbishing step for modern textiles and a conservation/restoration procedure for historic textiles. The bleaching process removes natural colorants, water-borne stains, and oil-borne soils. Throughout history, ubiquitous bleaching practices were carried out in direct sunlight because it became apparent that the sun had a catalytic effect. In the past, bleaching agents had been identified as indigenous acids, bases, and mineral salts. They were applied as soaks and sours to achieve high whiteness levels. Up until the eighteenth century it was common practice to bleach linens and cottons in the sun and woolens in the fumes of burning sulfur. These relatively primitive bleaching practices were not only cumbersome and arduously slow, but required extensive acreage. It was not until the introduction of chlorine-based materials in the late eighteenth century that bleaching became facile. In the late 1920s, hydrogen peroxide became the most prevalent bleach. Since that time, hydrogen peroxide continues to be the prominent bleaching agent for natural fibers and blends with synthetic fibers. Synthetic fibers generally require little bleaching except for size removal, where necessary. To heighten brightness, natural and synthetic fibers, yarns, and fabrics can be treated with fluorescent whitening agents (FWAs). These optical brighteners are designed for selective emission of blue light, which gives the illusion of 'whiter than white'.

Bleaching has been the source of consumer dissatisfaction and complaint because of diminished wear-life. The dissatisfaction has prompted a plethora of research reports on textile degradation from various bleaches and bleaching protocols. These reports are not only of academic interest; they have important practical implications. Most pertinent for wear-life estimation is the damage effected during the early stages of bleaching. In the early stages, permissive degradation should be kept to a minimum level when attempting to treat the colorant selectively while avoiding oxidation of the textile.

Reference to a brand or firm name does not constitute endorsement by the US Department of Agriculture over others of a similar nature not mentioned.

Comprehensive surveys on textile bleaching can be found in Marsh (1945), Peters (1967), Trotman (1968), Cegarra and Gacén (1983), Nettles (1983), Lewin and Sello (1984), and Duffield (1986). This chapter will cover developments in bleaching systems and processes from 1986 to the present and will examine the nature of the bleaching systems currently used.

2.2 Natural and synthetic fibers

Fiber processing commonly requires removal of accompanying impurities. Natural and synthetic fibers originate commonly in an impure state. Whiteness becomes masked during growth or manufacture, processing, and fabrication of fiber to yarn and fabric. In natural fibers, impurities arise from natural exudates, local vegetation, and indigenous contaminants. In synthetic fibers, discoloration results from the by-products of manufacture, which include processing auxiliaries, spinning oils in yarn-making, and formulations for yarn sizing in weaving.

It is generally recognized that the requirements for effective bleaching are specific to fiber type. A consideration of these requirements in terms of structure and reactivity of fibers and bleaching agents is relevant.

Fibers can be classified according to origin and structure. An abbreviated scheme is as follows.

Natural cellulosics
  cotton, flax, jute, ramie, sisal, coir
Regenerated cellulosic
  rayon
Synthetic modified cellulosic
  acetate
Natural polyamide
  wool, camel, cashmere, mohair, rabbit, silk
Synthetic polyamide
  nylon, aramid
Other synthetic fibers
  polyester, acrylic, polypropylene, spandex.

Much of the chemical reactivity of fibrous polymers is derived from that of the functional groups of the monomer units. Thus, all cellulosic fibers have similar chemical reactivities, though susceptibility to attack and physical and mechanical properties are determined principally by differences in fiber morphologies. Crystalline and amorphous character define morphology. The relative content of a fiber's crystalline and amorphous regions determines its properties of strength, extension, and reagent accessibility.

2.2.1 *Cellulosics*

Cellulose (Figure 2.1) reacts through primary and secondary hydroxyl groups and through the glycosidic bond of the monomer unit. Typical reactions of the alcohol groups are oxidation, esterification, and etherification. These reactions take place in a heterogeneous medium, that is, in the solid substrate, which contains absorbed water. This heterogeneity stems in part from the fibril nature of cellulose itself. In cotton, a seed-hair fiber, fibrils are organized in sheets of associated polymer chains, which stack laterally and longitudinally and then spiral across the fiber axis with periodic reversal of direction. These spiral reversals give rise to the characteristic convolutions in the fiber and are responsible for reduced strength compared to flax, a bast fiber, where fibril orientation is aligned along the fiber axis. These morphological features, along with the fact that the glycosidic bonds are not easily broken, give rise to the stability of cellulose under a broad range of conditions.

All cellulosic fibers can be damaged by oxidizing bleaches. Oxidation products are termed 'oxycelluloses', whereby alcoholic —OH groups are converted to aldehyde, ketone, and carboxyl groups (Figure 2.2; Nevell, 1987). Hydrolysis reactions can occur during bleaching with the formation of chain-shortened 'hydrocellulose' (Trotman, 1968). The presence of oxycelluloses and hydrocelluloses can cause a textile to fall apart. Hydrocellulose formation ruptures glycosidic bonds and produces one reducing (aldehyde) and one non-reducing end group. Qualitative and quantitative reagents and analyses for determining the extent of degradation are Fehling's solution, copper number, Turnbull's blue test, and methylene blue absorption (Merkel, 1984).

2.2.1.1 *Cotton.* A typical bleach for cotton has been sodium hypochlorite. It is inexpensive and can be processed at pH 9–11, where the bleach is stable and where there is minimum fiber damage at 13–21°C (55–70°F). The process involves pretreatments for singeing and desizing with sodium hydroxides or enzymes, washing, and souring at room temperature for 10–15 min with 0.5% sulfuric acid. The material is extracted with 4% sodium hydroxide, washed, soured with 0.55–1.0% sulfuric or hydrochloric acid at room temperature and then washed. Bleaching follows by saturating the material with sodium hypochlorite at pH 9.5 for 30 min at room temperature. This process is

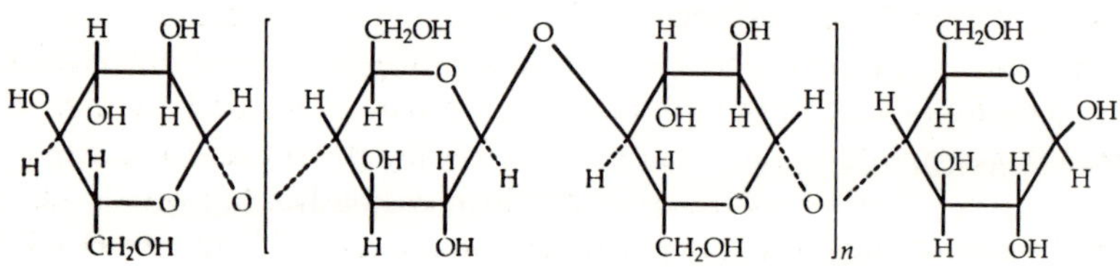

Figure 2.1 Cellulose.

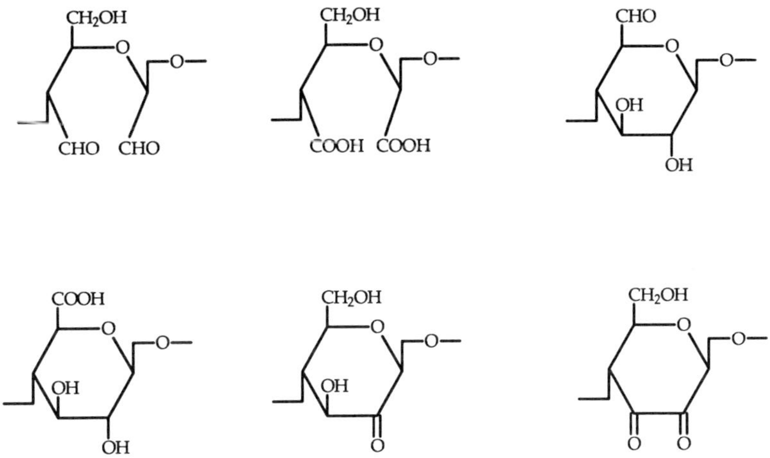

Figure 2.2 Oxycelluloses.

repeated as necessary. The material is then washed, antichlored with sulfur dioxide, and washed again at room temperature (Nettles, 1983).

Hydrogen peroxide is the most prevalent bleach. It is preferred over hypochlorite because it can be used more economically. Both hypochlorite and peroxide have been used together in one bath for scour/bleach formulations. Hydrogen peroxide serves as an antichlor and simultaneously releases oxygen for bleaching.

2.2.1.2 *Flax and lignified cellulosics.* Flax, ramie, hemp, and coir have higher crystalline content than cotton because fibril orientation is aligned along the fiber axis. These fibers are more resistant to chemical attack. The presence of lignin, although a drawback as a source of discoloration, is somewhat beneficial, because its completely amorphous character lends stress transfer when the relatively brittle fiber is placed under stress. Bleaches can be selected to remove lignin completely or to simply decolorize it. A processing consideration is the weight loss, which can be as high as 26% from hemicelluloses and lignin (Peters, 1967). The amount of weight loss is dictated by the textile's intended end-use. There are two approaches to removing color: bleaching with sodium chlorite to remove lignin or bleaching with sodium hypochlorite by acid chlorination of lignin to make it soluble for removal in an alkaline scour. Sodium chlorite is advantageous when less weight loss is required. It requires effective bleaching in 45 min at pH 3.8–4.2 at 80°C. A second bleaching with hydrogen peroxide may be required (Peters, 1967).

2.2.1.3 *Rayon.* Rayon lacks the highly ordered fibril character of cotton and flax. It is readily accessible and the effects of bleaching are more severe. In its preparation from xanthate-derivatized wood pulp, rayon is regenerated in an acid bath to cellulose before extrusion from a viscous solution as filament fibers. This process reduces the chain length of cellulose by one-third of the expected value for cotton and flax. Hence there are many more shorter chain lengths and thus a greater number of reducing ends. Rayon and acetate are extruded as filament fibers in the white state. Rayon is commonly cut to staple whenever bleaching may be required. Of the various forms of rayon, which include modal (rayon obtained by processes giving a high tenacity and a high wet modulus) and viscose, viscose is most vulnerable. It requires milder bleaching. Recommended procedures employ cold sodium hypochlorite at pH 10–11 with sodium carbonate. Peracetic acid can be used at 65°C for 1 h at neutral pH in the presence of sodium hexametaphosphate at neutral pH. Sodium chlorite can be used at 80°C for 30–45 min at pH 3.8–4.0. Other bleaching regimens for high wet modulus rayon and viscose have been reported elsewhere (Lewin and Sello, 1984).

2.2.1.4 *Acetate.* In the case of cellulose acetate, the reactivity of the acetate fiber is determined not by hydroxyl but by acetate groups. When cellulose is acetylated and extruded as filamentous acetate fiber, a different range of fiber properties results. Esterification conveys heat sensitivity, thermoplasticity, and solubility in acetone.

The conditions for bleaching cellulose acetate are influenced by its chemical reactions to acids and bases. Acetate is resistant to cold dilute acids, is decomposed by strong acids, and dissolves in acetic acids. It has good resistance to weak alkali. Strong alkali hydrolyzes the acetate groups and saponifies the fiber. Sodium hypochlorite and hydrogen peroxide can be used at pH lower than that used to bleach cotton. Safe bleaching can be carried out using paracetic acid in a mild alkali bath (Lewin and Sello, 1984).

2.2.2 *Natural polyamides*

2.2.2.1 *Wool.* Protein fibers are formed by the chemical combination of alpha-amino acids joined by peptide linkages. The amino acids can be liberated by alkaline hydrolysis, and this destroys the bonding between parallel polymer chains which are joined by the salt linkages of pendant amino acid groups.

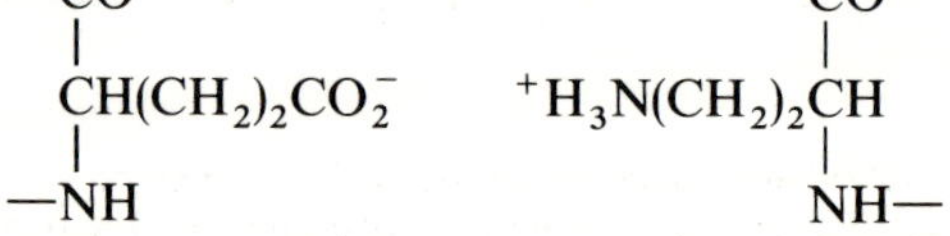

Wool is keratin protein, unique because covalent intermolecular bonding occurs through the disulfide linkages of the amino acid cystine. The salt links are a main consideration for fiber strength. They form from a weak acid and a weak base and the extent of their hydrolysis in water should increase with temperature. Fortunately, however, hydrolysis occurs at temperatures above normal processing.

Wool is not sensitive to weak acids. In fact, weak acids are used in dye applications. Moreover, dilute mineral acids are used for carbonizing wool, where the sorbed solutions do not weaken wool at the baking temperature of 120°C, at which temperature vegetable matter chars. Concentrated mineral acids do cause wool to swell and gelatinize.

Wool is sensitive to caustic sodium and potassium hydroxide, but not to sodium carbonate. Wool can be treated at the boil with ammonia without harm, but solutions of 5% sodium hydroxide cause considerable strength loss. In this case, the polymer chains are converted into sodium salts of amino acids with concomitant salt-link rupture, and cystine bonds cleave.

Attempts have been made to stabilize wool for exposure to alkali, and oxidizing and reducing agents (Speakman, 1941). Formaldehyde was used to strengthen wool by what was thought to be crosslinking of the basic amino group side-chains, represented simply as Wool—NH—CH—NH—Wool (D'yachenko and Shelpakova, 1939). In an early report, the optimum conditions for increasing the alkali-resistance of wool with formaldehyde were 0.5%, pH 6–8, 50°C, 6 h (Brown *et al.*, 1951).

Because of wool's particular bonding characteristics, the association of keratin protein takes the form of an alpha helix or coil crosslinked with dithiol linkages through cystine's —S—S— bonds at adjacent turns in the helix, as shown.

$$\begin{array}{lcr} —CO & & CO— \\ \ | & & | \\ CH—CH_2—S & —S— & CH_2—CH \\ \ | & & | \\ —NH & & NH— \end{array}$$

The particular challenge of bleaching wool involves the removal of dirt, vegetation, suint,* grease, and black hair impurities. The effects of oxidative and reductive bleaching processes on wool can be described as follows.

- Cleavage of —S—S— bonds by reducing agents, with the formation of thiol groups (equation 2.1) that can be readily reoxidized to the original form.

$$RCH_2—S—S—CH_2R' + 2[H] \rightarrow RCH_2—SH + HS—CH_2R' \qquad (2.1)$$

* Suint is a collection of sweat gland secretions which include potash salts of various fatty acids together with small quantities of sulphates, phosphate and nitrogenous materials (von Bergen, 1963).

- Cleavage at these sites with formation of thiol and cysteine sulfonate (Bunte salt) groups (equation 2.2).

$$RCH_2—S—S—CH_2R' + NaHSO_3 \rightarrow RCH_2—SH + NaO_3S—S—CH_2R' \tag{2.2}$$

- Reaction with alkali to form lanthionine crosslinks.

$$\rangle CH—CH_2—S—CH_2—CH\langle$$

- Cystine oxidation by hydrogen peroxide bleaching, resulting in cystine disulphide cleavage and complete conversion to cysteic acid (equation 2.3).

$$R—S—S—R' \rightarrow R—SO—S—R' \rightarrow R—SO_2—S—R' \rightarrow RSO_3H + R'SO_3H \tag{2.3}$$

Oxidation of cystine during bleaching diminishes wool fiber strength. Considering the inherent low tenacity of wool, this is to be avoided. During bleaching, the selective alteration of peptide bonds and/or —S—S— linkages can be controlled by the appropriate choice of bleaching system. Although oxidation intermediates are difficult to document quantitatively because they hydrolyze to cysteine and cysteic acid, they have been recorded by infrared spectrophotometry. Absorptions at 1070 and 1120 cm^{-1} have been assigned to —SO—S— and —SO_2—S— (Nettles, 1983; Merkel, 1984). These absorptions are diagnostic for oxidation damage. Wool degradation can be monitored by percentage alkali solubility, which correlates directly with percentage cysteic acid content (Duffield, 1986).

A traditional reducing bleach for wool had been sulfurous acid (SO_2) because it leaves wool relatively unaffected but removes colored matter. The white is known to lack permanency, however, so the bleach was supplemented by a hydrogen peroxide treatment.

Often, hydrogen peroxide baths are followed by reducing bleach baths for increased whiteness. Such dual bleachings are referred to as full bleaching. Arifoglu and Marmer (see references) have perfected a single bath variation of such dual bleaching. Another variation on bleaching wool includes combining peroxide bleaching with scouring, whereby wool emerges from the last scour bowl wet with dilute peroxide; slow bleaching ensues within the wool bale.

Peroxide bleaching sometimes may be carried out within the dye bath. When wool is dyed near the boil for approximately 60–90 min, yellowing can occur during the dyeing process. For many dyes that are stable to hydrogen peroxide, clear pastel shades can be achieved without prebleaching by incorporating a bleaching agent in the dye bath.

A hydrogen peroxide bleach formulation includes a phosphate or silicate stabilizer to control the rate of peroxide decomposition and subsequent rate of bleaching. Often, chelating agents are added to bind metal ions that can decompose the peroxide. For example, the system can include hydrogen peroxide, 35%, in the amount necessary to achieve a desired whiteness level, phosphate stabilizer, wetting agent, temperature 40–60°C, duration 1–8 h, and pH 7.5–8.5.

Because hydrogen peroxide in alkaline medium is present in excess to the amount of targetted chromophoric groups and susceptible keratin functionality, only 15–25% of the peroxide in the bleach bath is actually consumed during bleaching. Some mills therefore recycle the unspent peroxide.

2.2.2.2 *Silk.* Pure silk is fibroin protein. It is composed of simple amino acids and contains no disulfide (—S—S—) bonds. Without intramolecular crosslinking, silk's morphology results from the ionic bonding of amino acid groups. In contrast to wool's coil configuration, polypeptide chains of silk associate in ordered lateral sheet-arrays, which stack to build up fibril structure. These arrays convey greater strength and crystallinity to silk than does the coil configuration to wool. The permeability of silk is less than wool and there is relatively less tendency for chemical attack.

Like wool, silk has been traditionally bleached by peroxide. Because silk is not readily degradable, it can tolerate more severe conditions. Chlorine bleaching produces discoloration and is never prescribed. This discoloration has been attributed to the degradation of tyrosine residues (Earland, 1960).

When compared to wool bleaching, a typical silk bleaching system using hydrogen peroxide can be formulated for higher temperatures and alkalinity. A typical peroxide bleach bath includes alkaline hydrogen peroxide with sodium silicate and ammonia, pH 10, with tetrasodium pyrophosphate and EDTA, and processing at 40–60°C for 2–4 h.

2.2.3 *Synthetics*

Synthetic fibres can be engineered for different mechanical behaviors. They demonstrate variable crystallinities, crosslinking densities, and thermoplasticities. Trademarks within generic fibers represent a myriad of physical and mechanical properties that have been tailored to particular end uses. Generic fibers are classified by common chemical constitution and take their name from the identity of the polymer repeat unit comprising them. Some are listed below.

Nylon 6.10 (a polyamide):
$[\text{—NH(CH}_2)_6\text{NH—CO(CH}_2)_8\text{CO]}_n$
Polyester (polyethyleneterephthalate shown here):
$[\text{—OCH}_2\text{CH}_2\text{O—CO—}p\text{-C}_6\text{H}_4\text{—CO]}_n$
Polyolefin (polypropylene shown here):
$[\text{—CH(CH}_3)\text{—CH}_2]_n$

Acrylic (polyacrylonitrile shown here):

$[-CH_2-CH(CN)]_n$

Spandex (a polyurethane):

$[-O(CH_2)_4O-CO-NH(CH_2)_4NHCO]_n$

Synthetic fibers are generally manufactured with high levels of whiteness except when the conditions for fiber formation cause discoloration. In these cases, bleaching can include fluorescent whitening agents. Formulations are composed for safe temperatures and pH ranges that minimize acid and alkali degradation of the fiber. As an example, polyester fibers have been bleached with di- and tri-chloroisocyanuric acid at pH 3–9, at 60–65°C, for 30–45 min (Bel'tsov and Klaus, 1971).

2.2.3.1 *Nylon.* Heat setting temperatures for the various forms of nylon may cause yellowing from oxidation. Heat sensitivity is a special consideration for bleaching the nylon thermoplasts. Hydrogen peroxide is not used because the bleach bath processes at too high a temperature. Recommended bleaches are nonchlorinated types: sodium perborate, percarbonate, and peracetic acid. A peracetic acid bleach bath is processed with sodium pyrophosphate at pH 6–7. The material is introduced cold and the temperature is raised to 80–85°C over 30 min with subsequent bleaching at this temperature for 30 min. Where possible, lower temperatures and pHs are recommended (Trotman, 1968).

2.2.4 *Bleaching of fiber blends*

In 50/50 blends of natural and synthetic fibers, bleaching procedures are based on the sensitivity of the natural fiber component, provided that temperature and pH are not damaging to the synthetic fiber. When the natural fiber component represents a small percentage of the composition, it may withstand harsher conditions than would normally be prescribed for bleaching it alone. When a textile contains a fiber blend with fibers of equal sensitivity but unequal percentages, bleaching formulations are tailored to the safe requirements of the fiber present in the larger amount. For a review of bleaching methods reported on cotton/rayon, polyester/cotton, wool/cotton, wool/polyester and wool/acrylic, the reader is referred to Lewin and Sello (1984).

2.3 Bleaching agents

Natural and synthetic fibers are the essential elements of textile fabrics. The chemical, physical, and mechanical properties of fibers are directly related to yarn and to fabric behavior. When compared to fibers, yarn and fabric forms are less vulnerable substrates and after-effects of bleaching may be expected to be less pronounced.

Table 2.1 Categories of bleaching agents

Chlorine-containing oxidants	
Sodium hypochlorite	NaOCl
Sodium chlorite	$NaClO_2$
Sodium dichloroisocyanurate	An *N*,*N'*-dichloro compound
Inorganic peroxygen bleaches	
Hydrogen peroxide	H_2O_2
Ozone	O_3
Sulfur dioxide	SO_2
Sodium perborate	$NaBO_2 \cdot H_2O_2 \cdot 3H_2O$
Potassium permanganate	$KMnO_4$
Sodium percarbonate	$Na_2CO_3 \cdot 3H_2O_2$
Sodium bromate	$NaBrO_3$
Sodium bromite	$NaBrO_2$
Potassium peroxydiphosphate	$K_4P_2O_8$
Ammonium perdisulfate	$(NH_4)_2S_2O_8$
Ammonium hydrogen permonosulfate	NH_4HSO_5
Organic peroxygen bleaches	
Peracetic acid	$CH_3CO{-}OOH$
Peroxydodecanedioic acid	$HOO{-}CO{-}(CH_2)_{10}{-}COOH$
Tetra-acetylethylenediamine/H_2O_2	Generates $CH_3CO{-}OOH$
Reducing bleaches	
Sulfur dioxide	SO_2
Sulfurous acid	H_2SO_3
Sodium bisulfite	$NaHSO_3$
Sodium sulfite	Na_2SO_3
Sodium hydrosulfite (sulfoxylate, dithionite)	NaO_2SSO_2Na
Sodium formaldehyde sulfoxylate (sodium hydroxymethanesulfinate)	$HOCH_2SO_2Na$
Sodium borohydride	$NaBH_4$
Thiourea dioxide (formamidinesulfinic acid)	$H_2NC({=}NH)SO_2H$
Sodium sulfinate	HSO_2Na
Trisodium trithiocyanurate/H_2O_2	Generates HSO_2Na
Photosensitizing bleaches	
Fluorescent whitening agents	

Formulations for the major bleaches listed in Table 2.1 have traditionally been tailored for selective bleaching of coloring matter without diminishing a textile's structural integrity.

Bleach formulations usually contain buffering and sequestering agents such as silicates, phosphates, or oxalates. Some formulations call for metal salts, sodium hydroxide, ammonia, or sodium nitrate. Other additives to the bleach bath are surfactants, antideposition agents and optical brighteners.

The most common bleaching is oxidative in nature, and hydrogen peroxide is the most widely used agent for stain bleaching of proteinaceous fibers, and in the past, chlorinated bleaches for cellulosics. Reductive bleaching may be used alone as a substitute for oxidative bleaching, but its use is mainly confined to after-bleaching following peroxide treatment. The most widely used reductive bleach, though not the best, is sodium hydrosulfite (NaO_2SSO_2Na; also known as dithionite, hyposulfite, and sulfoxylate).

More favored as reductive bleaches are derivatives of sulfinic acid (H_2SO_2), all of which are thought to act by decomposition to a fleeting but reactive species, free sulfinate ion. The most common such derivative is sodium hydroxymethanesulfinate ($HOCH_2SO_2Na$; sodium formaldehyde sulfoxylate), but a newcomer now gaining favor is thiourea dioxide ($H_2NC({=}NH){-}SO_2H$; formamidinesulfinic acid).

The most effective bleaching regimens for stain incorporate both an oxidative and a reductive bleaching step and are referred to as 'full' bleaching. Such full bleaching is normally a two- or three-step process carried out in separate baths. Industry, however, typically avoids full bleaching, and relies on oxidative bleaching alone; the gains from extra whiteness, it concludes, do not offset the added expenses of preparing and heating an additional bleach bath.

2.3.1 *Chlorine-containing oxidants*

2.3.1.1 *Sodium hypochlorite.* Sodium hypochlorite ($NaOCl$) is a powerful oxidizing agent, the reaction product of chlorine gas and sodium hydroxide. $NaOCl$ is stable above pH 10. The rate of bleaching may be increased by adding acid in moderate amounts (to pH 5.0–8.5) to form hypochlorous acid ($HOCl$). Chlorine liberation occurs below pH 5.0. The oxidation of cellulose is greatest within the range pH 5–9.

2.3.1.2 *Sodium chlorite.* Sodium chlorite ($NaClO_2$) is a fairly stable solid source of chlorine dioxide (ClO_2) for bleaching. $NaClO_2$ can be purchased as approximately 81% solids and alkali stabilizers. The formation of ClO_2 occurs only on its dissolution in acidic medium (equations 2.4 and 2.5). ClO_2 is unstable and acts as an oxidative bleach (equation 2.6).

$$NaClO_2 + H^+ \rightarrow HClO_2 + Na^+ \tag{2.4}$$

$$5HClO_2 \rightarrow 4ClO_2 + HCl + 2H_2O \tag{2.5}$$

$$ClO_2 + 4H^+ + 5e^- \rightarrow Cl^- + 2H_2O \tag{2.6}$$

The formation of toxic and corrosive ClO_2 is a major industry concern. Bleaching should be carried out within the range pH 3.5–4.0 with buffer systems such as sodium dihydrogen phosphate. During cotton bleaching, when ClO_2 solutions were used at pH 1.76–6.05 at 20°C, oxycelluloses formed and were detected by increase in fluidity, copper number, and carbonyl content (Peters, 1967).

2.3.1.3 *Sodium dichloroisocyanurate.* Isocyanuric chlorine bleaches are the principal solid chlorine bleaching agents. Most common is sodium dichloroisocyanurate dihydrate, with 56% available chlorine (Farr *et al.*, 1992).

2.3.2 *Inorganic peroxygen bleaches*

2.3.2.1 *Hydrogen peroxide.* Hydrogen peroxide bleaching proceeds by dissociation into H^+ and perhydroxyl ions (HOO^-). This reaction is favored by alkaline conditions. At pH 11.5, the peroxide rapidly breaks down and molecular oxygen forms. Cotton would become severely damaged under these conditions. Activators and stabilizers are added to the hydrogen peroxide bath to control bleaching. Sodium hydroxide and sodium carbonate are the alkalis for cellulosics, and ammonia and tetrasodium pyrophosphate are used for wool. Sequestering agents are used to incapacitate heavy metal ions in the bleaching solution. These ions foster the decomposition of peroxide into radical species that attack and damage the fiber.

Attention is now being given to use of enzymes to assist peroxide bleaching. In wool, proteolysis and lipolysis open channels for better penetration of peroxide. Specific enzymes are selected for their stability in an alkaline peroxide medium (Fornelli, 1994).

Ironically, the same radical process is being used to advantage to bleach pigmentation in animal fibers. Bleaching for pigmentation involves the selective decomposition of melanin pigment granules in the interior of the fiber. Industry uses one variation or other of the process first reported by Laxer and Whewell (1955) and then perfected by Bereck (1985), which involves the selective mordanting of pigmented fibers by ferrous sulfate. The fiber or fabric is first treated with $FeSO_4$ in the presence of a reducing agent (e.g. phosphorous acid, H_3PO_3, and hypophosphorous acid, H_3PO_2) to prevent oxidation of Fe^{2+} to Fe^{3+}. Following rinsing (sometimes in the presence of a chelating agent such as ethylenenediaminetetraacetic acid, EDTA, or nitrilotriacetic acid, NTA) to remove the bulk of the $FeSO_4$, small amounts of Fe^{2+} remain adsorbed specifically to the melanin granules. Subsequent treatment of the fibers with H_2O_2 causes decomposition of the melanin pigment. The localized reaction on melanin is a free-radical one, superimposed on the non-radical and non-selective interaction of H_2O_2 with stain components. The bleaching of pigmented animal fibers was reviewed by Knott (1990).

2.3.2.2 *Sodium perborate.* Sodium perborate is a white crystalline powder that dissolves in water at room temperature to form hydrogen peroxide. In practice, it is used as a 1% solution at pH 10 (equation 2.7):

$$4[NaBO_2 \cdot H_2O_2] + 3H_2O \rightarrow NaB_4O_7 + H_2O_2 + 2NaOH \qquad (2.7)$$

2.3.2.3 *Potassium permanganate.* Potassium permanganate is unstable and decomposes in the presence of oxidizable substances; its manganese is converted into manganese dioxide, a brown solid. However, under controlled acidic conditions, soluble manganous salts form instead, and effective bleaching occurs. A formulation for permanganate bleaching to remove color from wool in acrylic or polyester blends was reported by Millson and von Bergen (1970).

2.3.2.4 *Sodium percarbonate.* Sodium percarbonate is a white powder with approximately 13% available oxygen. A 1% solution has a pH of 10.5. It decomposes above 20°C to form sodium carbonate and hydrogen peroxide.

2.3.2.5 *Sodium bromite.* Sodium bromite is used in cotton desizing as an oxidizing agent. Optimum bleaching takes place at pH 9.8–10.2 at 30°C. Without careful control of these conditions, depolymerization by glycosidic bond cleavage can occur.

2.3.2.6 *Others.* Sodium salts of peroxygens including diphosphate, disulfate and monosulfate are used as desizing compounds in an alkaline medium. Often they are used with hydrogen peroxide in a continuous bleaching process of desizing, scouring, and bleaching (Lewin and Sello, 1984).

2.3.3 *Organic peroxygen bleaches*

2.3.3.1 *Peracetic acid.* Peracetic acid is the acetyl derivative of hydrogen peroxide. It is available as a colorless 36–40% solution in dilute acetic acid. The stability of peracetic acid is less than that of hydrogen peroxide. Radical scission will result in hydroxyl radicals ($HO^{\bullet}$), and when ions such as iron are present, the effect is magnified (Lewin and Sello, 1984).

2.3.3.2 *Peroxydodecanedioic acid.* Unlike such hydrophilic peracids as peracetic acid, which are very soluble in water, function in the bulk phase for bleaching, and are useful at high temperatures, peroxydodecanedioic acid, a hydrotropic peracid, is oily by nature and less soluble. It has a well-defined critical micelle concentration, which suggests it operates from within the detergent micelle and as such is highly effective in low-temperature bleachings. It is important because of its stability as a solid organic peracid.

2.3.3.3 *Tetra-acetylethylenediamine* (*TAED*). TAED has been used in over half of European detergents as a hydrogen peroxide activator to meet the demands of the European market for low-temperature processing. It activates peroxide by forming peracetic acid at the low temperatures where peroxide would normally be an ineffective bleaching agent (equation 2.8).

$$(CH_3CO)_2NCH_2CH_2N(COCH_3)_2 + 2H_2O_2 \rightarrow 2CH_3CO\text{—}OOH + CH_3CONHCH_2CH_2NHCOCH_3 \quad (2.8)$$

This bleach was introduced into normal detergent formulations in 1990 (Farr *et al.*, 1992).

2.3.4 *Reducing bleaches*

2.3.4.1 *Sulfur dioxide, sulfites, and bisulfites.* Sulfur dioxide dissolves in water to form sulfurous acid, which on increasing alkalinity exists as bisulfite or sulfite. These are reducing agents, as illustrated by the half-reaction for sulfite (equation 2.9).

$$SO_3^{2-} + 2OH^- \rightarrow SO_4^{2-} + H_2O + e^- \quad (2.9)$$

A bleaching formulation for wool can be found in Millson and von Bergen (1970). These agents are inexpensive, but can impart fiber damage as well as bleaching activity.

2.3.4.2 *Hydrosulfite.* Hydrosulfite (sulfoxylate, dithionite) is used as an industrial bleach for bleaching pulp and paper. It has been used on wool for less damage than bleaching with hydrogen peroxide (Millson and von Bergen, 1970). At low pH or high temperatures, aqueous solutions are unstable and bisulfite is formed (equation 2.10).

$$2S_2O_4^{2-} + H_2O \rightarrow 2HSO_3^- + S_2O_3^{2-} \quad (2.10)$$

Following bleaching and rinsing, the fiber is treated with a little H_2O_2 to stem residual sulfurous odors. Full bleaching using sequential oxidation by hydrogen peroxide followed by reductive bleaching by hydrosulfite is discussed by Duffield (1986).

2.3.4.3 *Sodium formaldehyde sulfoxylate.* Sodium formaldehyde sulfoxylate (sodium hydroxymethanesulfinate) is formed from the reaction of hydrosulfite and formaldehyde. It can be processed at pH 3.2–3.5, is less reactive than hydrosulfite and is more stable to heat (Farr *et al.*, 1992).

2.3.4.4 *Sodium borohydride.* Sodium borohydride is a selective reducing agent for aldehyde and ketone groups. It is used for cellulose bleaching where oxycellulose formation and concomitant dehydration have produced yellowness. The reduction of aldehydic and ketone groups to alcohols reverses yellowness. Combination treatments with borohydride and ozone have been used to bleach and brighten cellulosic fibers (Wade, 1967).

2.3.4.5 *Thiourea dioxide.* Thiourea dioxide is an effective bleach when used alone or when used after hydrogen peroxide in a full bleaching process (Duffield, 1986; Cegarra *et al.*, 1988). Bleaching with thiourea dioxide is not common practice but it is effective when used alone, and the process compares favorably with hydrogen peroxide bleaching. A formulation can include a commercial thiourea dioxide product, wetting agent and EDTA sequestering agent. Reductive bleaching is carried out at pH 7.0 at 70°C for 60 min (Duffield, 1986).

Full bleaching, oxidative bleaching (e.g. with hydrogen peroxide) followed by reductive bleaching (e.g. with thiourea dioxide), can lead to superior whiteness and improved light fastness. Generally two-step full-bleaching processes incorporate rapid bleaching methods to maintain nearly similar exposure times to those of single-stage bleaching.

A series of recent reports (Arifoglu and Marmer, 1990a,b, 1992a; Arifoglu *et al.*, 1990, 1992; Cardamone *et al.*, 1992; Marmer *et al.*, 1993, 1994; Cardamone and Marmer, 1994), patents and patent applications (Arifoglu and Marmer, 1990c, 1991, 1992b,c,d, 1993a,b) disclose new approaches to full bleaching that integrate hydrogen peroxide bleaching and thiourea dioxide after-bleaching into a single-bath procedure. Residual hydrogen peroxide from the initial oxidative bleaching is utilized in a quick chemical reaction that converts the bath into a reductive medium. Thus, thiourea is added to the peroxide bath under controlled conditions, producing thiourea dioxide (equation 2.11). Thiourea dioxide is not a truly reductive agent; its hydrolysis produces sulfinate ion, the active species (equation 2.12), and reductive bleaching ensues (equation 2.13).

$$H_2NC(=NH){-}SH + 2H_2O_2 \text{ (pH 4.5–5.5)} \rightarrow H_2NC(=NH){-}SO_2H + 2H_2O \quad (2.11)$$

$$H_2NC(=NH){-}SO_2H + 2OH^- \text{ (pH 7–8)} \rightarrow H_2NC(=NH){-}OH \text{ (urea)} + SO_2^- \quad (2.12)$$

$$SO_2^- + 4OH^- \rightarrow SO_4^{2-} + 2H_2O + 3e^- \quad (2.13)$$

The proposed mechanism was supported by ^{13}C NMR spectroscopy (Arifoglu *et al.*, 1992) and the monitoring of redox potential.

The above studies confirmed the advantages of full bleaching. Whiteness was vastly superior to that from peroxide bleaching. Mechanical testing showed that damage to the fabric (a wool challis) was minimal, and the fabric handle, as assessed by using the Kawabata Evaluation System (KES-F; Kawabata, 1980), was softer, more flexible, and smoother than seen from conventional bleaching.

Full bleaching could be tied into the Bereck process for bleaching of natural pigmentation. A particular advantage is the elimination of traces of any orange-colored Fe^{3+} species (from the mordanting) by reduction to soluble

Fe^{2+} (Arifoglu and Marmer, 1990a,b, 1991, 1992b,d, 1993a) during the reduction step of full bleaching.

2.3.5 *Photosensitizing bleaches*

The photo-yellowing of wool is wavelength-dependent and is influenced by the wool's previous treatment and condition. Wool can be photobleached in air in the presence of water when exposed to blue light (380–475 nm). Reducing bleaches accelerate this effect (International Wool Secretariat, 1970). Ultraviolet light (less than 380 nm) causes yellowing, especially if the wool has been hydrogen peroxide bleached, and this effect is enhanced if wool is wet. It has been reported that hydrosulfite reductive bleaching inhibits the yellowing of peroxide-bleached fabrics (International Wool Secretariat, 1970).

The spectral distribution of sunlight varies by season and by time of day within a geographical location. The effect of sunlight, therefore, is not so predictable. Supposedly, the ratio of the exposures to 300 nm ultraviolet and to 400–475 nm blue light can determine whether the exposure leads to bleaching or yellowing.

Launer (1965) examined the broad band spectrum, 331–398 nm, to simulate the effects of blue light, ultraviolet, and sunlight on wool. The work was an extension of a study in 1949 in which he and co-workers found that the yellowing and bleaching of cellulose was wavelength dependent. They reported that wool could be effectively bleached at wavelengths above 398 nm, yellowed at less than 331 nm, and simultaneously bleached and yellowed when exposed to the wavelength range 331–398 nm. Thus, wool can be made to yellow for a given time period and subsequently bleached for an ensuing period of time. Apparently, the intermediate simultaneous yellowing and bleaching proceed by different pathways with consecutive reactions in the wool molecule. Maclaren (1963) found that one such yellowing reaction involved the destruction of cystine.

These phenomena imply that the yellowing effect of light, which is a common occurrence in sunlight exposure, can mask the bleaching effect. It is most desirable to select light of appropriate wavelength for bleaching — high enough in energy to whiten but too low to break down the fibrous components to yellow substances. It should be noted that cotton, too, has been bleached by near-ultraviolet light and by visible light (Launer, 1968); the deleterious effects of accompanying heat were noted.

To address the problem of lack of stability of the whiteness obtained by hydrogen peroxide bleaching when woolen fabrics are subjected to high heat in the dye bath, Simpson (1992a) investigated the application of photobleaching wool with blue light. Exposure of wool to blue light at 470 nm imparts 60.8 kcal/einstein of energy, which is lower than the energy required for covalent bond formation in wool. Thus, blue light in photobleaching is thought to react selectively with chromophores without attacking wool.

Qualifying concepts, however, include the possibility that the presence of catalysts could lower the activation energy barrier and that the presence of 'inherent structural strain' could cause some damage to —S—S— bonds (Simpson, 1992a).

Photobleaching is viewed as color-specific, that is, only colored moieties absorb the blue light until they are destroyed. Hydrogen peroxide bleaching has no such specificity, and it can damage the fiber while bleaching it.

Reports in Wool Science Review 39 (International Wool Secretariat, 1970) indicate that the results of photobleaching can be improved when carried out in the presence of reducing bleaches, such as sodium hydrosulfite or thiourea dioxide. More recently, Simpson (1992a) showed similar photobleaching in the presence of hydrogen peroxide. He applied blue light (420 nm) to bleach peroxide-impregnated wool fabrics and found that under this exposure, pH was the most important variable affecting the rate of peroxide bleaching. A pH range of 2–12 was chosen for 20-min bleachings in 2.5% (v/v) peroxide at 30°C. At pH 11, whiteness was optimum, and at such low temperatures and short exposure times, destruction of wool in this highly alkaline medium was within acceptable limits.

When compared to conventional thiourea dioxide bleaching, the most effective in terms of brightness was peroxide photobleaching. Relatively similar results were seen with reduction photobleaching, which gave results just slightly lower in brightness than thiourea dioxide-bleached samples not subjected to photoexposure. Overall optimum conditions for achieving brightness by oxidative photobleaching were established as 2.5% v/v peroxide, pH 10.5–11, 20–30 min.

No loss in tensile strength was seen on wool fabrics treated either by conventional bleaching with alkaline hydrogen peroxide or by high-pH oxidative photobleaching. All bleached samples were within the acceptable limits for alkali solubility (24%). This value was approximately the same for conventional peroxide and photobleached (18.9% and 17.9%, respectively), and this represents an increase of 8–9% over the untreated fabrics. Weight loss for both bleaching systems was within acceptable limits. Blue light may have a catalytic bleaching effect, which allows the circumvention of the high temperatures and long exposure times generally associated with conventional alkaline peroxide bleaching (Simpson, 1992a).

2.3.6 *Fluorescent whitening agents*

'Blueing agents' had traditionally been used on cotton. The phenomenon of blueing involves the ability of the blueing agent to absorb yellow light, thereby selectively emitting blue light. When yellow chromophores are present, they are selectively absorbing blue light. When the blueing agent is added to a yellowed textile, the emitted blue light combines visually with the yellow light. The textile will be perceived as gray–white from the additive light mixing of the complementary colors, blue and yellow.

An extension of this phenomenon is seen with fluorescent whitening agents (FWAs). These optical brighteners absorb invisible ultraviolet light and emit blue to violet light. For an effective FWA, there must be affinity of the agent for the textile substrate. The FWA must be wash-fast, except if it is a component of a laundry detergent and can thus be replenished with every laundering.

Use of FWAs, however, has a major drawback. Their presence on fabrics induces yellowing reactions by light. Fabrics that initially appear bright white may in time revert to a severely yellowed condition. This, of course, is most critical to white fabrics and fabrics dyed a pastel color.

Wool that has been brightened to FWAs tends to photo-yellow, and the effect is accelerated in sunlight. By the fact that FWAs absorb in the ultraviolet spectral region, high energy radiation can lead to degradation unless, in the design of the agent, this energy can be dissipated without causing damage. Thiourea and thiourea–formaldehyde, as well as other reducing agents, have been found to offer some protection (International Wool Secretariat, 1970). The most effective FWAs are those that are applied after peroxide bleaching but the most efficient are those that have no negative effect on the light-fastness of the dyed textile.

Many studies have addressed the design of FWAs for whitening wool without sensitizing it. Typically, a FWA is included in the last quarter of a reducing bleach bath, which follows peroxide bleaching. After the reducing bleach is consumed, acidifying the bath ensures that the last traces of brightening agent will be exhausted. Caution must be exercised, however, because pH change can cause a hue shift in some dye systems and this might be difficult to correct with after-treatment.

FWAs are used on cotton with mechanically-held finishes and with those finishes to which cotton has crosslinked. The applications of FWAs are versatile; for example, when they are added directly before or after printing, the color is brightened.

FWAs have the same specificity for fibers that dyes have. The individual application processes are similar to dyeing except that much less FWA is used than dye. A broad range of FWAs has been invented to cover the broad spectrum of substantivities found in untreated and pretreated textiles and their blends as the result of various wet-processing conditions. Other prevailing prerequisites are bath-compatibility of the FWA and penetration into the fiber.

The anionic FWAs which are substantive to cellulose and rayon are derived from stilbene. Their planar structures are similar to the planar structures of direct dyes so that the same mode of attachment applies. The incorporation of sulfonate groups, as in direct dye molecules, provides the necessary water solubility for wet-processing. Synthetic fibers can be treated with FWAs at the melt–spin stage of filament fiber development. Cellulose acetate can be brightened with pyrazole-derived agents.

Hydrophobic polyester requires a hydrophobic FWA. Usually naphthalimides are applied as aqueous dispersions. Like disperse dyes, such FWAs can require high temperatures and carriers. Optical brighteners are applied to nylon in the manner of exhaust dyeing with acid dyes. Tailored FWAs have been developed for the various nylons. For example, some nylons can be dyed with water-insoluble disperse dyes, which suggests that insoluble FWAs can be supplied as suspensions. Acrylic textiles have affinity for cationic dyes and are thus substantive to cationic optical brightening agents of the amino or substituted amine group variety.

A comprehensive review of FWAs for textiles can be found in Sarkar (1971) and Levene and Lewin (1984).

A survey of some of the textile bleaching literature since 1986 is contained within the appendix to this chapter. Included are descriptions and evaluations of the various bleaching processes along with information on the after-effects of treatments.

2.4 Energy and the environment

Since about 1990, the textile industry in the USA and Western Europe has stepped up support for process design modifications and relevant basic and applied research designed to improve bleaching systems. These are ongoing movements to bring current wet-processing methods into compliance with those regulations governing effluents and occupational health and safety. These movements are intensifying with the ever-increasing demand by consumers for energy efficient and environmentally friendly products.

Studies on wet textile processing, including bleaching, have been directed toward conserving water, energy, and chemicals while limiting wastewater and wastewater contamination. The many reports that are summarized in the appendix show new proposed protocols for bleaching systems and processes that modify the traditional methods for increased efficiency and enhancement of textile properties with little or no loss of fiber integrity.

A current trend is to combine several process stages into a single automatically controlled continuous line with high production speed. Increased speed may require greater bleach concentration or more powerful bleaches, more effective catalysts, tailored stabilizers and activators, and higher temperatures. Some reports show the use of solar radiation and photolysis for circumventing the problem of fiber damage, which is affected by heat application (Gulrajani *et al.*, 1990; Simpson, 1990, 1992a,b; Collins and Davidson, 1993).

Various reports of investigations to modify traditional bleaching systems such as hypochlorite (HOCl, NaOCl), peroxide (H_2O_2), and chlorite ($NaClO_2$) are described in the appendix. For example, the use of hypochlorite prevails because of its relatively low cost, processing ease, and high efficiency for whiteness achieved at fairly low temperatures. Drawbacks include the formation of toxic chlorinated organic byproducts and the tendering of cellulosics, so modifications of the system have been examined (Nevell and Singh, 1986; Shenai and Narkhede, 1987).

An alternative to hypochlorite bleaching is chlorite. It is used on rayon, modal, acetate, polyester, acrylic, and nylon. It is preferred over peroxide for bleaching cotton because, unlike peroxide, chlorite is not affected by heavy metal ions. Strict control is essential to maintain the narrow useful pH range, 3.5–4.0, and this is difficult because liberated $HClO_2$ can lower the pH to the level where cotton is tendered. Advantages for the use of chlorite over hypochlorite include significant reduction in the organic halogen compound content of wastewater streams. This problem, however, is not eliminated entirely with chlorite bleaching. An added disadvantage is the production and volatility of chlorine dioxide (ClO_2), formed in solution, which contributes to corrosion of equipment and to biological irritation.

Newer systems have been designed for more energy efficient and safer usage of chlorite bleach, for example, the combined scouring/bleaching with sequential chlorite–hypochlorite–peroxide (Goswami and Mukherjee, 1993). Another approach is single-stage scouring and bleaching by padding with chlorite activated by TEA at 60°C (e.g. Gulrajani and Venkatraj, 1986; Gulrajani *et al.*, 1987; Petrick, 1988).

Generally, hydrogen peroxide is not considered a water pollutant. In wool bleaching, however, phosphate stabilizers for the peroxide do present pollution problems. There is sustained interest in reducing phosphate discharges by recycling the bleach bath. Other studies are related to phosphate replacements, and various pH buffers and amides to activate peroxide and reduce bleaching time (e.g. Sukumar and Gulrajani, 1985; Das *et al.*, 1986). In the case of cotton, another application of enzymes is seen. Enzymatic desizing is followed by hypochlorite padding and storing before actual bleaching using persulfate-activated hydrogen peroxide (Sarma *et al.*, 1989).

Peracids are stronger oxidizing agents than peroxide and consequently they are more effective in bleaching at lower temperatures. For example, peracetic acid, formed *in situ* from peroxide and acetic anhydride, has an adequate bleaching activity at 60°C. By incorporating 2,2′-bipyridine with peracetic acid, bleaching can be made effective as low as 30°C (Rucker, 1989).

Peroxide systems incorporating the enzyme peroxidase have the advantage of low peroxide concentration without high pH (Pedersen and Reesens, 1992).

Reductive bleaching is an alternative to peroxide/phosphate bleaching for wool. Although thiourea dioxide, hydrosulfite, and bisulfite bleach to an inferior whiteness level when compared to conventional peroxide, this can be compensated for by adding a surfactant. Surfactants, furthermore, offer the added benefit of protecting wool's disulfide crosslinks (Gacén *et al.*, 1989, 1992, 1993).

The design of single-bath full bleaching, incorporating sequential oxidative (peroxide)/reductive (thiourea dioxide) steps, saves time, reagents, water and equipment when operated over the same time-frame as oxidative peroxide bleaching alone. Process parameters were chosen to reach comparable and superior whiteness to the conventional method (Arifoglu *et al.*, 1990; Cardamone *et al.*, 1992).

The following survey includes other process designs for bleaching systems which address energy and environmental concerns.

Survey of bleaching protocols

Bleaching system
Textile substrate
Bleaching protocol
Remarks and reference

HOCl, catalysts: Mn(II), Co(II), Cu(I) hydroxides

Cotton staple, combed sliver

- Pretreatment of cotton by precipitation with metal salts
- Oxidative bleaching
- 30°C
- pH 4–11

Oxidation rate curves of exposure time versus cellulose oxygen consumption from hypochlorite decomposition showed a maximum rate for bleaching in the presence of manganese and cobalt at pH 7 and in the presence of copper at pH 7.5. The catalysis effect increased in the order: Co ≫ Cu > Mn. Faster reactions produced more aldehyde groups than other oxycellulose forms. Aldehyde groups formed with copper oxidize more readily with subsequent hypochlorite bleaching (Nevell and Singh, 1986).

HOCl, cobalt sulfide

Cotton yarn

- Pretreatment of cotton by impregnation with cobalt chloride and sodium sulfide
- Oxidative bleaching at 30°C at pH 3–10

Cobalt sulfide accelerated hypochlorite oxidation of cellulose with maximum rate at pH 6.1–7.5. Maximum extent of oxidative degradation occurred at pH 7 where oxygen consumption, copper number, carboxyl value and alkali solubility (%) were greatest. Degradation increased with oxidant concentration. Oxycellulose groups were characterized by treatments with chlorous acid and sodium borohydride (Shenai and Narkhede, 1987).

NaOCl, soda ash, ammonium persulfate, sodium hyposulfite, sodium metasilicate, sodium percarbonate, hydrogen peroxide

Dyed fabric, especially blue jeans

- Two-step
- Continuous or batch processes
- Paste-form application for decolorization
- Acidic whitening
- Oxidative bleaching
- Whitening agent

Decolorization was applied to specific areas to introduce a white color area. The four-step chemical treatment process involved paste application, washing with an acidic whitening agent, oxidation bleaching, and washing in a bath containing acidic whitening agent. For sharp definition of pattern, process parameters for each step must be strictly regulated. The entire process was reported as requiring 20 min (Lin, 1989).

Bleaching system
Textile substrate
Bleaching protocol
Remarks and reference

$HOCl$, H_2O_2
Cotton hosiery knit
- $HOCl/H_2O_2$ separate bleaching after caustic boil, or
- H_2O_2 boil after caustic boil
- HOCl followed by cold H_2O_2 after caustic boil

Caustic boil followed by single-stage peroxide bleaching or hypochlorite bleaching followed by peroxide was recommended for required whiteness. Hypochlorite bleaching alone was not recommended to produce white hosiery (Tendulkar, 1993).

NaOCl, $NaClO_2$, H_2O_2
Jute fabric, grey and scoured
- NaOCl, 2 h, 30°C, 5–15 g/l, pH 11
- $NaClO_2$, 2 h, 95°C, 5–30% owf, pH 4.2
- H_2O_2, 30 min, 80°C, 0.5–4.0 g/l, pH 11

Bleaching jute fabric with NaOCl produced poor whiteness but this effect was reduced if scouring preceded bleaching. Scouring before NaOCl bleaching increased whiteness index by 6 units (from 40 to 46). With H_2O_2 bleaching, a value of 42.98 was obtained for whiteness index, and scouring before bleaching reduced this value. H_2O_2 bleaching was recommended as most satisfactory (Pandey et al., *1993).*

NaOCl, H_2O_2, $NaClO_2$
Linen, grey fabric
- Bleaching sequence: NaOCl, H_2O_2, $NaClO_2$
- Modified sequence: $NaClO_2$, NaOCl, H_2O_2
- pH 4.5–11, 30–90°C, 60–180 min

Scouring/bleaching was more effective than grey-state bleaching. Fabrics scoured with 10 g/l sodium carbonate and 2 g/l surfactant, 30 min, 95°C, followed by bleaching with the combination NaOCl—H_2O_2—$NaClO_2$ showed 29.4% less weight loss and required 90 min less time. Whiteness index of these samples was comparable to more severe scour/bleach conditions. Fluidity was less and light fastness was the same. The modified combination $NaClO_2$—NaOCl—no rinse—H_2O_2 was cost-efficient and produced higher, more uniform whiteness (Goswami and Mukherjee, 1993).

Bleaching system
Textile substrate
Bleaching protocol
Remarks and reference

$NaClO_2$ activated by triethanolamine (TEA)
Cotton cambric fabrics

- Single stage scouring/bleaching by padding
- pH 4.6
- 20–60°C
- 1–9 h
- Vary concentrations of scouring agent (emulsified perchloroethylene), $NaClO_2$, TEA

Optimum treatment conditions for good whiteness, absorbent fabrics, and minimum strength loss were: scouring agent 4.5% owf, 60°C, 5.5–6.5 h. At 60°C, less ClO_2 is liberated (Gulrajani and Venkatraj, 1986).

$NaClO_2$, TEA, HC, (triethanolamine hydrochloride-activated sodium chlorite), scouring agent (emulsifier, perchloroethylene solvent, pine oil wetting agent)
Polyester/cotton fabrics 67/33

- Pad, store
- Combined scour and bleach
- 40–60°C
- pH 4.6
- 1–9 h
- $NaClO_2$, 0.0–0.8% owf

Canonical analysis from second-order polynomial equations showed that whiteness index was affected by concentration of bleaching agent, treatment time, and temperature. Bleaching agent concentration had the maximum effect on whiteness index. Sodium chlorite had a marginal effect on fabric strength. Optimal bleaching conditions: $NaClO_2/TEA \cdot HCl$, 0.8/0.4% owf, treatment time 7–8 h, temperature 55°C, starting pH 4.6. Under these conditions evolution of chlorine dioxide fumes was not observed (Gulrajani et al.*, 1987).*

$NaClO_2$, formic acid, disodium phosphate, bleaching range to process 50 tons per day
Cotton knit, cotton tubular knits

- Select concentration temperature, pH to keep ClO_2 at a minimal level
- Continuous bleaching range
- Fabric in rope form
- 85°C
- 4–5 h
- Acidic

Sodium chlorite bleaching caused the least damage to cotton fiber: less weight loss, less grease, less wax from cotton in the effluent, and a fuller and softer hand. This system is recommended for bleaching fibers that are sensitive to alkali (rayon, acetate, silk) (Petrick, 1988).

Bleaching system
Textile substrate
Bleaching protocol
Remarks and reference

$NaClO_2$, formic acid activator, non-ionic emulsifier, perchloroethylene
Polyester/cotton (67/33)

- Combine scour, bleach
- Padding
- Continuous high pressure steaming combined with foam
- $NaClO_2$, 0.2–1.8% owf
- 90–130°C

Whiteness increased with concentration of bleaching agent. This effect was greater than the effect of temperature and processing time. Temperature marginally affected whiteness. Optimal conditions: scouring agent concentration 2.0% owf; bleaching agent concentration 1.8% owf; temperature 120°C; processing speed, 68 s inside pressure chamber. Whiteness achieved was 79.88 units which was comparable to the conventional three-stage sequence: desize, scour, bleach. Wettability after treatment was satisfactory (Venkatraj and Gulrajani, 1989).

$NaClO_2$, formic acid, solar oven, scouring agent
Polyester/cotton suiting (67/33)

- Pad-batch
- Combined scour, bleach
- $NaClO_2$ concentration 0.25–1.5% owf
- pH 4.0
- 65–94°C
- 30–180 min

A solar oven was designed as a circular trough solar concentrator with a cylindrical absorber, which is a copper tube into which fabric rolls were placed. Solar radiation came directly to this tube. A quadratic polynomial was used to express the relationship among time, temperature, % $NaClO_2$ consumed, fabric whiteness and wetting time. Whiteness was most correlated to $NaClO_2$ consumed. Temperature and time made smaller contributions to whiteness. The same relationship or process variables applied to wettability. Fabric strength was retained and this represented an improvement over conventional bleaching. Optimal conditions: $NaClO_2$, 0.64–0.84% owf; scouring agent, 1.0% owf; pH 4.0; 150–180 min; 80–85°C (Gulrajani et al., *1990).*

Bleaching system
Textile substrate
Bleaching protocol
Remarks and reference

$NaClO_2$, H_2O_2 activator, wetting agent, Na_2HPO_4
Cotton, loomstate; cotton/polyester, loomstate
- Activation of $NaClO_2$ by H_2O_2 at pH 8 or 10 before bleaching
- Bleaching at 95°C at pH 8 or 10
- Concurrent desize, scour, bleach

Bleaching occurred by H_2O_2 induced activation of $NaClO_2$ by a free-radical mechanism. The bleaching effect of $NaClO_2$ increased with higher H_2O_2 concentrations. Whiteness increased with exposure time and pH, and in the absence of Na_2HPO_4. Higher H_2O_2 concentrations increased whiteness and wettability. Better results were obtained for all-cotton textiles. Copper number (aldehyde content) and carboxyl content were not greater than for unbleached textiles. Tensile strength loss was less at pH 10 but overall loss was not a detriment to recommending this system (El-Sisi et al., *1990).*

$NaClO_2/K_2CrO_4$, non-ionic wetting agent
Cotton, loomstate; cotton/polyester blend
- Pad, batch
- 90°C
- pH 4–8
- Concurrent desize, scour, bleach

Bleaching was based on activation of $NaClO_2$ by K_2CrO_4 through free-radical formation on cellulose by the interaction of Cr(VI) with OH groups. Optimum conditions for cotton: pH 6, 90 min, 90°C; optimum conditions for cotton/polyester: pH 6, 60 or 90 min, 90°C; in both cases, $[NaClO_2] = 20$ g/l and $[K_2CrO_4] = 3$ g/l. Whiteness indices of > 75 in cotton and in the blend were achieved in the presence of K_2CrO_4. Wettability improved by this treatment. Optimum conditions of time, temperature, and concentration could be chosen so that tensile strength would not fall below that of the loomstate fabric. Copper number was greater at low pH (Hafiz et al., *1991).*

$NaClO_2/KMnO_4$, non-ionic wetting agent
Cotton, loomstate; cotton/polyester blend
- Pad, batch
- 90°C
- pH 4–10
- Concurrent desize, scour, bleach

Bleaching occurred by permanganate-initiated hydroxyl free radicals. Whiteness was greater with $KMnO_4$ *present and by increasing pH from 4 to 10. Wettability increased with* $NaClO_2/KMnO_4$ *concentration, exposure time, and it was greatest at pH 4. Tensile strength decreased with concentration and time and was not influenced by pH. Copper number and carboxyl group content increased with concentration and was greater at lower pH. These values were not lower than untreated fabric. Results suggested selective, primary attack of oxidant on noncellulosic colorants before attack on cellulose itself (El-Rafie* et al., *1990, 1991).*

Bleaching system
Textile substrate
Bleaching protocol
Remarks and reference

$NaClO_2$, sodium hydrosulfite prescour, non-ionic wetting agent
Linen, loomstate

- $[NaClO_2] = 2–10$ g/l
- 65–95°C
- 1–3 h
- pH 4–10

A reductive scour with Na_2CO_3 resulted in fabrics more susceptible to bleaching with $NaClO_2$. Maximum whiteness was with 8 g/l $NaClO_2$, 2.5 h, 95°C, pH 4–6. $NaClO_2$ bleaching after reductive scour resulted in the best bleaching effect when compared to conventional scouring that was followed by bleaching using NaOCl, $NaClO_2$, and H_2O_2 in succession. The cessation of bleaching at pH 8–10 was taken as evidence of a free-radical mechanism for inducing $NaClO_2$ activation whereby the redox system 'fiber reductant/$NaClO_2$, oxidant' prevails (Abou-Zeid et al., *1991).*

$NaClO_2$/HCHO, non-ionic wetting agent
Cotton, loomstate

- Pad, batch
- 70°C
- pH 10
- Consecutive desize, scour, bleach

Bleaching was based on the activation of $NaClO_2$ by formaldehyde. The incorporation of formaldehyde with $NaClO_2$ avoided the problem of chlorine dioxide formation from an acid $NaClO_2$ bleach bath and increased the rate and the amount of sodium chlorite decomposition. Raising the bleaching temperature enhanced this effect. The presence of formaldehyde contributed to better wettability, higher whiteness, and greater weight loss with increasing concentration of formaldehyde. Copper number and carboxyl content of the bleached fabrics were lower than those of the unbleached fabrics. Tensile strength loss could be controlled by regulating formaldehyde concentration, time, and temperature of bleaching (Hebeish et al., *1993).*

Bleaching system
Textile substrate
Bleaching protocol
Remarks and reference

$NaClO_2$, additives, with and without buffer (ammonium sulfate/tetrasodium pyrophosphate), H_2O_2

Synthetic fibers, cellulosic fibers, viscose, linen yarns

- $[NaClO_2] = 0.75$ g/l
- pH 3.8
- 90°C
- 30 min
- Component streams: J-box, washing machine, circulating-liquor machine
- Production unit: pretreatment, dyeing, finishing

AOX content studies of $NaClO_2$ bleaching liquor showed that $NaClO_2$ is ecologically acceptable. Without buffer, bleaching liquor from bleaching synthetic fibers was 2.6–2.8 ppm AOX, and it was 0.6–1.2 ppm AOX with buffer. These values were below the 3.0 mg/l (ppm) allowable in Germany's component streams. A bleaching $NaClO_2$ liquor could be selected for viscose to limit AOX to 1.3 ppm. AOX bleaching liquors from $NaClO_2$ bleaching of linen yarns in stage 1 were 8.0 ppm, and were 1.2 ppm in H_2O_2 stage 2 bleaching. Combined wastewaters produced a component stream below the limit AOX value. Alkaline process by boil-off before bleaching drastically reduced AOX loading and this process produced acceptable whiteness levels (Kleber, 1993).

H_2O_2/NaOH, $MgSO_4 \cdot 7H_2O$, EDTA, wetting agent, Fe^{3+}, Cu^{2+}

Polyester/cotton blend, (68/32) fabric, loomstate

- Simultaneous desize, scour, bleach (DSB)
- Winch-beck
- 95°C
- 60 min

Increasing $[MgSO_4]$ decreased H_2O_2 decomposition and led to higher whiteness, less wettability and no strength loss. [NaOH] 4–6 g/l was preferred. Above this concentration H_2O_2 decomposition was favored, whiteness and wettability did not improve, polyester hydrolyzed, cellulose degraded, and tensile strength of the fabric was lost. $[H_2O_2$, 35% w/v] 2–14 ml/l did not affect H_2O_2 decomposition. There was no improvement in whiteness above 6 ml/l. Within this range of $[H_2O_2]$, increasing concentration led to strength loss. Recommended was 4–6 ml/l $[H_2O_2]$. The effect of Cu^{2+} ions to catalyze H_2O_2 decomposition was greater at higher metal ion concentrations and resulted in whiteness loss but better wettability. Mg^{2+} and EDTA stabilized H_2O_2 decomposition in alkaline medium (Hebeish and El-Bazza, 1985).

Bleaching system
Textile substrate
Bleaching protocol
Remarks and reference

H_2O_2

Cotton cambric fabrics

- Pad fabric with H_2O_2/sodium silicate and 0.1 M buffer
- Buffer: borax (pH 9), sodium carbonate/sodium bicarbonate (pH 10), sodium hydroxide and disodium hydrogen phosphate (pH 11)
- 70–95°C
- Vary exposure time
- Sizes: acrylic and starch

pH buffers increased the rate of H_2O_2 decomposition. Fabrics with acrylic size showed a slower rate of H_2O_2 decomposition but size caused more damage to cellulose. Copper number and carboxyl content for acrylic-sized fabrics were higher than for starch-sized fabrics. In the buffered system, increase in alkalinity (pH 9–11) formed oxycelluloses of lower aldehyde content (lower copper number) and higher carboxyl content. By adding pH buffers to the bleach bath either bleaching time or amount of peroxide could be reduced (Sukumar and Gulrajani, 1985).

H_2O_2, amides: (urea, benzamide, formamide, *N*,*N*-dimethylformamide (DMF))

Cotton poplin fabric, desized, scoured

- Pad/batch
- Room temperature, 60°C or 90°C
- 3–24 h
- pH 8

Amides were incorporated into the conventional hydrogen peroxide bleach bath to facilitate the controlled decomposition of peroxide in order to shorten bleaching time. Decomposition of peroxide was faster with amides in the bleach bath. The sequence in rate of peroxide decomposition resulting from the presence of amides was formamide > DMF > benzamide > urea. The best whiteness was obtained with benzamide, thereby indicating that the whiteness of the bleached fabrics was not the result of peroxide decomposition. The increase in whiteness index with increase in time and temperature, when bleaching in the presence of amides was most pronounced for bleaching at 20°C (increase of 5–6 units). This effect was only marginal at 60°C or 90°C. Amide activation of peroxide was recommended in cold bleaching processes (Das et al., *1986).*

Bleaching system
Textile substrate
Bleaching protocol
Remarks and reference

H_2O_2, NaOH, borate–silicate
Jute fibers, raw and demineralized

- 80°C
- 60–240 min
- pH 8.0–10.0

Bleaching was most effective at pH < 9.0 in terms of whiteness, with 8.5 the best for demineralized fibers. At pH 8.0 there was minimum H_2O_2 decomposition. Proposed from these findings: bleaching with H_2O_2, in the borate silicate system, is through perhydroxyl ions from H_2O_2 dissociation. Brightness increased up to 40 min bleaching time, and beyond 40 min there was strength loss. H_2O_2 consumption is minimized and brigher jute is obtained by demineralizing jute before bleaching. At pH 8.0–8.5, there was increased bleaching efficiency of H_2O_2 in the borate–silicate system. This system reduced bleaching time by 33% (Sikdar et al., *1987).*

H_2O_2, sodium silicate, sodium carbonate, emulsifying system (non-ionic detergent, trichlorethylene, pine oil)
Coir fibers

- Sequence: prewet, hypochlorite bleaching, peroxide bleaching, wash, sun dry
- H_2O_2 (50%): 6%–8%
- 5–22 h
- 80°C

Both hot and cold bleaching could be used. Subsequent peroxide bleaching nearly doubled the brightness achieved after hypochlorite bleaching. Hot bleaching required 5 h; cold bleaching 20–22 h. More strength was retained after cold bleaching. Hot bleaching caused 50% reduction in strength and cold bleaching 40% (Mandavawalla and Tendulkar, 1988).

H_2O_2, NaOH
Cotton fabrics, grey

- Single stage cold-pad-batch for desize, scour, bleach
- 4–24 h
- 20–80 ml/l of 30% H_2O_2

Fabric weight loss did not increase when [NaOH] = 30–40 g/l and [H_2O_2] = 40 ml/l. Tensile strength decreased as NaOH and H_2O_2 concentrations increased. At 40 ml/l H_2O_2 and 40 g/l NaOH, strength loss was 6.0%, weight *loss was 5.3%, and absorbency was 5.9 s. Batching time of 8 h was minimum for good whiteness and bleaching efficiency but this time could be extended to 12 h. Adequate wettability and whiteness for dyeing pastel shades were achieved in the system without silicate stabilizer (Jamadagni* et al., *1988).*

Bleaching system
Textile substrate
Bleaching protocol
Remarks and reference

H_2O_2, sodium silicate, perchloroethylene scour solvent
Cotton fabric, grey

- Combined desize, scour, bleach
- Solar energy accelerated
- Pad-batch
- pH 10.5
- 20–110°C
- 0–90 min
- H_2O_2: 0.05, 1.8, 2.15%

Temperature was the most significant variable. Maximum whiteness was achieved at 80–95°C. Solar radiation led to high rates of H_2O_2 decomposition at temperatures above 103°C. Optimal conditions: 4% scouring agent, 1.8% H_2O_2 (100%), 2% sodium silicate, <100°C, 90 min, pH 10.5. When fabrics bleached by solar energy were compared to those bleached conventionally, all parameters (weight loss, %; strength, kg/g; whiteness index and wetting times) were approximately the same. Solar energy provided chemical and thermal energy savings (Gulrajani and Gupta, 1989).

H_2O_2, NaOH, persulfate, silicate
Cotton materials

- Pretreatment: steaming, 5 min
- Treatment: alkaline H_2O_2, room temperature, 1–10 h
- Variables: [H_2O_2], [NaOH], [persulfate], time

Breaking strength was influenced by [NaOH] and treatment time. High breaking strength was found with 10 g/l and 5 h treatment. Increased water absorption was found at increasing [H_2O_2] and [persulfate]. Optimal values for breaking strength were found from the multiple regression equation of the dependence of water absorption on these variable pairs: [NaOH]–[H_2O_2], [NaOH]–[persulfate], [NaOH]–time, [H_2O_2]–[persulfate], [H_2O_2]–time, [persulfate]–time. Whiteness index was determined similarly, and maximum whiteness were achieved with 70 g/l H_2O_2 and 14 g/l persulfate. Optimal conditions for whiteness index were: [H_2O_2], 40 g/l; [persulfate], 10 g/l; [NaOH], 20–40 g/l. Compared to the classical two-stage hypochlorite/alkaline peroxide boiling treatment, low temperature alkaline H_2O_2/persulfate bleaching improved breaking strength by 21% but whiteness was lowered by 4%. Subsequent optical whitening increased whiteness to levels comparable to those achieved by the classical bleaching treatment (Muresan et al., *1989).*

Bleaching system
Textile substrate
Bleaching protocol
Remarks and reference

H_2O_2, persulfate activator
Cotton fabrics
- Preparation treatment: HOCl
- Pad/batch H_2O_2, alkaline
- Vary [H_2O_2], [NaOH]
- 25–45°C

Peroxide consumption could be reduced by enzyme desizing followed by HOCl and sequential H_2O_2-activated potassium persulfate. Whiteness increased with increasing [NaOH], [H_2O_2], and temperature. Fluidity increased with increase in [NaOH]. Wicking time was in general less than after conventional H_2O_2 scouring/bleaching when done at high temperatures. Absorbency was better at 25°C with [NaOH] = 3.25 owf, [H_2O_2] = 0.6% owf. Conditions for freeing fabric from motes (immature seeds covered with fuzzy fibers): [NaOH] > 1.5%; [H_2O_2] > 0.6; T > 35°C. The H_2O_2/persulfate system was described as a one-step, low temperature scour/bleach process for enzymatically desized and 'chemicked' fabrics (Surma et al., *1989).*

H_2O_2/acetic anhydride (*in situ* formation of peracetic acid), 2,2′-bipyridine (bipy), sodium lauryl sulfate (SLS), transition metal cations, sodium tetraborate (borax)
Cotton plain weave, desized, scoured
- [Peracetic acid], constant
- Vary [bipy] and [SLS]
- pH 4.75–10.0
- 20–60°C

Bleaching was catalysed by tris-chelate formation from bipy/metal ion complexation with Fe^{2+} in cotton. The presence of sodium lauryl sulfate reduced peracetic acid decomposition by its association with the tris-chelate. Maximum whiteness, 73.08 WI units, was achieved at pH 7 and it decreased to 55.30 at pH 10. This indicated that increasing pH caused decomposition of peracetic acid. Higher WI was achieved with the tris-chelate at all bleaching temperatures. Maximum whiteness was obtained at 50°C (79 WI). Higher bleaching temperatures contributed to peracetic acid decomposition. The most effective metal ions used were cobaltous, ferrous, and manganous. The system was effective at 30°C (Rucker, 1989).

Bleaching system
Textile substrate
Bleaching protocol
Remarks and reference

H_2O_2, metal salts, $FeSO_4$, $CuSO_4$, $CoSO_4$, sequestering agents: $Na_4P_2O_7$, sodium oxalate, Na_2EDTA

Wool jersey knit fabric, dark pigmented Karakul fiber

- Two-step mordant/bleach
- Mordant: 70°C, 90 min
- Bleach: pH 8.6–8.8, 55°C, 2 h

Order of increasing fiber damage: copper(II) > iron(II) > cobalt(II). Optimal conditions for mordanting and bleaching pigmented fibers were as follows: Mordanting: $CoSO_4$, pH 3, 80°C, 60 min; bleaching: H_2O_2/EDTA, pH 9, 65°C, 60 min. Cobalt(II) showed less damage to wool than the conventional iron(II) system. The cobalt bath was more stable, but more H_2O_2 was needed to achieve similar bleaching effects. Reductive after-treatment was not necessary because this bleaching did not discolor wool (Trollip, 1990).

H_2O_2, thiourea, thiourea dioxide, or sodium hydroxymethanesulfinate

Woolen flannel, scoured

- Single-bath sequential oxidative/reductive bleaching
- [thiourea] 3.07–7.69 g/l
- H_2O_2/alkaline, 60°C, 1 h followed by thiourea addition, pH 7.4–7.6 and continued bleaching at 60°C, 25 min

The variables [thiourea], oxidative peroxide bleaching time, reductive bleaching time and temperature were examined to optimize bleaching efficiency. Whiteness index was greatest at [thiourea] $\geqslant$ 4.61 g/l and after 15 min beyond thiourea addition. Maximum temperature for oxidative/reductive bleaching with the acceptable alkali solubility. (%) of 20.12 ± 0.34 was at 60°C. Bleaching efficiency was markedly improved with the H_2O_2 bleaching alone. In situ *conversion of thiourea by reaction with residual H_2O_2 from oxidative bleaching in step 1 produced thiourea dioxide, which hydrolyzed at pH 7.5 to form the reductive species, sulfinate anion (Arifoglu* et al., *1990).*

Bleaching system
Textile substrate
Bleaching protocol
Remarks and reference

H_2O_2, Hercosett cationic resin, anionic optical brightener (OBA)
Wool knit fabric

- Pad/dry/cure
- Ternary dispersions of resin, peroxide, OBA
- 20–80°C
- 1–20 min
- pH 7.0
- Vary OBA from 1.5 to 3.5 g/g resin

Resin/OBA/H_2O_2 aqueous systems showed acceptable loss in ball penetration resistance (BPR, %), cysteic acid content and alkaline solubility. Least damage was at lowest temperature, 20°C, and shortest exposure time, 1 min. BPR (%) after 17 min 15 s at 71°C was within the acceptable 10% level. Whiteness levels above 100% were achieved at 80°C in 10.5 min. Shrinkage increased with temperature and was highest at 71°C, 3 min 45 s (De la Maza et al., *1992).*

H_2O_2, urea, nonionic wetting agent
Linen fabrics scoured and unscoured

- Bleach grey fabric
- Separately bleach scoured fabric
- Vary H_2O_2 and urea concentrations
- pH 4–10
- 30–150 min
- 40–95°C

Optimum bleaching conditions: H_2O_2 (7 g/l), urea (8 g/l), nonionic wetting agent (1 g/l), pH 6, 95°C, 2 h. Mechanism for peroxide oxidation: free-radical, involving hydroxyl and perhydroxyl free radicals, and perhydroxyl ions, depending upon pH. This mechanism was promoted by the presence of urea. Within the concentration range for urea of 2–12 g/l, whiteness increased with urea concentration, irrespective of pH. Whiteness increased nevertheless in the order pH 6 > pH 8 > pH 4 > pH 10 (El-Rafle et al., *1992).*

H_2O_2 with peroxidase enzyme and O_2, with oxidase enzyme, bleach boosters (metal or halide ion or organic compound), cellulose, pumice
Dyed textiles; cotton denim, indigo-dyed

- Simultaneous bleaching and formation of localized color variation
- pH 4–10

The system was effective for concentrations of H_2O_2 < 100 mM. An advantage *was that the wastewater was almost colorless. The bleaching process was effective for textiles dyed with synthetic azo dyes, natural dyes or synthetic dyes that were nature-identical (such as indigo). Optimal conditions: enzyme protein per liter of bleach liquor, 0.1–10 mg/l; pH 5–9; 55°C; 60 min (Pedersen and Reesens, 1992).*

Bleaching system
Textile substrate
Bleaching protocol
Remarks and reference

H_2O_2 (alkaline), thiourea (thiourea dioxide *in situ* formation), hydrolysis of thiourea dioxide to form sulfinate reducing species and urea

Wool fabric

- Sequential oxidation/reduction
- [H_2O_2] 22, 16 or 8 g/l
- 50°, 55° or 60°C
- 85 min

ARS patented bleaching showed greater whiteness than conventional alkaline bleaching. Equal whiteness was found with 16 g/l peroxide in the ARS process where thiourea was 70% ow 100% H_2O_2 and with 22 g/l conventional alkaline H_2O_2 bleaching over the same time and temperature settings. NMR spectroscopy was used to determine the amount of thiourea to add in the reductive step for complete conversion to sulfinate and urea. Multiple regression equations were predictive for establishing temperature and [H_2O_2] values to achieve certain whiteness levels. By the Kawabata system (Kawabata, 1980) ARS bleached fabrics showed greater softness, flexibility, and smooth-feeling than conventional H_2O_2 bleached. Comparisons were made with conventional bleaching with alkaline H_2O_2 (22 g/l, 60°C) within the same time limits. Better whiteness was achieved using ARS bleaching under the following conditions of [H_2O_2] for the oxidative step, [H_2O_2] for reaction with thiourea, and run temperature: 22, then 16 g/l at 50°C; 22, then 8 or 16 g/l at 60°C. ARS bleaching resulted in some strength loss (3–14%) compared to conventional peroxide bleaching (Cardamone et al., *1992; Marmer* et al., *1994).*

H_2O_2, NaOH

Cotton fabric, sized and polyester/cotton fabric, sized

- Continuous one-stage, desize–scour–bleach
- pH 12.5
- 23–95°C
- [NaOH] 0.02–2.8 g/l
- [H_2O_2, 35%] 0.1–2.8 g/l

Degree of desizing was time, temperature and [NaOH] dependent. At [NaOH] 2% owf, increasing the peroxide concentration reduced the rate of desizing. H_2O_2 catalytically initiated hydrolysis of size, most effectively at 0.1% owf [H_2O_2]. [H_2O_2] was the most significant factor in whiteness; whiteness increased with time, temperature, and [H_2O_2]. Only fabrics treated at 65–95°C were absorbent. Increasing time, temperature, and [H_2O_2] increased rate of *size removal, which was complete at temperatures below 65°C. Fluidity tests showed no chemical damage from treatments (Levene and Prozan, 1992).*

Bleaching system
Textile substrate
Bleaching protocol
Remarks and reference

H_2O_2, enzymes (xylanase and cellulase)
Jute fibers

- Enzyme pretreatment
- pH 5
- 60 min
- Room temperature

Lignin and hemicellulose content decreased by 3.9% after enzyme/bleach treatment. Loss in dry weight was 5.23%. A proper balance of xylanase and cellulase enhanced bleaching. Xylanase contributed to whiteness and cellulase to enhanced swelling or softening (Kundu et al., *1993).*

H_2O_2, Na_2CO_3, $NaHCO_3$, urea, trisodium pyrophosphate (TSPP), ethylene diaminetetraacetic acid (EDTA), NaCl
Silk yarn, mulberry, 20/22

- Combined degumming and bleaching in one bath
- Sodium lauryl sulfate (22 g/l)
- EDTA, 1 g/l
- TSPP, 3 g/l
- Nonionic detergent, 1 ml/l
- H_2O_2 (30%), 13 ml/l
- pH 10.1
- 1 h
- 90°C

Addition of urea up to 50 g/l increased whiteness; from 50 g/l to 150 g/l, whiteness decreased. Breaking load was not affected up to addition of 100 g/l urea, but at 150 g/l urea breaking load decreased by 28%. The improvement in whiteness by 15% by urea addition was attributed to amide complexation with H_2O_2 to facilitate faster but controlled peroxide decomposition. Increased effectiveness, less time and chemicals are benefits from using combined degumming and bleaching with H_2O_2 and EDTA and TSPP. Conditions can be optimized for controlling weight loss and enhancing whiteness by the addition of urea and NaCl (Saligram et al., *1993).*

Bleaching system
Textile substrate
Bleaching protocol
Remarks and reference

H_2O_2, alkali
Cellulose and rayon textiles (CS_2-free and modal)
- Alkaline boiling off and cold pad-batch bleaching
- Alkaline boiling off and hot bleaching
- Alkali treatment and cold pad-batch bleaching
- Alkali treatment and hot bleaching

Carbon disulfide-free rayon had higher strengths and whiteness than cotton, viscose, and modal rayons. Abrasion resistance was comparable to modal, less than cotton, and higher than viscose. Absorbency was similar to modal and better than viscose and cotton. Least damage was with alkali/H_2O_2 cold pad-batch bleaching. No excessive strength losses were found (Hellwich et al.*, 1994).*

Ozone, steam
Cellulosic garments, dyed and wet; applicable to vapor and cellulosic blends
- Batchwise or continuous
- Ozone: 10–100 mg/l
- 40–100°C
- Recycle excess ozone to treat effluent

This method was proposed as an alternative method to stone washing for decolorizing or fading dyed garments. Requirement: the dye must be ozone reactive. Blocking agents on the garment determined the desired effect. Ozone-treated fabrics retained more abrasion resistance and strength than NaOCl-bleached fabrics. Shading effects were possible with ozone treatment that were not possible with stone washing. Ozone treatment corrected for yellowed background areas that result from conventional discoloration methods (Wasinger, 1992).

Bleaching system
  Textile substrate
    Bleaching protocol
Remarks and reference

$KMnO_4$
  Cotton cloth, desized, scoured, cotton hanks
- $KMnO_4$, 0.5% owf
- H_2SO_4, $\frac{1}{4}$ part $KMnO_4$
- HOAc, $\frac{3}{4}$ part $KMnO_4$
- Room temperature
- 30 min

Tensile strength and copper number losses were less than after conventional hypochlorite bleaching. Carboxyl content was lower for $KMnO_4$ bleaching of cotton cloth but higher for cotton hanks. Whiteness of permanganate-bleached fabrics was comparable to fabrics bleached by hypochlorite. When considering energy and time expenditure, the permanganate process was less costly than conventional hypochlorite bleaching, followed by hydrogen peroxide bleaching. Estimate production cost saving was 16.0% and saving in process time 30% (Lokhande and Wasif, 1991a).

$KMnO_4$, H_2O_2
  Polyester/cotton fabric (48/52) (desized, scoured, and mercerized)
    Formulation I: bleach
- $KMnO_4$
- H_2SO_4
- HOAc
- Oxalic acid
- NaOCl
- H_2O_2
- Sodium silicate
- Sodium hydroxide
- 85°C, 2 h

    Formulation II: scour, mercerize, bleach
- NaOCl
- $KMnO_4$
- Oxalic acid
- H_2O_2
- Sodium silicate
- Sodium hydroxide
- 85°C, 2.5 h

When compared to conventional $NaClO_2/H_2O_2$ *bleaching formulations, formulations I and II produced less strength loss with comparable whiteness values. Slightly lower copper number values were obtained from* $KMnO_4$ *bleaching when compared to conventional bleaching. Bleaching by formulations I and II cost substantially less than conventional bleaching in terms of chemicals, energy comparable and time (Lokhande and Wasif, 1991b).*

Bleaching system
Textile substrate
Bleaching protocol
Remarks and reference

Peracetic acid, H_2O_2
Cotton fabric

- Ultrasonic radiation
- Apply peracetic acid (15%) with H_2O_2 (30%)
- Adjust pH with caustic soda
- 50°C
- 1 h
- pH 1–7

Addition of peracetic acid, 1% owf at 50°C, pH 6 with ultrasound, produced fabric brightening. This brightness was not exceeded above 2% owf concentration peracetic acid. Bleaching by peracetic acid was markedly increased by ultrasound at pH 1–4. A higher degree of whiteness was achieved at pH 6. Damage to fibers within the pH range was less than damage from conventional hydrogen peroxide bleaching. Peracetic acid with concomitant ultrasound could be used at low temperature and without activators to achieve comparable whiteness, and less damage when compared to conventional peroxide bleaching (Poulakis et al., *1991).*

Peracetic acid (PAA) generated from the equilibrium solution of H_2O_2, and HOAc with subsequent H_2O_2/alkaline bleaching
Cotton jersey knit fabric and yarns; linen

- Continuous bleaching
- Fabric impregnation
- Multistage
- Rope form; J-Box
- Open width, winchbeck
- pH 3–9
- 20–80°C
- 20–80 min
- [PAA] = 1.75–8.75 g/l
- With and without metal ions

High brightness for cotton was achieved with [PAA] 1.5–2.5 g/l at room temperature, followed by H_2O_2/alkaline treatment at 90°C. For linen yarns, the recommendation was scour, alkaline H_2O_2 bleaching and a PAA bleaching stage at pH 5–6, 80°C. In all cases, PAA bleaching produced less fiber damage than NaOCl bleaching in the optimum pH range 6–7, 50–80°C, 20 min. Brightness increased with [PAA] and with temperature. The PAA system *showed overall less fiber damage than NaOCl systems. Unlike H_2O_2 systems, where metal ions can catalytically degrade fibers, with PAA Fe ions have no detrimental effect and there is no loss in brightness. Continuous scour and PAA bleaching at room temperature produced whiteness values equivalent to NaOCl/H_2O_2 bleaching with less damage to fiber (Steiner, 1993).*

Bleaching system
Textile substrate
Bleaching protocol
Remarks and reference

$NaHSO_3$, $Na_2S_2O_4$, $NaHSO_2 \cdot CH_2O$, $KMnO_4$, Na_2O_2 (sodium peroxide), H_2O_2, $NaBO_2$

Silk, mulberry and Tasar varieties, degummed

- $NaHSO_3$: 5% owf, 1 h, 60°C
- $Na_2S_2O_4$: 3% owf, 2 h, 50°C
- $NaHSO_2 \cdot CH_2O$: 24%, 30 min, boil
- $KMnO_4$: 0.25% owf, 30 min, room temperature
- Na_2O_2: 0.5% owf, pH 8–8.5, 4–5 h, 65°C
- H_2O_2, EDTA: 13 ml/l, 35%, pH 8–8.5, 2–3 h, 70°C
- H_2O_2, Na_2SiO_3: 1 h, 90°C
- $NaBO_2 \cdot H_2O_2$: 1% owf, pH 9.2, 1 h, 80°C

Both oxidative and reductive bleachings including sodium formaldehyde sulfoxylate, $HOCH_2SO_2Na$ or $NaHSO_2 \cdot CH_2O$, were necessary to whiten silk varieties. Degumming increased whiteness, decreased yellowness, and caused loss in moisture by ~5% and loss in tenacity by 7%. Highest whiteness was found by bleaching with $H_2O_2 \cdot Na_2SiO_3$, followed by H_2O_2, EDTA. Bleaching with $H_2O_2 \cdot Na_2SiO_3$ showed lowest loss in moisture regain. Average tenacity loss from all bleachings was 27%, but degradation was least by H_2O_2. Average loss in moisture regain was 16% from all bleaching. $KMnO_4$ and $NaBO_2 \cdot H_2O_2$ gave only negligible increase in whiteness. $Na_2S_2O_4$ and $NaHSO_2 \cdot CH_2O$ were unstable and not recommended. The bleaching agent recommended for mulberry and Tasar silk was H_2O_2 with sodium silicate stabilizer (Shukla et al., *1993).*

$NaHSO_3$, sodium lauryl sulfate (SLS)

Wool tops

- $[NaHSO_3] = 3$ g/l
- [SLS] = 5 g/l
- 50°C
- 2 h
- pH 3.0–6.5
- [SLS] = 0–10 g/l at pH 5.5

The presence of SLS imparted protection of wool's —S—S— bonds. When compared to the untreated sample, SLS contributed to increased whiteness and this increase was constant within the pH range 3.0–6.5. Best results were obtained at pH 5.5–6.5, where cysteine content was lowest and whiteness was highest. Optimal conditions: $[NaHSO_3]$, 4.5 g/l; [SLS], 5 g/l; bleaching time, 1 h; bleaching temperature, 50°C; pH 5.5 (Gacén et al., *1993).*

Bleaching system
Textile substrate
Bleaching protocol
Remarks and reference

Sodium hydrosulfite, sodium laurylsulphate
Wool tops

- [Hydrosulfite], 5 g/l
- 2.5 h
- 60°C
- pH 3, 4, 5, or 6
- [SLS] = 0, 2.5, 5, or 10 g/l

Optimal conditions for hydrosulfite bleaching; [hydrosulfite], 3.5–5 g/l; [SLS], 2.5–5 g/l; pH 5–6; 1–2 h; 55–60°C. Under these conditions, the presence of SLS reduced cysteine content by 27% and increased whiteness by five units (Gacén et al., *1989).*

Thiourea dioxide (TD)
Wool

- [TD] = 1.5–3.0 g/l
- 55–85°C
- 1–4 h

Increased temperature, [TD] and bleaching time improved whiteness. Temperature increase most highly influenced cysteine content. There was a linear dependence of [TD], temperature, and treatment time on cysteine content. At higher temperatures, strength losses were greater. A maximum acceptable cysteine content of 1.5% was established at 65°C after 60 min of bleaching, or at 80°C after 30 min (Cegarra, 1988).

Thiourea dioxide (TD), sodium laurylsulfate (SLS)
Wool top

- [TD] = 1.5 g/l
- 70°C
- 2.5 h
- [SLS] = 0.3 or 5 g/l
- Removal of SLS at pH 6.5, 8 or 9

Most efficient removal treatment for SLS after TD bleaching was at pH 9 and 45°C, which reduced SLS retention to 1.6–1.7%. SLS removal resulted in decreased whiteness but no change in yellowness. Even after SLS removal, wool bleached with SLS present was whiter than when bleached without it. The presence of SLS with or without TD increased whiteness (Gacén et al., *1992).*

Bleaching system
  Textile substrate
    Bleaching protocol
Remarks and reference

Photobleaching, H_2O_2, sodium formaldehyde sulphoxylate, thiourea dioxide
  Wool fabric
- Blue light illumination
- Dry heat photobleaching, 105°C, 1 h
- Simulated sunlight photobleaching, 40°C, 48 h

Samples, untreated and bleached by several processes (including peroxide, sulphoxylate, thiourea dioxide, dry photobleaching, and catalyzed photobleaching) were evaluated for brightness after subjecting them to the following aftertreatments: boiling, Xenotest exposure, sunlamp and dry heat exposures. In each after-treatment, reversion of brightness was measured. Xenotest and catalyzed photobleaching increased brightness of all samples to the greatest extent. When compared to conventional bleaching, photobleaching did not produce a more stable brightness. When the reflectance of wool was increased by bleaching, degradative processes clearly resulted in subsequent reflectance losses (Simpson, 1990).

Visible radiation, UV radiation, H_2O_2/blue light, thiourea dioxide, sodium formaldehyde, sulphoxylate, sodium dithionite
  Wool fabric
- Dry photobleaching
- Wet photobleaching

Reflectance properties of photobleached and conventionally bleached fabrics were reported. Some treatments were in the presence of blue light illumination. Dry photobleaching at ambient temperature for 36 h using fluorescent tubes did not alter brightness but the presence of reducing agents increased brightness. Alkaline peroxide in the presence of blue light rapidly brightened the samples in 20 min. Maximum reflectance over the 400–650 nm range was found for fabrics bleached by peroxide-catalyzed photobleaching. Emission curves of bleached samples with excitation at 290 nm showed that conventional bleaches increased fluorescence at 340–350 nm. Photobleaching in the presence of oxidizing or reducing agents caused a 10–15 nm red shift in the wavelength of maximum absorption. Excitation with 350–360 nm radiation showed that all bleached samples except those that were dry photobleached had greater fluorescent emissions. High pH oxidative bleaching in the presence of blue light produced greater whiteness than bleaching by conventional means and this altered wool's fluorescence pattern. These studies followed the changes in tryptophan with wool degradation (Simpson, 1992b).

Bleaching system
Textile substrate
Bleaching protocol
Remarks and reference

Photobleaching: argon ion laser 480 nm and 351 nm; north light fluorescent lamps

Wool fabric; raw Merino wool; normal yellow color wool

- Wet and dry states
- Monochromatic and filtered light
- Measurements by diffuse reflectance and fluorescence spectroscopy
- Ambient temperature

UV radiation produced a rapid increase in the natural fluorescence of wool. Dirty wool showed smaller increases in fluorescence upon irradiation. Wet fabrics irradiated by filtered blue light became progressively whiter but did not increase in fluorescence intensity. The enhancement of whiteness from photobleaching was not attributed to increased fluorescence but to a possible oxidation mechanism for destroying yellow chromophores. Blue light irradiation did not result in a difference in fluorescence intensities when clean and dirty wool staples were compared. Dry and wet fabrics yellowed from UV irradiation at 351 nm and whitened from exposure to 488 nm irradiation. This effect was greater for wet fabrics (Collins and Davidson, 1993).

FWA, silicone shrink-resistant resin, acrylic resin, chlorinated organic solvent

Woolen cloth, bleached

- FWA incorporated into acrylic resin
- Sequential or consecutive application of silicone and acrylic resins with FWAs
- Padding
- FWA, 2.5% owf resin

Highest whiteness was achieved with satisfactory rate of photo-yellowing when silicone was applied before applying the acrylic emulsion that contained the suspended FWA. This result could be achieved only with high add-on levels. The improvement in whiteness was moderate when compared with conventional FWAs. Up to 5% resin application did not affect fabric hand and improved brightness by 3%. The best results for area shrinkage resulted from applying the silicone finish first. The drawback is that 10–14 days were needed for full curing of the silicone. The treatment resulted in 1% loss in brightness upon storage but the rate of photo-yellowing was significantly reduced. When silicone was applied over the acrylic resin, insufficient shrinkproofing was found. Applying silicone and acrylic resins together showed an improvement in fabric hand (Levene, 1987).

References

Abou-Zeid, N.Y., Higazay, A. and Hebeish, A. (1991) Reductive scouring of linen fabrics — a key for rapid bleaching. *Melliand Textilberichte*, **72** (5), 362–365 (English translation in *International Textile Reports* version, E146–E147).

Arifoglu, M. and Marmer, W.N. (1990a) Novel approaches to the bleaching of stained and pigmented wools. *Proceedings of the 8th International Wool Textile Research Conference, Christchurch, New Zealand, February 1990*, vol. IV, pp. 330–339.

Arifoglu, M. and Marmer, W.N. (1990b) Sequential oxidative and reductive bleaching of stained and pigmented wool in a single bath. *Textile Research Journal*, **60**, 549–554.

Arifoglu, M. and Marmer, W.N. (1990c) Sequential oxidative and reductive bleaching in a multicomponent single liquor system. US Patent 4,961,752, October 9, 1990.

Arifoglu, M. and Marmer, W.N. (1991) Sequential oxidative and reductive bleaching of pigmented and unpigmented fibers. US Patent 5,017,194, May 21, 1991.

Arifoglu, M. and Marmer, W.N. (1992a) Sequential oxidative/reductive bleaching and dyeing in a single bath at low temperatures. *Textile Research Journal*, **62**, 123–130.

Arifoglu, M. and Marmer, W.N. (1992b) Sequential oxidative and reductive bleaching of pigmented and unpigmented fibers. US Patent 5,084,066, January 28, 1992.

Arifoglu, M. and Marmer, W.N. (1992c) Sequential oxidative and reductive bleaching in a multicomponent single liquor system. US Patent 5,103,522, April 14, 1992.

Arifoglu, M. and Marmer, W.N. (1992d) Sequential oxidative and reductive bleaching of pigmented and unpigmented fibers. Australian Letters Patent 618,874, May 4, 1992.

Arifoglu, M. and Marmer, W.N. (1993a) Sequential oxidative and reductive bleaching of pigmented and unpigmented fibers. New Zealand Letters Patent 232,161, February 4, 1993.

Arifoglu, M. and Marmer, W.N. (1993b) Sequential oxidative and reductive bleaching and dyeing in a multicomponent single liquor system. US Patent 5,264,001, November 23, 1993.

Arifoglu, M., Marmer, W.N. and Carr, C.M. (1990) Sequential oxidative and reductive bleaching of wool in a single bath. *Textile Research Journal*, **60**, 319–328.

Arifoglu, M., Marmer, W.N. and Dudley, R.L. (1992) Reaction of thiourea with hydrogen peroxide: ^{13}C NMR studies of an oxidative/reductive bleaching process. *Textile Research Journal*, **62**, 94–100.

Bel'tsov, V.M. and Kalaus, I.V. (1971) Oxidative degradation of polyester and cellulose acetate fibers during their bleaching with chlorine derivatives of isocyanuric acid. *Sbornik Nauchnykh Trudov, Leningradskogo Instituta Tekstil'noi Legkoi Promyshlennosti im. S.M. Kirova*, **12** (1), 158–161 (*Chemical Abstracts*, **78**, 44898p).

Bereck, A. (1985) Bleaching of dark fibres in wool. *Proceedings of the 7th International Wool Research Conference, Tokyo*, vol. IV, pp. 152–162.

Brown, A.E., Hornstein, L.R. and Harris, M. (1951) The chemical modification of wool-treatment with formaldehyde solutions. *Textile Research Journal*, **21** (4), 222–227.

Cardamone, J.M. and Marmer, W.N. (1994) Bleaching efficiency and photostability of wool, wool/cotton blends and all-cotton fabrics. *Textile Chemist Colorist*, **26** (6), 19–22.

Cardamone, J.M., Marmer, W.N., Arifoglu, M. and Barndt, H.J. (1992) Optimum conditions for sequential oxidative/reductive bleaching of wool in a single bath. In *Book of Papers, 1992 International Conference & Exhibition, American Association of Textile Chemists & Colorists (AATCC), Atlanta*, AATCC, Raleigh, North Carolina, pp. 197–202.

Cegarra, J. and Gacén, J. (1983) The bleaching of wool with hydrogen peroxide. *Wool Science Review*, **59**, 3–42.

Cegarra, J., Gacén, J., Caro, M. and Pepio, M. (1988) Wool bleaching with thiourea dioxide. *Journal of the Society of Dyers and Colourists*, **104** (7/8), 273–279.

Collins, S. and Davidson, R.S. (1993) Aspects of the photobleaching and photoyellowing of wool. *Journal of the Society of Dyers and Colourists*, **109** (5/6), 202–209.

Das, T.K., Mandavawalla, A.K. and Datta, S.K. (1986) Amide activated bleaching processes. *Textile Dyer and Printer*, **19** (21), 21–28.

De la Maza, A., Manich, A.M., Parra, J.L. and de Castellan, M.D. (1992) Optimizing Hercosett/optical brightener agent/hydrogen peroxide systems applied to untreated wool for shrinkproofing. *Textile Research Journal*, **62** (3), 162–168.

Duffield, P.A. (1986) Review of wool bleaching. *IWS Technical Information Bulletin*, International Wool Secretariat, Ilkley, England.

D'yachenko, P.F. and Shelpakova, K.F. (1939) Tanning of casein by means of aldehydes. *Journal of the Society of Leather Trade Chemists*, **23**, 12–22.

Earland, C., Stell, J.G.P. and Wiseman, A. (1960) The oxidative insolubilization of proteins. *Second Quinquennial Wool Textile Research Conference, Harrogate, 18–28 May, 1960, Journal of the Textile Institute, Transactions*, **51** (12) T817–T827.

El-Rafie, M.H., Abdel Hafiz, S.A., El-Sisi, F.F., Helmy, M. and Hebeish, A. (1990) A fast desizing/scouring/bleaching system for cotton-based textiles (Part 1). *American Dyestuff Reporter*, **79** (12), 49–51.

El-Rafie, M.H., Abdel Hafiz, S.A., El-Sisi, F.F., Helmy, M. and Hebeish, A. (1991) A fast desizing/scouring/bleaching system for cotton-based textiles (Part 2). *American Dyestuff Reporter*, **80** (1), 45, 48–49.

El-Rafie, M.H., Higazy, A. and Hebeish, A. (1992) Bleaching of linen fabrics using a hydrogen peroxide/urea system. *American Dyestuff Reporter*, **81** (3), 48–67.

El-Sisi, F.F., Abdel Hafiz, S.A., El-Rafie, M.H. and Hebeish, A. (1990) Development of a one-step process for desizing/scouring/bleaching cotton based textiles. *American Dyestuff Reporter*, **79** (10), 39–40, 42.

Farr, J.P., Smith, W.L. and Steichen, D.S. (1992) Bleaching agents. In *Kirk-Othmer Encyclopedia of Chemical Technology*, 4th edn (Eds J.I. Kroschwitz and M. Howe-Grant), John Wiley & Sons, New York, pp. 279, 285.

Fornelli, S. (1994) Towards an IQ of enzymes — enzymatic treatment of protein fibres. *Melliand Textilberichte*, 75 (2), 120–125 (English language translation in *International Textile Reports* version, E33–E35).

Gacén, J., Cegarra, J. and Caro, M. (1989) Wool bleaching with reducing agents in the presence of sodium lauryl sulfate. Part II — bleaching with nonstabilised hydrosulphite. *Journal of the Society of Dyers and Colourists*, **105** (12), 438–441.

Gacén, J., Cegarra, J., and Caro, M. (1992) Wool bleaching with reducers in the presence of sodium lauryl sulphate. Part IV — treatments to remove the surfactant absorbed by wool. *Journal of the Society of Dyers and Colourists*, **108** (7/8), 325–327.

Gacén, J., Cegarra, J., Caro, M. and Cayuela, D. (1993) Wool bleaching with reducing agents in the presence of sodium lauryl sulphate. Part V — bleaching with sodium bisulfite. *Journal of the Society of Dyers and Colourists*, **109** (9), 301–303.

Goswani, K.K. and Mukherjee, A.K. (1993) Bleaching of linen (*Linum usitatissimum*). *Indian Journal of Fibre and Textile Research*, **18**, 82–86.

Gulrajani, M.L. and Gupta, S. (1989) A solar energy accelerated process for combined desizing, scouring, and bleaching of cotton fabrics. *Textile Research Journal*, **59** (4) 217–226.

Gulrajani, M.L. and Venkatraj, R. (1986) A low temperature scouring/bleaching process or cotton using sodium chlorite. *Textile Research Journal*, **56** (8), 476–483.

Gulrajani, M.L., Venkatraj, R. and Chandrasekaran, P. (1987) A process for the combined scouring and bleaching of polyester/cotton fabrics at low temperature. *Journal of the Society of Dyers and Colourists*, **103** (9), 299–303.

Gulrajani, M.L., Mittal, R.M. and Gupta, S. (1990) A process for combined scouring and bleaching of a polyester/cotton blend in a solar oven. *Textile Research Journal*, **60** (6), 361–367.

Hafiz, A., El-Sisi, F., El-Rafie, M.H., Helmy, M. and Hebeish, A. (1991) Sodium chloride/potassium chromate cooxidant induced concurrent desizing, scouring and bleaching of cotton and cotton/polyester blend fabrics. *American Dyestuff Reporter*, **80** (3), 37–47.

Hebeish, A. and El-Bazza, S. (1985) Novel method for desizing, scouring, and bleaching polyester/cotton blends. *American Dyestuff Reporter*, **74** (10), 33–42.

Hebeish, A., El-Sisi, F., El-Bisi, M.K. and El-Rafie, M.H. (1993) Bleaching of loomstate cotton fabric using a sodium chlorite/formaldehyde system. *American Dyestuff Reporter*, **82** (6), 40-43; and **82** (7), 31–34.

Hellwich, B., Weidaver, G. and Müller-Litz, W. (1994) Studies on viscose fibres and other cellulose fibres produced without carbon disulphide in selected pretreatment and bleaching processes. *Melliand Textilberichte*, **72** (1), 46–49 (English translation in *International Textile Reports* version, E16–E17).

Huson, M.G. (1992) The mechanism by which oxidizing agents minimize strength losses in wool dyeing. *Textile Research Journal*, **62** (1), 9–14.

International Wool Secretariat (1970) Yellowing of wool. *Wool Science Review*, **39** (August), 27–39.

Jamadagni, R., Joshi, G. and Shah, R. (1988) Use single stage cold-pad-batch process for bleaching. *Journal of the Textile Association*, **49** (5), 183–187.

Kawabata, S. (1980) *The Standardization and Analysis of Hand Evaluation*, 2nd edn, The Textile Machinery Society, Osaka, Japan, p. 20.

Kleber, R. (1993) Sodium chlorite bleaching in its ecological context. *Melliand Textilberichte*, **74** (5), 395–397 (English translation in *International Textile Reports* version, E182–E183).

Knott, J. (1990) *Fine Animal Fibres and their Depigmentation Process*, EEC COMETT Program, Universidade do Minho (Portugal) (available from DWI, D-5100 Aachen, Germany).

Kundu, A.B., Ghosh, B.S. and Chakrabarti, S.K. (1993) Enhanced bleaching and softness of jute pretreated with polysaccharide degrading enzymes. *Textile Research Journal*, **63** (8), 451–454.

Launer, H.F. (1965) Effect of light upon wool: Part V, Yellowing and bleaching by ultraviolet and visible arc light. *Textile Research Journal*, **35** (9), 813–819.

Launer, H.F. (1968) Photobleaching—a common basic phenomenon. *Nature*, **218**, 160–161.

Laxer, G. and Whewell, C.S. (1955) Some physical and chemical properties of pigmented animal fibres. *Proceedings of the International Wool Research Conference* (*Australia*), vol. F, pp. 186–200.

Levene, R. (1987) Surface whitened, shrink resistant wool cloth of improved lightfastness. *Textile Research Journal*, **57** (5), 298–306.

Levene, R. and Lewin, M. (1984) The fluorescent whitening of textiles. In *Handbook of Fiber Science and Technology, Vol. 1: Chemical Processing of Fibres and Fabrics—Fundamentals and Preparation*, Part B (Eds M. Lewin and S.B. Sello), Marcel Dekker, Inc., New York, pp. 257–304.

Levene, R. and Prozan, R. (1992) Preparation procedures for woven cotton and polyester/cotton fabric. Part 2—oxidative desizing with hydrogen peroxide. *Journal of the Society of Dyers and Colourists*, **108** (9), 375–382.

Lewin, M. and Sello, S.B. (eds) (1984) *Handbook of Fibre Science and Technology: Vol. 1, Chemical Processing of Fibres and Fabrics—Fundamentals and Preparation*, Part B, Marcel Dekker, Inc., New York, pp. 64–65, 228–234.

Lin, P.T. (1989) Method for the decolourization of dyed fabric. UK Patent Application 87300717.4, August 23, 1989.

Lokhande, H.T. and Wasif, A.I. (1991a) Environment-friendly and economical potassium permanganate bleaching process: Part I—Bleaching of cotton goods. *Indian Journal of Fibre and Textile Research*, **16** (3), 218–222.

Lockhande, H.T. and Wasif, A.I. (1991b) Environment-friendly and economical potassium permanganate bleaching process: Part II—Bleaching of polyester/cotton fabric. *Indian Journal of Fibre and Textile Research*, **16** (3), 270–274.

Maclaren, J.A. (1963) Studies in wool yellowing. *Textile Research Journal*, **33** (10), 773–778.

Mandavawalla, A.K. and Tendulkar, S.R. (1988) Economic process for bleaching of coir using hydrogen peroxide. *Textile Dyer and Printer*, **21** (14), 17–21.

Marmer, W.N., Cardamone, J.M., Arifoglu, M. and Barndt, H.J. (1993) Single-bath full-bleaching of wool: Optimization of bleaching conditions and subsequent dyeing in the same bath. *DWI Reports*, **111**, 211–228 (*Proceedings of the 1992 Aachen Textile Conference*).

Marmer, W.N., Cardamone, J.M., Arifoglu, M. and Barndt, H.J. (1994) Optimizing process conditions in sequential oxidative/reductive bleaching of wool. *Textile Chemist Colorist*, **26** (5), 19–24.

Marsh, J.T. (1945) *An Introduction to Textile Bleaching*, 3rd edn, B.I. Publications, Madras.

Merkel, R.S. (1984) Methods for analyzing damage in textile materials. In *Analytical Methods for a Textile Laboratory*, 3rd edn (Ed. J.W. Weaver), American Association of Textile Chemists and Colorists, Raleigh, North Carolina, pp. 34–55.

Millson, H.E. and von Bergen, W. (1970) Dyeing, bleaching and printing. In *Wool Handbook*, Vol. 2, 3rd edn (Ed. W. von Bergen), Interscience Publishers, New York, pp. 766, 769.

Muresan, A., Butnaru, R., Marcu, M. and Muresan, R. (1989) Preparation–cleaning treatment of cotton materials with reduced energy consumption. *Cellulose Chemistry and Technology*, **23**, 443–453.

Nettles, J.E. (1983) *Handbook of Chemical Specialties, Textile Fiber Processing, Preparation and Bleaching*, John Wiley & Sons, New York, p. 400.

Nevell, T.P. (1987) Degradation of cellulose by acids, alkalis, and mechanical means. In *Cellulose Chemistry and its Applications*, (eds T.P. Nevell and S.H. Zeronian), John Wiley & Sons, New York, p. 249.

Nevell, T.P. and Singh, O.P. (1986) Effect of manganese (II), cobalt (II), and copper (I) hydroxides on the kinetics of the oxidation of cellulose by hypochlorite. *Textile Research Journal*, **56** (4), 270–280.

Pandey, S.N., Chattopadhyay, S.N., Pan, N.C. and Day, A. (1993) Bleaching of jute fabric—a comparative study. *Textile Dyer and Printer*, **26** (10), 23–26.

Pedersen, S. and Reesens, E. (1992) Process for bleaching of dyed textiles. International Patent Application PCT/DK92/00119, October 29, 1992.

Peters, R.H. (1967) *Textile Chemistry*, Vol. II, Elsevier Publishing Company, New York, pp. 249–250.

Petrick, M. (1988) The application of sodium chlorite bleach for tubular cotton fabric as used on a large scale bleaching range in Turkey. *International Textile Bulletin Dyeing, Printing, Finishing*, **34** (3), 58–63.

Poulakis, K., Buschmann, H.J., Denter, U. and Schollmeyer, E. (1991) Bleaching of cotton with peracetic acid assisted by ultrasonic radiation. *Textile Praxis International*, **46** (4), 334–335.

Rucker, J.W. (1989) Low temperature bleaching of cotton with peracetic acid. *Textile Chemist and Colourist*, **21** (5), 19–25.

Saligram, A.N., Sawant, P.B, Shukla, S.R. and Nadiger, G.S. (1993) One-stage degumming and bleaching of silk using peroxide in the presence of urea. *American Dyestuff Reporter*, **82** (11/12) 47–49, 58.

Sarkar, A.K. (1971) *Fluorescent Whitening Agents*, Merrow Publishing Co., Ltd., Watford, UK.

Sarma, T.S., Mittal, R.M. and Narasimham, T. (1989) Preparation of cotton fabrics at low temperatures using persulfate activated hydrogen peroxide. *Textile Research Journal*, **59** (12), 748–753.

Shenai, V.A. and Narkhede, A.S. (1987) Hypochlorite oxidation of cellulose in the presence of cobalt sulphide. *Textile Dyer and Printer*, **20** (10), 17–22.

Shukla, S.R., Mathur, M. and Narayan, A. (1993) Comparative studies on bleaching silk. *American Dyestuff Reporter*, **82** (7), 42–46.

Sikdar, B., Adhikari, D. and Das, N.N. (1987) Bleaching of jute by hydrogen peroxide in borate buffer. *Indian Journal of Textile Research*, **12** (2), 93–96.

Simpson, W.S. (1990) Comparison of chemical and photochemical bleaches for wool. In *Proceedings of the 8th International Wool Textile Research Conference* (*Christchurch, New Zealand, February 1990*), vol. IV, pp. 279–287.

Simpson, W.S. (1992a) Novel wool bleaching process utilising blue light. *Wool Research Organisation of New Zealand Communications*, No. C121, WRONZ (Inc.), Christchurch, NZ.

Simpson, W.S. (1992b) Interaction of wool with visible and uv radiation and its relationship to bleaching and appearance. *Wool Research Organisation of New Zealand Communications*, No. C120, WRONZ (Inc.), Christchurch, NZ.

Speakman, J.B. (1941) The chemistry of wool and related fibres. *Journal of the Textile Institute*, **32**, T106.

Steiner, N. (1993) Evaluation of peracetic acid as an enviromentally safe alternative for hypochlorite in textile bleaching. In *Book of Papers, International Conference and Exhibition, American Association of Textile Chemists and Colorists* (*AATCC*), *Montréal, Canada*, AATCC, Raleigh, North Carolina, pp. 214–219.

Sukumar, N. and Gulrajani, M.L. (1985) Kinetics of bleaching agent decomposition in a single-stage preparatory process. *Textile Research Journal*, **55** (6), 367–371.

Tendulkar, S.R. (1993) Bleaching of hosiery fabric by usage of hydrogen peroxide. *Textile Dyer and Printer*, **26** (1), 24–26.

Trollip, N.G. (1990) The bleaching of pigmented animal fibers using cobalt(II) as the peroxide catalyst. In *Proceedings of the 8th International Wool Textile Research Conference* (*Christchurch, New Zealand, February 1990*) vol. IV, pp. 307–319.

Trotman, E.R. (1968) *Textile Scouring and Bleaching*, Charles Griffin and Co. Ltd., London, pp. 16, 116.
Venkatraj, R. and Gulrajani, M.L. (1989) Bleaching of polyester/cotton blended fabrics with a foamed composition. *Indian Journal of Textile Research*, **14**, 35–38.
von Bergen, W. (ed.) (1963) *The Wool Handbook*, Vol. 1, 3rd edn, Interscience Publishers, New York, p. 143.
Wade, R.C. (1967) Method of bleaching cellulose fibres. US Patent 3,318,657, May 9, 1967.
Wasinger, G. (1992) Ozone decolorization of garments. US Patent 5,118,322, June 2, 1992.

3 Fire-retardant textiles

D.P. OULTON

3.1 The problem

Fire takes a steady toll of both human life and property. The results are often tragic and in many cases (at least potentially) avoidable. The problem has many components which can be summarized as follows.

1. A wide range of commonly used materials is flammable.
2. A range of situations, in theatres, restaurants, clubs and aircraft, can cause heavy loss of life in the event of fire.
3. Use of particular materials, garments and upholstery structures can lead to sharply increased personal hazard.

Use of reduced flammability materials, testing of both materials and complete products, regulations and legislation have been applied to the problem. Each has contributed to reducing the risk of accidental injury, death or loss.

Progress in reducing the hazard has been made, but is constrained by a series of factors.

1. The risk is not widely appreciated by the general public, except when a major disaster occurs.
2. Any solution which restricts personal freedom is likely to be avoided by a significant number of people.
3. Any product with reduced aesthetic qualities will be less successful even if it has fire-retardant properties.
4. Any solution which adds significantly to product cost will tend to be resisted.

3.1.1 *The risk*

About 16 000 fires a year result from ignition of textile materials in the UK, most of them domestic fires.

In the UK, fires in dwellings produce about 6000 casualties per year, 650 of them fatal (Home Office, 1985). A major proportion of deaths (at least 50%) are attributable to toxic fumes and smoke, but in clothing fires the major hazard is burns.

In the USA one person is burned to death every 44 minutes in clothing and other fires.

If fire happens in public places horrific loss of life has occurred, as the following examples illustrate.

At Boston, Massachusetts, USA, 750 customers were present at a nightclub in 1942. The fashion at the time included stiffened lace party dresses. A common finishing stiffener was, unbelievably, nitrocellulose. In the flash fire which engulfed the customers, 432 died and 214 were injured.

The notorious 'Ring Theatre Disaster' in Vienna, Austria, caused 450 deaths when a back-stage fire spread rapidly to the auditorium.

A relatively minor fire at a store in Manchester, UK, killed ten people. It is however important in that it highlighted the danger of inflammable upholstery foam.

A whole range of aircraft, train and auditorium fires have been influential in both highlighting dangers and generating legislation to control risks.

3.2 The combustion process

In order to understand how flammability hazards can be reduced, it is necessary to study the nature of the combustion process. Combustion can very easily develop into a runaway self-feeding event producing enormous heat, suffocating and toxic fumes, and total destruction. Figure 3.1 shows the essential components: fuel, an oxidant (usually air) and an ignition source.

Heat and fumes feed back readily burned gases and give increased ignition. The rate of combustion increases, and can reach explosive violence in so called 'flashover' events. Such an event caused major loss of life in the Kings Cross Underground station disaster in London, UK, in 1989.

A major contribution to saving life is to 'retard' the combustion process. By slowing down the developing situation, crowds can escape, fires can be extinguished, and garments removed. Time is available to formulate and execute the necessary measures.

3.3 Developing a fire-retarding solution

Over many years a fourfold attack has been mounted on loss of life and injury by fire.

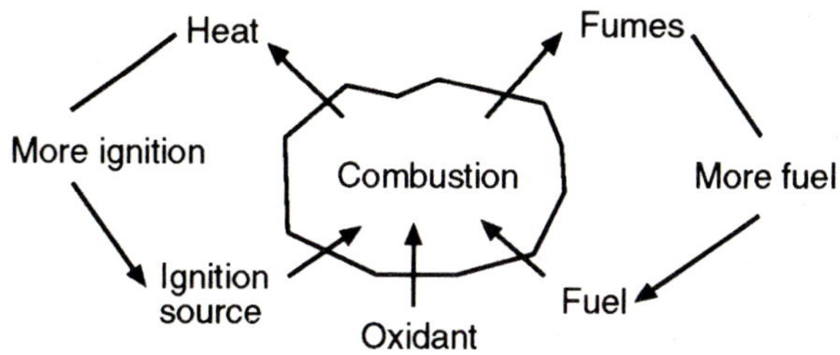

Figure 3.1 The combustion process.

(*a*) *Design for fire retardancy and escape.* This covers every aspect of garment, furniture, architectural, and transport design. It has resulted in fire fighters' uniforms capable of protecting against 2000°C exposure for 30 s. Buildings are designed to prevent spread of fire and smoke, and theatres have a safety curtain between stage and auditorium.

(*b*) *Legislate to eliminate known risks.* The UK Children's Nightwear (Safety) Regulations, SI 616 (HMSO, 1959), prohibit the sale of nightwear which does not pass appropriate flammability tests. The UK 1980 Upholstered Furniture (Safety) Regulations were the result of much detailed assessment of fires in the home. The Manchester store fire, long predicted by those aware of hazards, gave the final impetus to legislation in the UK on foam-filled upholstery.

(*c*) *Develop adequate test methods.* It is often not clear precisely what the nature of the hazard is, until similar events can be reproduced in a laboratory. Flammability testing has been continuously improved in recent years. It has also been realized that testing materials and components on their own is inadequate. Modern testing often involves complete structures, furniture, windows, drapes, etc., tested in carefully designed settings (Horrocks, 1986).

(*d*) *Develop fire-retardant materials.* It is probably true to say that fire-retardant materials have been developed to meet every conceivable hazard. In theory nobody should be injured if the best available fire-retardant material were always used, and an optimum garment or furnishing structure were chosen. It is, however, also equally true that only space programmes and the like can afford the really exotic materials. Many cheaper solutions give reduced comfort, colour, or aesthetic properties. There are no solutions at all which do not add significantly to the cost of products.

In the following sections we will concentrate on the chemistry of fire retardancy, fire-retardant fibres, and fire-retardant finishes for flammable materials.

3.3.1 *Terminology and standards*

It is important, when discussing fire retardancy, and when selling fire-retardant products, to use precise terminology. If you make an assertion that a product 'has low flammability' or 'is flame-resistant' or 'does not burn', you may have to defend the statement in a court of law. Terminology, testing and labelling must be clear and unambiguous.

1. Use approved and defined terminology.
2. Refer to the BS or ISO tests by which the terminology is established (e.g. BS 6336: 1982)
3. Ensure that the products and materials pass the appropriate tests.

The terms 'flameproof' and 'fire-proof' should be avoided. 'Fire-retardant' is a useful general term. 'Does not sustain flame' is the approved term (BS 6336: 1982). 'Self-extinguishing' is widely used to describe materials which only burn in contact with an ignition source. 'Reduced fire hazard' or 'resists ignition' are also acceptable terms but should only be used in connection with a specified test. A variety of carefully designed tests is available by which the property can be measured.

Test methods and their execution and interpretation are a highly specialized area. Individual testing houses set up appropriate apparatus, train staff and receive validation and certification. It is common practice to use such certified testing procedures to ensure product integrity.

Relevant British and international standards include ISO 9000, BS 5750 and 7750, and NAMAS certification. These cover validation and accreditation. All British standards in this field make a clear distinction between methods of test and performance in testing. BS 5438, 6807, 5852 for example, describe test methods. Performance in these tests is classified in standards such as BS 5722, 7177, 6249. Both the test and the performance criteria met need to be quoted when describing a product; for example, 'Flame-retardant finish durable to repeated washing BS 5438: 1976 Test 2a to BS 5722 performance for sleep wear'. Table 3.1 lists some important British Standards. Some equivalent US standards include:

DOC FF 3-71 & 5-74	Children's sleepwear
CS191-53	
ASTMD 1230-1983	Tests for flammable fabrics
NFPA 702-1975	
CPSC NBBIR 76-1072 (1976)	Burn simulation (mannequin) test
US FAR 25.85b	Fire resistance of aircraft seating

Table 3.1 Some important British Standard tests for flammability

British Standard	Purpose
BS 2963: 1958	Test for flammability of fabrics
BS 3119: 1959	Tests for flameproof materials
BS 3120: 1959	Performance requirements for flameproof clothing
BS 3121: 1959	Performance requirements for fabrics of low flammability
BS 5438: 1976	Test methods for vertically orientated fabrics and assemblies (small flame ignition)
BS 5722: 1974	Performance of fabrics and assemblies for sleepwear and dressing gowns
BS 5852: 1979	Fire tests for furniture, textile and foam interior combinations: pt 1 ignition by smokers' materials; pt 2 ignition by flaming sources
BS 5867: 1980 (pt 2)	Flammability requirements for curtains and drapes
BS 6307: 1982	Ignition of textile floor coverings (small ignition source)
BS 6357: 1983	Tests for resistance to molten metal splash for protective clothing

3.4 Flammability of materials

Materials can be classified quite clearly into those which will burn in air once ignited, and those which are self-extinguishing if not in direct contact with an ignition source. Amongst those materials which will burn freely, two other properties may add significantly to the hazard:

(*1*) *The heat of combustion.* Any material with a high heat of combustion provides increased ignition for surrounding material. Heats of combustion can vary from 11.6 kcal/g for polypropylene down to 3.4 kcal/g for viscose rayon.

(*2*) *Thermal properties.* A series of thermal transition temperatures cover important events when a material is exposed to an ignition source.

- T_g The glass transition or softening temperature, particularly in clear rigid platics
- T_m The melting point (thermoplastics only)
- T_p Pyrolysis temperature: rapid thermal degradation of the polymer
- T_c Combustion temperature: spontaneous ignition occurs

Some materials such as wool and modacrylic have a T_c as high as 600–700°C and are difficult to ignite.

3.4.1 *Limiting oxygen index*

The atmosphere contains about 21% oxygen. The rest (nitrogen, CO_2, water, etc.) does not support combustion. It has been found by experiment that increasing the ratio of oxygen to nitrogen makes some materials burn that are self-extinguishing in air. Conversely, materials which burn readily in the air are self-extinguishing in nitrogen-rich atmospheres.

This leads to a useful and general test for classifying flammability. The limiting oxygen index (LOI) test seeks to establish the minimum oxygen content atmosphere in which a material will continue to burn.

Those materials with a LOI greater than 25 are generally self-extinguishing in air. Those with the lowest LOI values burn most readily. Very few materials indeed refuse to burn in atmospheres approaching 100% oxygen.

A full description and discussion of the LOI test is given elswewhere (Lewin and Sello, 1984), and Table 3.2 lists a range of materials and their LOI values.

3.4.2 *Insulation and retardancy*

An important feature of fire retardancy is heat insulation. Any structure or material which prevents heat reaching further combustible material, or the

Table 3.2 Limiting oxygen indices (LOI): minimum percentage oxygen contents for sustained combustion

Substance	LOI
Polyethylene oxide	15.0
Poly(methyl methacrylate) (Perspex)	17.0
Polyacrylonitrile	18.0
Polyethylene	18.0
Polystyrene	18.5
Polypropylene	18.6
Cellulose	19.0
Poly(ethylene terephthalate)	21.0
Poly(vinyl alcohol)	22.0
Nylon 6.6	23.0
Wool	25.2
Modacrylic	26.8
Polycarbonate	27.0
Nomex (Aramid)	28.5
Phenol–formaldehyde resin	35.0
Neoprene	40.0
Poly(vinyl chloride)	60.0
Carbon	60.0
Polytetrafluoroethylene	95.0

human body, is valuable. Several types of material achieve this:

1. Non-thermoplastic materials (wool, carbon fibre)
2. Carbonaceous char
3. Silica and silicon-based polymers

A further by-product of insulating properties is the limitation of access of oxygen to the combustion centre.

3.5 Which materials are best?

This is a typically ill-defined question. We must distinguish between end uses. We must solve the often contradictory demands of performance, cost, and aesthetic effects, and we must assess the effects of combining materials into products. For example, a very different solution is called for in hazardous environments such as those a firefighter may encounter or a racing driver may suffer in a crash.

We will leave a fuller discussion until we have reviewed what can be done to enhance the flame-retardant properties of popular cellulosic fibres, polyester and wool.

3.5.1 *Wool as a low-flammability material*

Wool has many of the required properties. Its LOI at 25.2 is, however, marginally above the atmospheric oxygen content. It will thus burn in air under favourable conditions. It is non-thermoplastic and has a very high T_c (600°C). Furthermore, its heat of combustion is low at 4.9 kcal/g. In dense tightly woven structures wool is an excellent insulator, and produces dense black char when burnt.

From the above it is possible to predict that wool drapes, furnishings, and carpets will not be a major hazard. Any fire will burn, if at all, only slowly. In a carpet, heat generated rises away from the fuel and the ignition source is lost. In a curtain, heat rises towards fresh fuel, so burning behaviour is quite different.

3.5.2 *Flammability of polypropylene*

With an LOI of 18.6 and a heat combustion of 11.6 kcal/g, polypropylene burns readily and fiercely in air. As a thermoplastic with T_m 165°C it shrinks and melts away, exposing more material or skin to heat. Molten flaming polymer readily starts fires elsewhere. Its T_c at 550°C indicates some resistance to casual ignition, e.g. by a smouldering cigarette.

3.5.3 *Flammability of cellulosic materials*

Cellulosics have a similar LOI to polypropylene but lower heat of combustion (4.3 kcal/g). A T_c value of 350–420°C indicates relatively easy ignition. As a result, cotton and viscose textiles burn freely if ignited. A small flame is enough to ignite the hem of a nightdress. Combustion spreads rapidly, and can cause fatal burns within 15 s of ignition. By contrast, flame-retardant treated cotton is self-extinguishing and completely safe for use in nightwear.

3.5.4 *Thermoplastic fibres as low-flammability materials*

Both polyamide (LOI 23.0) and poly(ethylene terephthalate) (LOI 21.0) support combustion in air. Thermoplasticity, however, has an important effect on burning behaviour. Synthetic fibres shrink considerably just below their melting point, T_m. This causes materials to shrink away from ignition sources making established combustion less likely. Further, molten flaming polymer falls away and removes the heat likely to spread combustion. Thus, in free-hanging garments such as nightdresses, brushed nylon provides an acceptable risk and good aesthetic properties.

3.5.5 *Testing flammability*

The tests established in BS 5438 and ASTM D1230 provide controlled reproducible ignition systems, and measurement of following combustion. This may be either as flame or as glowing char. Each test has been carefully compared with the burning behaviour of garments, furnishings, and carpets.

In addition, appropriate test methods lay down standardized washing and solubility tests. These are used to assess durability of flame-retardant finishes.

Several special hazards have been identified and allowed for in testing. Both soap and fabric softeners are flammable materials. In hard-water conditions, significant deposits can build up, masking the flame-retardant effect.

Bleaching of garments with sodium hypochlorite, which is widely used in the USA, particularly for children's clothes, can modify the finish, or produce oxycellulose. This renders an otherwise durable finish ineffective.

3.5.6 *Durability*

A dilemma faces a garment maker who wishes to describe a finish as 'durable'. How long must the protection last? Detailed research suggests that if the protection is still intact after 200 domestic wash–dry cycles, it is likely to outlast the garment. Both extended testing and accelerated testing have been used and compared in establishing suitable standards, e.g. BS 3121, BS 5722, and the American DOC FF 3.71.

3.5.7 *Tests for furniture*

The dramatic and powerful combustion resulting from ignition of foam-filled upholstery has caused serious loss of life. House fires can become uncontrollable in under 3 min. Tests (e.g. BS 5852) have been established for such composite structures. The tests use a series of ignition sources of increasing intensity. They simulate ignition sources from a smouldering cigarette, through a lighted match, to burning debris. The latter simulates situations such as train and aircraft fires, or a domestic fire which is already in progress.

In each case, a controlled reproducible ignition source is used. Time, temperature, duration and size of ignition event are the critical factors.

The outcome of the attempted ignition will depend mainly on the materials involved. However, 'barrier layer' structures are important. Figure 3.2 shows such a structure.

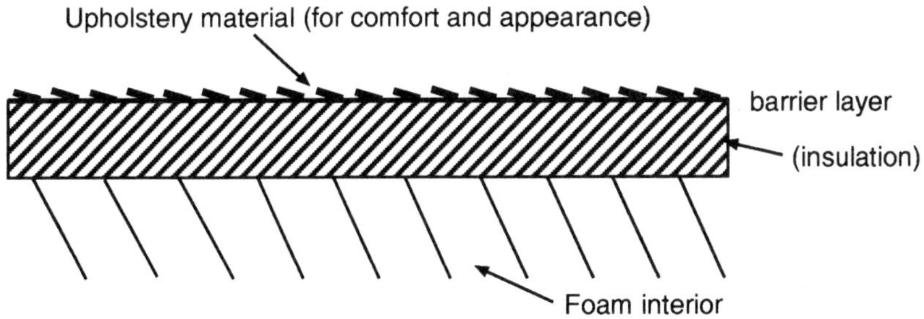

Figure 3.2 A barrier-layer fire-retardant structure.

3.5.8 *Barrier-layer fire-retardant structures*

The barrier layer must:

(a) Stay substantially intact even after pyrolysis
(b) Minimize heat transfer to the interior
(c) Limit the supply of oxygen to any combustion occurring underneath.

A variety of materials have been used successfully as barrier-layer materials. Carbon fibre is widely used, so is treated cotton.

Back coating of upholstery with polymer-bound fire-retardant chemicals is widely used for less demanding domestic applications.

3.6 The chemistry of fire-retardant finishing

A range of strategies can be used to modify the combustion characteristics of textile materials.

3.6.1 *Exclusion of oxygen*

If it is possible to separate the fuel (fibres and flammable combustion gases) from the oxidant the combustion will collapse and be extinguished.

Vapour phase flame retardants achieve this by generating large volumes of nonflammable gases on pyrolysis, usually HBr and HCl. This is the mechanism which gives PVC its high LOI. It is also the basis of several back-coating formulations.

Fibre coating flame retardants use a different strategy. A medium–high melting-point glass-forming material such as borax–boric acid is used. This has the added advantage of not producing toxic fumes on pyrolysis.

Chlorinated hydrocarbon fire extinguishing systems and to some extent water-based fire extinguishers also work in part by exluding oxygen. However, removal of heat (ignition source removal) is also an important strategy.

3.6.2 *Modification of the pyrolysis route*

Given that combustion is maintained mainly by heat generation, any system that reduces heat of combustion is likely to be helpful. The two major natural fibre classes, cellulosic and protein, can be made self extinguishing (LOI up to 30) by this strategy.

The detailed chemistry of pyrolysis is highly complex. Several texts, and indeed multivolume publications, are devoted to it (Fenimore, 1970; Kuryla, 1979; Lewin and Sello, 1984; Horrocks, 1986).

Pyrolysis of cellulose. Fortunately the overall process can be expressed more simply. Cellulose (**1**) can be considered for this purpose as a polymeric carbohydrate.

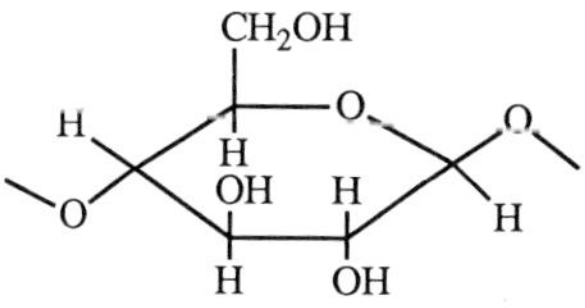

OR $(C_6H_{10}O_5)_n$

(1)

On pyrolysis the chains break up, giving a wide range of low molecular weight aldehydes (**3**), ketones (**2**) and alcohols (**4**), which are highly volatile and flammable. In addition, less volatile but equally flammable higher molecular weight tars are produced. Some material is also turned into solid carbonaceous char.

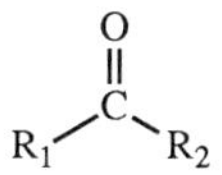

(2)

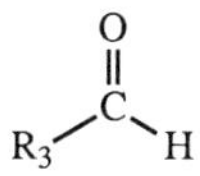

(3)

R_4—OH

(4)

Given a free supply of oxygen and an ignition source cellulose oxidizes completely with a substantial evolution of heat.

$$(C_6H_{10}O_5) + 6O_2 \xrightarrow[350°C]{} 6CO_2 + 5H_2O + \text{Heat} \tag{3.1}$$

The mechanism of oxidation is largely a free-radical one:

$$\text{Ignition} \quad R\text{—}H + 2O_2 \rightarrow R\text{—}OOH + RO^{\bullet} + HO^{\bullet} \tag{3.2}$$

$$\text{Combustion} \quad R\text{—}H + HO^{\bullet} \rightarrow R^{\bullet} + H_2O \tag{3.3}$$

Chain branching:

including $R\text{—}H + HO_2^{\bullet} \rightarrow RO^{\bullet} + HO^{\bullet}$

or $O_2 \rightarrow 2O^{\bullet}$

or $R\text{—}H + O^{\bullet} \rightarrow RO^{\bullet} + H^{\bullet}$

or $H^{\bullet} + O_2 \rightarrow HO^{\bullet} + O^{\bullet}$

Cellulose may also decompose by dehydration:

$$(C_6H_{10}O_5) \xrightarrow[\text{Catalyst}]{240°C} 5H_2O + 6C \tag{3.4}$$

The result is a mass of carbonaceous char, little if any flammable volatiles, and much less heat.

Under normal combustion conditions a mixture of equations (3.1) and (3.4) occurs. Reaction (3.1) is favoured by lightweight, open, free-hanging materials, which burn fiercely. Heavily folded tight-woven materials will often smoulder and char, mainly by equation (3.4).

It is also possible to influence pyrolysis by using suitable chemicals in favour of equation (3.4). By favouring dehydration:

(a) Pyrolysis starts at a lower temperature
(b) Less heat is evolved
(c) Much less volatile flammable gas and tar is produced

The combustion of cellulose will then be inhibited successfully:

(a) By reducing the heat and flame which spread the fire
(b) By promoting char formation which may limit access to air

The result is a self-extinguishing material with an LOI of 28 or above.

Wool can be raised to a fire-retardant standard suitable for any environment by appropriate finishing. The strategy is similar to that of cellulose, but the chemistry is more complex due to the proteins involved.

Many of the commercial finishes available promote dehydration and char formation rather than oxidation.

3.6.3 *The role of Lewis acids in promoting dehydration*

The action of Lewis acids is central to the fire-retardant finishing of cellulose, by favouring dehydration.

A Lewis acid is a chemical which, under appropriate conditions, is strongly electron-absorbing. It is paired in the reaction with a Lewis base, which donates electrons (equation 3.5).

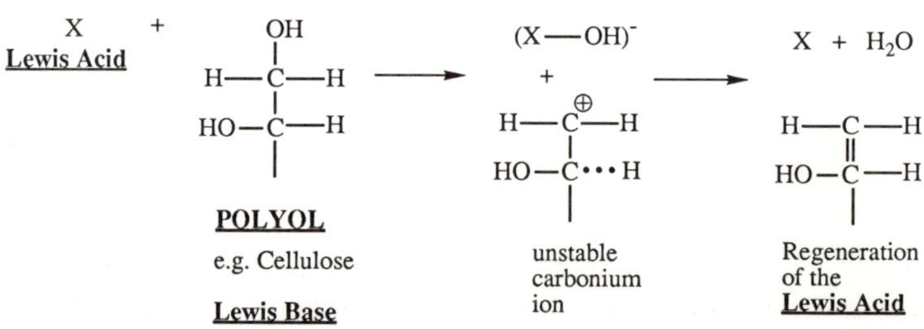

(3.5)

Thermogravimetric analysis of cotton treated with monoammonium phosphate (MAP; a good Lewis acid) confirms a fundamental change in the pyrolysis route (Table 3.3).

A wide range of water-soluble phosphorus-containing chemicals are active Lewis acids and give good temporary flame retardancy to cotton. The problem

Table 3.3 Percentage loss in weight after 10 min

Temperature	250°C	280°C	300°C
Untreated	3	5	7 (of dry wt)
MAP (1.7% P add-on)	50*	70%	80

* Carbonization virtually complete

is to combine this with:

(a) Durability lasting the life of the garment
(b) Minimal effect on the softness and drape
(c) Minimal effect on dyes
(d) Low cost
(e) Low emission of toxic fumes
(f) Environmentally friendly application
(g) No health hazard to wearer or finisher

A large number of commercial finishes have met most of these criteria but none has met all of them.

3.6.4 *Vapour-phase fire-retardant action*

The majority of chemicals active by the vapour-phase mechanism are chlorinated hydrocarbons. On pyrolysis, these release large quantities of nonflammable gases such as HBr and HCl and Cl_2. A good example is the now notorious 'tris' – tris-(2,3-dibromopropyl) phosphate ($CH_2Br \cdot CH \cdot BrCH_2O)_3P(O)$ (**5**).

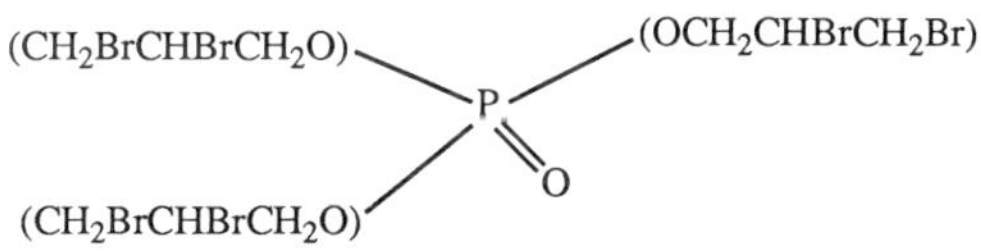

Tris-(2,3-dibromopropyl) phosphate (**5**)

A great advantage of vapour-phase retardancy is that it works on any substrate and does not depend on the chemical nature of the material. Tris was used extensively on polyester and polyester cotton until it was found to be potentially carcinogenic, and banned from further use.

It should be noted that tris combines both chlorinated hydrocarbon and Lewis acid entities. On the approriate substrate, e.g. polyester/cotton, two separate mechanisms come into action. The result is often better than that obtained by either single mechanism. This is called a synergistic effect, and is common in flame-retardant finishing.

3.6.5 *Metal oxides and salts as flame retardants*

Good flame retardancy can be achieved on both cellulose and wool by incorporating metal oxides and/or salts in an appropriate finish.

The literature does not give many references to mode of action, but one author suggests they might act as free radical chain terminators, e.g.:

$$SbX_n + H^{\bullet} \rightarrow HX + SbX_{n-1} \tag{3.6}$$
$$HX + HO^{\bullet} \rightarrow H_2O + X^{\bullet}$$
$$X^{\bullet} + X^{\bullet} \rightarrow X_2$$

Titanium and zirconium are used successfully on wool as their citrates and fluorides. Antimony oxide and chloride find use on cellulosics.

3.6.6 *Synergistic flame-retardant effects*

Synergism is the term used when the sum of the effects is greater than the individual effects of the components. It seems mysterious and almost magical, but can be understood if the complex nature of combustion and retardant action are understood.

Many of the mechanisms of combustion are interdependent. Generation of volatiles precedes the spread of flame. Oxidation produces volatiles. Dehydration reduces material for volatile generation. It thus follows that simultaneous attack on more than one mechanism can produce a collapse of the combustion process.

In practice, metal oxides are often combined with vapour-phase retardants. Nitrogen-containing materials are combined with phosphorus. This results in some remarkable 'witches' brews' which are sold as commercial flame retardants. The picture is further complicated by the need to bind the chemicals to the fibre to make them durable.

The above account of combustion and pyrolysis mechanisms is necessarily over-simplified. In practice, the pyrolysis of cellulose and protein follow a wide range of routes between the two extremes of equations (3.1) and (3.4). Hundreds of intermediate chemicals, short-lived free radicals, tars and volatile products have been identified.

It follows that no clear equations or specific routes for fire retardant action can be stated with certainty. The mechanisms and reactions given above illustrate some of the possible routes.

3.7 Commercial solutions for fire retardancy

3.7.1 *Design*

Substantial progress has been made in the design of buildings, aircraft, trains, upholstery and garments to avoid flammability and larger-scale fire risks.

Alongside this, progress has been made in testing. The emphasis in testing is on complete structures rather than individual materials, which are adequately characterized by existing tests.

The key to designing low-risk structures is spread of combustion and combustion products (toxic fumes, smoke, etc.). If a fire remains small, it can be dealt with easily, and does not cause panic.

For a given material, e.g. cellulose, the rate of combustion varies enormously. This is illustrated (with suitable precautions) by igniting two sheets of newspaper. The first is held open and ignited at the bottom edge. (Be prepared to extinguish substantial flames within seconds.) The second sheet is rolled tightly. This burns very much more slowly, and controllably.

It is clear from this and similar observations that fabric and garment structure can play a substantial role in fire retardancy. Fine nets, free hanging edges, streamers, and open structures with free access to air all present maximum combustion rates, and likely points for casual ignition. By contrast, heavy close-woven materials, multilayer structures, and freedom from easily ignitable features reduce the risk.

In many design briefs it is not possible to avoid the risks described above for aesthetic or comfort reasons. In such cases designers return to flame-retardant materials to minimize the risk. In structures for particularly hazardous environments such as firefighters' uniforms, aircraft interiors, and protective overalls, both flame-retardant materials and fire-resistant structures are used.

3.7.2 *Intrinsically fire-retardant materials*

Few natural materials, with the exception of mineral fibres, are intrinsically self-extinguishing. Most synthetic polymers are flammable, unless specifically modified to make them self-extinguishing, by physically or chemically including fire-retardant chemicals.

Substantial research effort has been invested in the manufacture of intrinsically fire-retardant textile fibres. All the major polymer types with the exception of polyolefins have fire-retardant versions. The chemical nature of the modifications is not generally revealed. It is however reasonable to assume that the fire-retardant properties are obtained by addition of one or more fire-retardant chemicals to the polymer mass prior to spinning the fibre.

Examples chosen at random from the many commercial offerings include:

Viscose F.R. made by Lenzing of Austria. This has normal textile properties associated with viscose and is promoted for upholstery, domestic and industrial protective markets.

FIDION F.R. is a flame retardant polyester produced by Enichem Fibre Spa of Italy, who also produce SIRONIL F.R. and PANOX described below.

Trevira C.S. (comfort and safety) is another fire-retardant polyester by Hoechst of Germany. It is widely used and promoted for domestic and public auditorium upholstery and curtain applications. It is described as a phosphorus-containing copolymer.

Acrylic co-polymers: a wide range of flame retardant acrylic fibres is on the market including SIRONIL F.R. The comonomers used, up to 30% in some cases, include vinyl chloride, vinyl alcohol, and vinylidene chloride.

All the above are very similar in both appearance and properties to their generic fibre types.

Poly(vinyl chloride) fibres, such as Rhovyl and Clevyl, have a share of the fire-retardant market.

Oxidized acrylics such as PANOX (Enichem) and PANOTEX Universal. Carbon fibres are highly fire-resistant and widely used as barrier-layer materials.

The oxidized acrylics are essentially carbon fibres and are naturally black. They are non-thermoplastic, and resist even an intense blow torch flame for a significant time.

Aramid fibres, such as Nomex and Kevlar, have high enough LOI values to be self-extinguishing. In addition, they are strong and have high melting points. They find use in applications such as racing drivers' overalls.

Silicon-based polymers, unlike carbon-based materials, are generally incombustible. The Kemira company of Finland have recently announced their new fibre VISIL 33 A.P. (The Textile Institute, 1993). It is a viscose fibre with a silicic acid backbone. Smoke emission is low and it is free from toxic fumes.

The potential hazards from the use of phosphorus-containing and chlorine-based products add to the cost of fire retardant solutions containing them. In an environmentally conscious era it is vital to test the chemicals used, the effluents produced (both air and water), and the products themselves. No responsible company can ignore the safety of its workers or its customers.

Tests of both chemicals and products for toxicity, carcinogenic activity, teratogenic (damage to unborn foetuses) activity, and dermatitis are used. Concern is growing over the hazards from combustion products as well as from materials themselves. It will no doubt be an interesting problem for future chemists to isolate hazards, devise tests and invent new fire-retardant solutions avoiding the hazards.

3.7.3 *Commercial fire-retardant treatments for cotton*

Good commercial fire-retardant treatments successfully combine durability with low flammability. It is necessary to deliver the appropriate active chemical in a washfast form. The resulting material should not lose strength, abrasion resistance, lightfastness or general durability, so the finish needs to be chemically inert and neutral in pH.

THPC

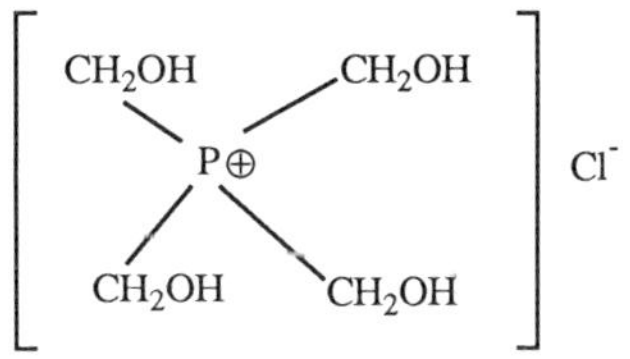

Tetrakis-hydroxymethylphosphonium chloride **(6)**

For upholstery fabrics, back coating with a polymeric binder containing metal oxide and chlorinated hydrocarbon flame retardants is commercially successful. Water-based rather than solvent-based application systems are preferred for environmental reasons.

Two different approaches are used to obtain durable phosphorus-containing finishes on cotton.

(*1*) *Use of highly cross-linked polymeric resins.* These are applied as low molecular weight chemicals and polymerized inside the fibre. This locks the resin permanently into the interior structure of the fibre.

(*2*) *Use of reactive chemicals which bond covalently with cellulose.* A chemical bond with cellulose ensures high durability, provided the chemical bond is not easily attacked by conditions in use or washing. Domestic and industrial washing can involve both bleaches and alkali so the choice of chemical link must be made carefully.

3.7.3.1 *Systems based on THPC (tetrakis-hydroxymethylphosphonium chloride* **(6)** THPC has several important properties. The *P*-methylol groups are quite reactive due to the electron withdrawing effect of the phosphorus atom:

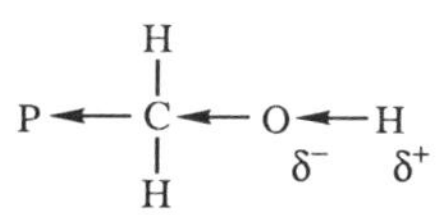

They thus readily undergo condensation reactions, e.g. with other *P*-methylol groups or with amine groups such as urea or ammonia (Scheme 3.1).

Compound **(7)** is a widely used resin precondensate. It has six reactive *P*-methylol groups, which react to produce highly cross-linked insoluble resins. Proban 210 from Albright & Wilson is based on this precondensate.

Resin formation based on condensation reactions of both *P*-methylol and *N*-methylol are widely used in textile finishing. In almost all cases the condensation polymerization is carried out at elevated temperatures with an acid catalyst in a pad–dry–bake sequence.

The Proban process is unique in using a room temperature ammonia

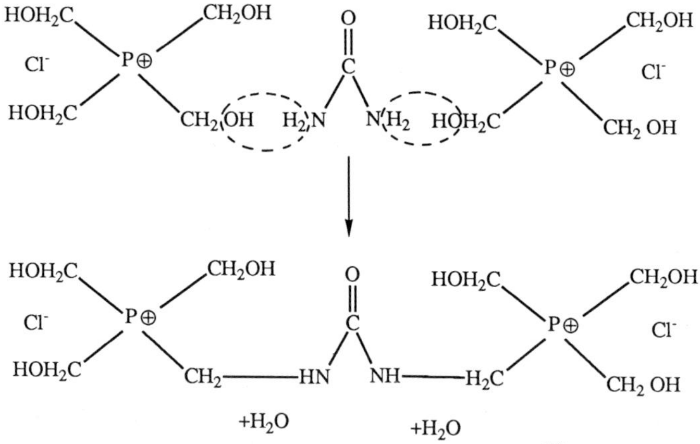

(7)

Scheme 3.1

gas-based polymerization. The central phosphorus atom can be in one of three states:

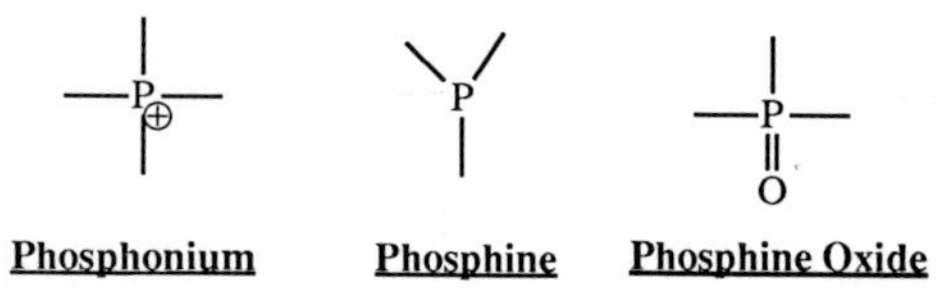

The phosphonium salt precondensate is buffered to pH 6 in a pad–dry application using sodium acetate; at pH 6 most of the salt has been converted to the phosphine form. At this pH, reaction rate with ammonia is at a maximum.

$$\underset{\text{Phosphonium}}{-P^{+}(CH_2OH)_3} \rightarrow \underset{\text{Phosphine}}{-P-(CH_2OH)_2} + HCHO + H_2O + Cl^{-} \quad (3.7)$$

Full polymerization is very rapid and exothermic, giving a three-dimensional cross-linking polymer (**8**). It should be noted that stoichiometrically the atomic ratio N:P has risen to 2:1 in the polymer from 1:1 in the precondensate.

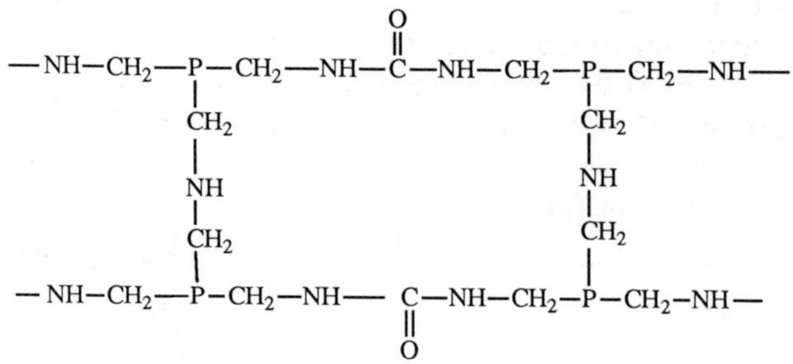

(8)

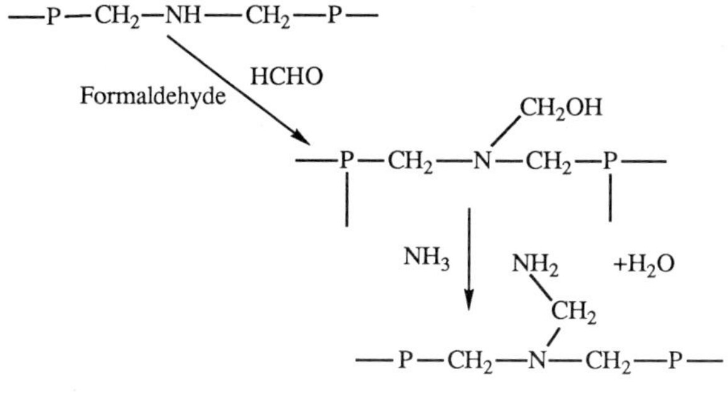

Scheme 3.2

This ratio can be monitored to ensure efficient curing. Side reactions of secondary amine groups may increase the ratio (Scheme 3.2 and equation 3.8).

$$6HCHO + 4NH_3 \rightarrow \underset{\text{Hexamethylene tetramine}}{N_4(CH_2)_6} \qquad (3.8)$$

If the pH rises significantly above 6 the reactivity of the precondensate falls due to phosphine oxide formation (equation 3.9).

$$—P—(-CH_2OH)_2 + OH^- + H_2O \longrightarrow H_2 + —\overset{\displaystyle O}{\overset{\|}{P}}—(CH_2OH)_2 + OH^- \qquad (3.9)$$

The THPOH process. In a similar finishing process known as the THPOH process, THPC (the phosphonium salt) is converted to the phosphonium hydroxide form (equation 3.10).

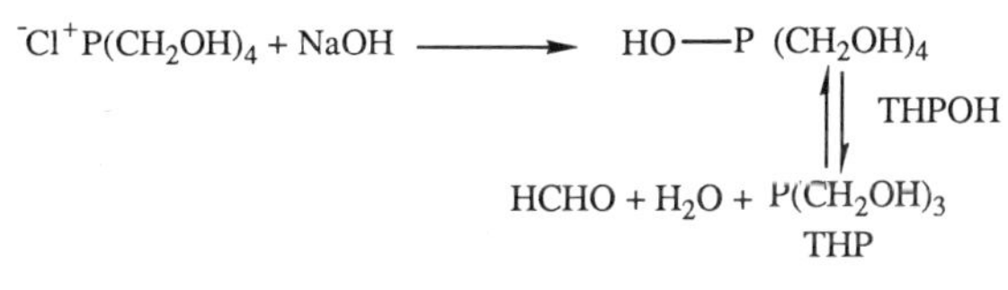

(3.10)

THPOH/THP show high reactivity with ammonia. The reaction rate peaks very sharply at pH 6. The phosphine form of the resin smells quite strongly, It also has a tendency to reduce lightfastness of dyes. It is thus necessary to follow polymerization by an oxidative wash-off with hydrogen peroxide, which converts the phosphine form to the phosphine oxide (equation 3.11).

(3.11)

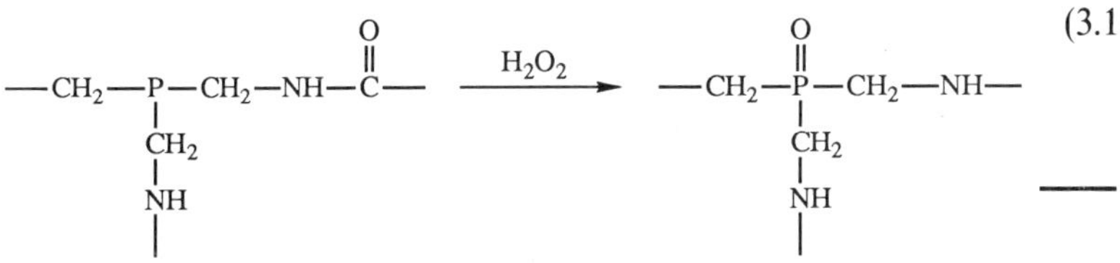

Over-oxidation can lead to loss of durability due to the formation of amine oxides and subsequent rupture of CH_2—NH_2 bonds (equation 3.12).

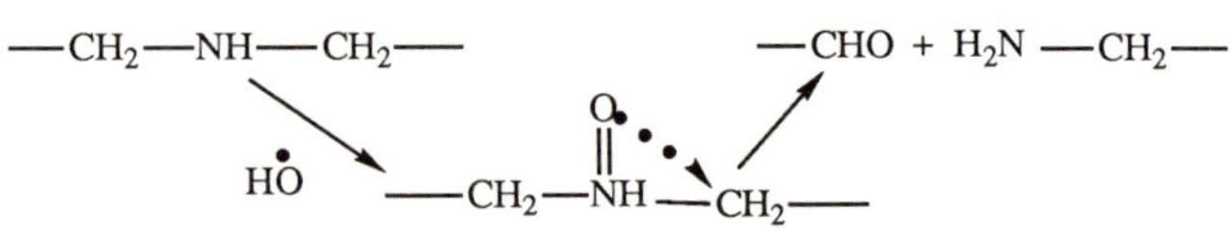

(3.12)

3.7.3.2 *Flame retardancy by phosphonylation of cellulose.* The general formula for phosphonic acid is:

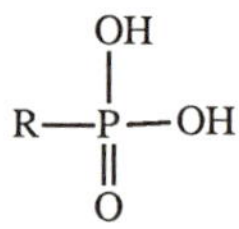

where R is an organic group.

The *N*-methyloldialkyl phosphonopropionamides, e.g. compound (**9**), form stable covalent bonds with cellulose and give good wash fastness in a typical *N*-methylol condensation reaction (Scheme 3.3). The commercially successful Pyrovatex CP (Geigy) finish is based on this method.

(9)

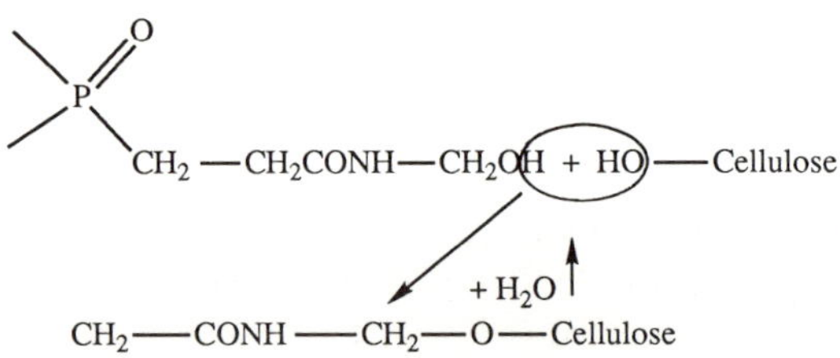

Scheme 3.3

3.7.4 *Finishes for wool*

Wool is only weakly flammable (LOI 25), has a high ignition temperature, and low heat of combustion. Considerable effort has been invested, however, into making it thoroughly flame-retardant, for use in high-risk environments.

THPC and halogenated hydrocarbons have been used effectively, but cheap effective durable finishes are now widely used based on titanium and zirconium complexes; 0.5% Ti and 1.9% Zr add-on are required to raise the LOI to 27. Citric (**10**) acid is a useful addition to the formulation. The alpha hydroxyl (* in compound **10**) stabilizes the complexes (**11**). Titanium hexafluoride $[TiF_6]^{2-}$, zirconium hexafluoride $[ZrF_6]^{2-}$ and zirconium difluorocitrate $[ZrF_2(\text{citrate})]^{2-}$ are also used in various versions of Zirpro wool finishing (International Wool Secretariat). Flame-retardant finishes can be applied to wool simultaneously with shrink-resist finishes in some instances.

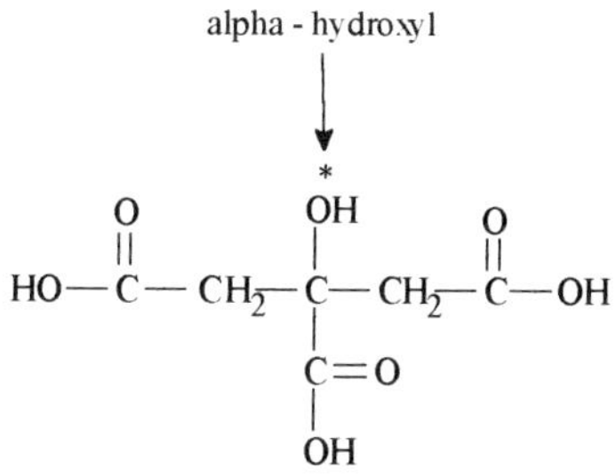

Citric Acid

(10)

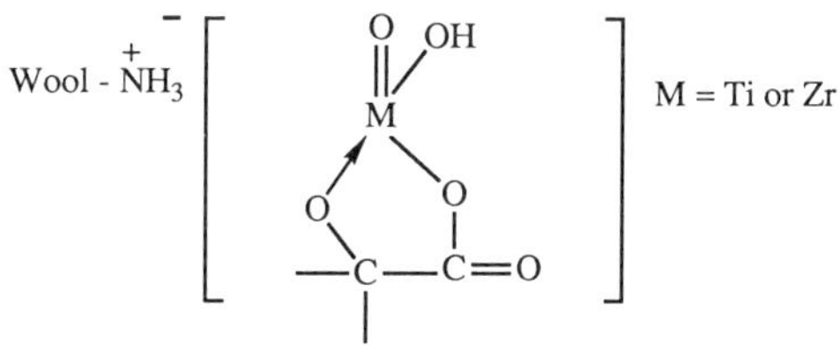

Citrate

(11)

3.7.5 *Finishes for polyester/cotton*

Finishes for 100% polyester have been largely unsuccessful, mainly for durability. Polyester is too hydrophobic for penetration by precondensates, and has too few sites for covalent bonding. The necessary phosphorus-containing chemicals can, however, be incorporated during fibre manufacture, as in Trevira CS and DuPont's Dacron 900F.

Polyester/cotton with up to 50% polyester can be successfully finished by using either THPC or phosphonate-type flame retardants. Rohringer *et al.* (1975) give a graph relating LOI to percentage blend composition at various phosphorus add-on levels. They report LOI greater than 30 for all blends above 50% cotton using a phosphonopropionamide finishing (Pyrovatex) with 3% phosphorus content.

Several attempts have been made to include bromide, either in the precondensate or by post-treatment, when using phosphorus-based retardants.

Caliban F.R. P.44 (White Chemical Co.) has been shown to be a safe and effective coating retardant. It is bound with a latex to give durability and combines aromatic bromide and antimony oxide. Further improvement is given by using a brominated latex.

Both these approaches use vapour-phase as well as condensed-phase mechanisms to increase the retardant effect synergistically.

3.8 Premature failure of flame retardancy

The possibility that an otherwise durable finish might suddenly lose its effect is a serious matter. Several cases of this have been identified. In most cases specific causes have been revealed, and can be avoided.

3.8.1 *Ion-exchange effects in hard water*

Phosphate, phosphonate and/or phosphoric acid groups perform well as Lewis acids, providing condensed-phase flame retardancy. Unfortunately, the calcium and magnesium salts are too stable and have much reduced effect.

Free phosphate groups can pick up Ca^{2+} or Mg^{2+} by ion exchange from hard water (equation 3.13), causing failure of appropriate tests after as few as 20 domestic laundering cycles. This compares with the 50-plus of normal durability.

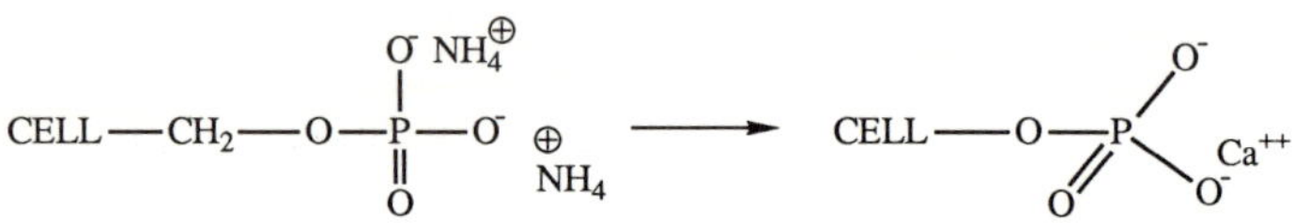

(3.13)

Cellulose diammonium Phosphate **Stable Ca Salt**

Phosphonates, as used in Pyrovatex C.P., are less susceptible to this problem; however, some ion-exchange activity has been demonstrated.

3.8.2 *Sensitivity to chlorine-containing bleaches*

All the phosphorus-containing flame retardants suffer premature loss of durability if laundered with hypochlorite-type bleaching agents. In laboratory tests THPOH–amide and THPC–urea finishes failed after only ten launderings (Horrocks, 1986). Failure appears to be due to direct loss of phosphorus by attack on the resin. As a result of the above findings, such fire-retardant finishes are sold with the explicit warning 'use no bleach' clearly labelled on the garment.

3.8.3 *Detergent and softener effects*

Laundering with soap in hard-water areas can lead to a build-up of insoluble lime soaps on the fabric. These are flammable and can cause premature failure of flammability tests.

Hospitals often use soap/bleach laundering which combines two possible causes of failure.

The recent popularity of domestic fabric softeners gives rise to concern. Many of the cheaper and most effective softeners are either poly(ethylene oxide)- or hydrocarbon-based. Both materials are highly flammable, and if present in significant quantities, can mask flame retardancy.

3.8.4 *Steam sterilization*

The practice of steam sterilizing hospital laundry has an unexpectedly strong effect on finishes which depend on covalent linking to celulose. THPC-based finishes are reasonably resistant, but some phosphonate-based finishes have been shown to lose up to 60% of added phosphorus after only five steam sterilizations with complete loss of flame retardancy. The probable cause is hydrolysis of the ether link to cellulose (equation 3.14).

$$\text{CELL—CH}_2\text{—O—CH}_2\text{NH—} \xrightarrow[\text{Heat}]{H_2O} \text{CELL—CH}_2\text{—OH} + \text{HO—CH}_2\text{—NH—} \qquad (3.14)$$

3.9 The cost of fire-retardant solutions

Even before the extra cost of fire-retardant materials is taken into account, some additional costs are incurred when offering flame-retardant products. These include testing, certification, labelling and promotional costs. It is also likely that product liability insurance will cost more.

The extra cost of materials with reduced fire hazard is also significant. In the upholstery field, back-coating finishes are available which give durable fire retardant action. They add about 10–20% to material costs. At the other end of the scale, specialist fire-retardant fibres can cost three to five times as much as equivalent general textile fibres.

In barrier-layer applications, the choice of barrier fabric is between oxidized acrylic (e.g. Panotex by Universal Carbon Fibres) and fire-retardant treated cotton. Oxidized acrylic has reduced fume generation, and costs up to four times the price of an equivalent cotton fabric.

Fire-retardant treated cotton and wool costs 20–50% more than equivalent untreated fabric. There is also some loss of aesthetic properties.

In view of the above, it is not perhaps surprising that fire-retardant solutions tend to be confined to products that meet specific hazards. Good examples are children's nightwear, aircraft and train interiors, public auditoriums, and protective clothing. Some progress has also been made in domestic upholstery. Here, elimination of highly flammable interior foam has also been a major contributor to reducing the risk.

References

BS 6336: 1982 (1982) Development and presentation of fire tests, and their use in hazard assessment, BSI, London.

Fenimore, C.P. (1970) The mechanisms of pyrolysis. In *Proceedings of 4th Materials Research Symposium*, Gaithersburg, NBS Special Publication 357, National Bureau of Standards, Gaithersburg, Maryland, USA.

HMSO (1959) *The Nightwear (Safety) Regulations*, Statutory Instrument SI616, HMSO, London.

Home Office (1985) *Fire Statistics: United Kingdom 1983*, HMSO, London.

Horrocks, A.R. (1986) Flame retardant finishing of textiles, *Review of Progress in Coloration*, **16**, 90.

Kuryla, W.C. (ed.) (1979) *Flame Retardancy of Polymeric Materials*, Vols 1–5, Marcel Dekker, New York.

Lewin, M. and Sello, S.B. (eds) (1984) *Handbook of Fibre Science and Technology*, Vol. 2, Marcel Dekker, New York.

Rohringer, P., Stensby, P. and Adler, A. (1975) Mechanistic study of flame inhibition, *Textile Research Journal*, **45**, 586–592.

The Textile Institute (1993) *Flammability 93, Proceedings of The Textile Institute Conference*, The Textile Institute, Manchester, UK.

4 Physical and chemical effects of domestic laundering processes

D.P. BISHOP

4.1 Introduction

The manufacture and marketing of household laundry products constitutes a large, profitable and highly competitive business. In 1990 the 'world' market for these products (excluding China and the former USSR) was estimated to be about 16 million tonnes per annum with a value of approximately £15 billion (Karsa, 1990). More recently, the European market alone was stated to be worth £6 billion and the US market $10 billion (Markus, 1994). Substantial investment in research and development is essential for the major manufacturers, if they are to maintain, or improve, their individual positions in this competitive market. It is therefore not surprising that much research investment has been directed towards gaining a fundamental understanding of detergency processes, with a view to developing products with improved cleaning performance. As a consequence there are numerous publications, including several up-to-date textbooks, which deal comprehensively with aspects of surfactant science relating to the theory and technology of detergency (Cutler and Davis, 1972, 1975, 1981; Cutler and Kissa, 1987).

During the past twenty years or so, manufacturers of domestic detergents have also invested heavily in improving the environmental acceptability of their products and manufacturing processes. The debate over the replacement of phosphates and the need for continuing improvement in the rate and completeness of biodegradation of the surfactants being used has been widely publicised (Brussels, 1987; SDIA, 1989a,b). It is not therefore the purpose of this chapter further to discuss detergency, or the environmental issues surrounding fabric washing products. The aim is rather to consider the physical and chemical effects of the washing process, and washing products, on some of the most frequently washed textile substrates.

From the textile manufactuer's point of view, domestic washing is similar to a mild scouring process which is often carried out in machines which, although smaller, are similar in design to those used for processing piece goods in the textile industry. The physical and chemical effects of such a process, whether applied for preparation of grey goods, or for washing-off dyed and finished articles, do not normally give cause for concern. The cumulative effects, however, of repeating the process, perhaps 25 to 50 times in the first year of use, can be seriously damaging to some textiles. The investigation of these long-term effects is costly, and does not usually form a part of the quality control procedures employed by textile manufacturers or retailers. Nevertheless, a better understanding of the effects of repeated domestic washing, and of the concerns of consumers for the care of washable textiles, may pay dividends in terms of improving perceived garment quality, for little or no on-cost in manufacture.

The manufacturers of fabric washing products have always tried to formulate their products well within the constraints imposed by textile substrates. Clearly, any product found to cause damage to textiles — no matter how good its cleaning performance — will be pilloried by consumer organisations and rejected by the consumer. For these reasons, products with the potential to damage carry appropriate warning labels. For example, brands containing proteolytic enzymes usually warn the consumer not to soak articles made of wool or silk.

Over the years, the diversity of washable textiles, and in particular machine-washable textiles, has increased dramatically. Not only are there increasing numbers of dye class, fibre type, and fabric finish combinations to wash, but some developments — microfibres for example — are changing the washing characteristics of previously well-known dye–substrate combinations. Worldwide fashion trends and changing lifestyles have led to increasingly full shades and bright hues in new types of casual clothes and sportswear. These are predominantly of knitted constructions in which dimensional stability, or the preservation of elastomeric properties, is essential to continued 'fitness for purpose'. Furthermore, many textiles now being imported into Europe are dyed to washfastness standards that are adequate only for countries where domestic washing is done at ambient water temperatures of about 10–25°C. When these textiles are washed under typical Northern European conditions (40–60°C) the resulting dye desorption and cross-staining causes considerable problems.

This continual change in the characteristics of typical domestic wash loads has provided new challenges — and opportunities — for the manufacturers of household detergents. Their response is to be seen in the proliferation of brands on supermarket shelves which now offer superlative cleaning at low wash temperatures, care for coloured fabrics, and special care for delicate fibres such as wool and silk.

These are, however, comparatively early days in the development of more caring, specialist domestic laundry products, and it is to be expected that continuing efforts to understand the cumulative effects of repeated washing, on a wider variety of textile substrates, will lead to further innovation in this competitive product arena. It is therefore the aim of the following discussion not only to acquaint the textiles student with some of the more important physical and chemical consequences of domestic laundering processes, but also perhaps to stimulate new thoughts on improved textile aftercare in the minds of the manufacturers of textiles and household detergent products.

4.2 Domestic laundry processes—Effects of water, mechanical action and tumble drying

4.2.1 *Introduction*

Whenever and however domestic washing is done, be it on a river bank, in a sink, bowl or bucket, or in an automatic washing machine, the principles of the process are the same. Fabrics are saturated with water and agitated or beaten in the presence of products which are designed to aid wetting, and the breakdown and removal of soils. After squeezing out dirty water, rinsing in clean water (usually several times) and finally squeezing out as much water as possible, the fabrics are dried. Usually at least some of the dry fabrics are finished by pressing or ironing.

In the final rinsing stages, fabric finishing or conditioning products such as softeners, or starch, may be applied, sometimes to selected articles following a common wash process. In most countries, drying is still done mainly by hanging the clothing in the open air, but in Western countries, drying is increasingly being done in tumble driers, where softeners and antistatic agents are often also applied. Thus it is in rinsing, drying and pressing that the domestic launderer seeks to improve the final appearance and handle of the clean washing by applying a variety of finishing techniques.

It is recognised by domestic consumers (as well as in commercial laundries) that the wet rubbing and beating operations essential to effective cleaning are at least partly responsible for the deterioration in textiles that have been washed, and used or worn many times. There is in fact good evidence that washing processes generally contribute more to fabric damage than do use or wear. For example, data from a typical wash and wear trial (Mohamed, 1982) showed 43% tensile strength loss in hospital uniforms (made of 50/50 polyester/cotton) after 25 wash/wear cycles, and 39% strength loss in the 'washed only' control articles. Thus 90% of the total damage was caused by the wash process in this particular case.

The fabric changes that occur during washing, such as shrinkage, distortion, fibre damage, fabrics becoming stiff and harsh, colour fading, and cross-staining by fugitive dyes, are all highly dependent on fibre type, fabric construction, dye class and the fabric finishing processes applied, as well as on the wash process and product. For the purpose of the following discussion, an attempt has been made to separate fabric changes according to:

(i) The physical effects of water, temperature and agitation on different fibre types, with some reference to fabric construction, and drying method.
(ii) The chemical effects of individual ingredients of fabric washing products on different fibre types, with some reference to common fabric finishes.

(iii) The combined physicochemical effects of the wash process and washing product ingredients, on the colourfastness and cross-staining properties of some of the most commonly used dye class/fibre type combinations.

4.2.2 *Physical effects of water, temperature and mechanical action on textile fibres*

4.2.2.1 *Previous history.* The ways in which water, and variables such as temperature and mechanical action, affect fabrics during domestic washing processes, especially during the first few washes, often depend strongly on the previous history of the fibres, yarns and fabrics, as well as on fabric geometry and the physicochemical properties of the fibres themselves. Consequently, factors such as:

(i) yarn structure and the nature of stresses built into yarns during spinning
(ii) fabric structure and the nature of stresses built into fabrics during knitting or weaving
(iii) physical and chemical effects of scouring, bleaching, mercerising, dyeing and chemical finishing processes
(iv) drying methods, heat setting and mechanical finishing

may need to be considered when studying the performance of individual fabrics or garments. Since the combination of these variables is different for almost every finished fabric, it is only possible here to identify and discuss those factors that are known to be most relevant to the behaviour of important fibre types and fabric constructions under typical laundering conditions.

4.2.2.2 *Combined effects of water and wash temperature: glass transition temperatures, fibre swelling and diffusion rates.* The extent to which water affects the second-order or glass transition temperatures (T_g) of textile fibres, depends on their chemical and physical structures. Some typical dry and wet T_g values for common washable fibres are given in Table 4.1, together with their recommended washing temperatures. For cellulosic and proteinaceous fibres in which the secondary forces between molecules result largely from polar or semipolar interactions such as salt linkages and hydrogen bonds, T_g may be lowered by 200°C or more when the accessible polar sites are hydrated by imbibed water. In these cases, the wet T_g is below 0°C and the fibres are in the rubbery state at all possible wash temperatures.

The 50–60°C reduction in T_g that occurs on wetting typical polyamide fibres is attributed to disruption of hydrogen bonding between the amide hydrogen and carbonyl oxygen atoms in adjacent polymer chains. The resultant wet T_g is close to 0°C and hence polyamides are also always washed in the rubbery state.

The wet T_g values for other fibres listed in Table 4.1 fall within the normal range of washing temperatures, but their recommended wash temperatures

Table 4.1 Principal dry and wet glass transition temperatures for common fibres and their recommended wash temperatures

	T_g (°C) in dry air	T_g (°C) in water	Recommended wash temperature range
Cotton	None before thermal decomposition at ~200°C	<0	All temperatures
Viscose rayon	None before thermal decomposition at ~200°C	<0	Up to 60°C
Cellulose triacetate	~180	~90	Up to 40°C
Cellulose diacetate	~180	~60	Up to 40°C
Polyester	~100	~85	Up to 60°C
Nylon 6	~56	Close to 0	Up to 60°C
Nylon 6.6	~62	Close to 0	Up to 60°C
Acrylics	70–90	50–70	Up to 40°C
Wool	None before thermal degradation	<0	Up to 40°C

are all below T_g. The consequences of washing these fibres above the recommended temperatures are discussed in sections 4.2.3.3, 4.2.3.5 and 4.2.3.6. It should, however, be noted here that while the rates of diffusion (in fibres) of water, water hardness ions, and other species present (either in washing products, or in the fibres themselves) increase with temperature according to the Arrhenius equation:

$$D = D_0 \exp\left(-\frac{\Delta E}{RT}\right)$$

there is, at T_g, a step change in typical diffusion coefficients of approximately two orders of magnitude. This clearly has implications not only for the colourfastness and cross-staining properties of dyed fabrics, but also for the location, and reaction of species diffusing into the fibre from washing product solutions.

The extent to which fibres imbibe water and swell is highly correlated with their wet T_g lowering. Fibres containing many hydrophilic sites become heavily hydrated; the polymer becomes rubbery; and swelling continues until the free energy of mixing is balanced by the elastic restraining forces imposed by 'fixed' points in the swollen network such as cross-links, entanglements, or crystalline regions not penetrable by water. Equilibrium swelling increases with increasing temperature, and with applied stress, according to well-established relationships (Flory, 1953). For typical fibre diameters of 10–20 μm, equilibrium swelling in water is achieved within less than 30 s given a typical diffusion coefficient of the order of $10^{-7}\,\mathrm{cm}^2\,\mathrm{s}^{-1}$. Thus, fabrics washed at above their wet T_g, achieve equilibrium fibre swelling at any chosen wash temperature, but since the stresses imposed by the washing action are likely to change rapidly throughout the wash process, water may be continually 'pumped' in and out of the fibres as the network tries to maintain equilibrium. In highly swollen structures this may accelerate the rate of fibre breakdown, although fibre-on-fibre, and fibre-on-metal abrasion and beating must be considered the prime causes of mechanical degradation.

4.2.2.3 *Mechanical action in washing and tumble drying.* The nature of the mechanical action applied in washing processes varies widely across the range of handwash methods used and the washing machine types available in different countries around the world. It seems reasonable, however, to assume that for any given combination of fibre type and fabric construction, similar changes must occur, but at different rates, according to the mechanical severity of the wash method employed. There have been few comparative studies between different handwash and machine wash methods, but there is certainly sufficient evidence to show that handwashing cannot be assumed to be generally less damaging than machine washing. Indeed, some handwash techniques, such as rubbing fabrics on the ribbed concrete surfaces of the typical wash 'tanques' found in many South American countries, are known to be particularly damaging. Similarly the beating of wet fabrics against rocks or walls, or with heavy wooden sticks or bats, as practised in parts of India and Southeast Asia, is also more severe than any washing machine action.

The mechanically induced changes occurring in textiles during washing and tumble drying processes may result from:

(i) movement of yarns relative to one another
(ii) migration of fibres within yarns
(iii) damage to individual fibres by abrasion and beating

These types of change may be expected to lead to:

(i) changes in garment shape and size, and in fabric thickness
(ii) fuzzing, pilling, felting and fibre shedding, and
(iii) fibre splitting, fibrillation and breakage

Mechanical property changes, and associated visual and tactile changes accompany each of these types of change. Their relative importance for different fibres and fabric constructions are discussed in section 4.2.3.

4.2.2.4 *Effects of washing product formulation on mechanical action.* The chemical effects of individual ingredients of fabric washing products are discussed in section 4.3, but it is relevant here to note that washing products do significantly modify the physical effects of the mechanical action applied during washing. This occurs in two ways.

(i) Foam, generated by agitating surfactant solutions, cushions fabrics against the beating and rubbing action, thereby reducing fabric damage. The volume and quality (bubble-size distribution and stability) of the foam generated is of course dependent on the surfactant and antifoam types and levels used in a given formulation, as well as on product dosage, and the mode of mechanical action. It should be borne in mind that excessive foam cushioning reduces soil removal as well as minimising fibre damage. Furthermore, horizontal axis washing machines require particularly low-foaming products to avoid the inconvenience and potential electrical hazards of foam overflow, and to prevent the formation of foam-locks in pumps and pipework.

(ii) Surfactants adsorbed on fibre surfaces, as either monolayers or multilayers, reduce the frictional forces operating at fibre–fibre and fibre–metal contacts. In the case of water-swollen fibres, the frictional forces that operate at fibre–fibre contacts are particularly high. This is because the true area of contact, for a given normal force, increases as the fibre becomes rubbery and more easily deformed. In addition, the ploughing component of the frictional force also increases as the fibre becomes more deformable. Consequently, the lubrication provided by adsorbed surfactants plays an important role in mitigating against the damaging effects of beating or abrading water-swollen fibres.

4.2.3 *Physical effects of washing on fibres and fabrics*

4.2.3.1 *Cotton and regenerated cellulose fibres*

(*a*) *Swelling, shrinking and wrinkling.* Cotton fibres swell about 40% by volume in water. This is almost completely accounted for by radial swelling; longitudinal swelling accounting for only about a 1–2% increase in fibre length. This markedly anisotropic swelling behaviour is explained by the fact that the crystalline, microfibrillar structures in cellulosic fibres are not penetrated by water. Swelling therefore occurs only between microfibrillar structures, and consequently their orientation in the fibre determines the swelling anisotropy.

For regenerated cellulose fibres with their generally lower level of crystallinity, but nevertheless high degree of orientation achieved by the drawing process, swelling is typically in the range of 70–130% by volume with only a 3–5% increase in fibre length.

When fibres are disposed in a spiral configuration—as they are in spun yarns—it can be shown that radial fibre swelling with no significant increase in fibre length must lead to a reduction in spiral length, and hence to yarn shrinkage. The shrinkage of wet ropes is a familiar manifestation of this phenomenon; when it occurs in textiles, the changes in fabric geometry are complex.

In woven structures, 'wet relaxation' caused by fibre swelling leads to an increase in yarn crimp (weave angle) and a thickening of yarn cross-section.

Yarns that have been flattened to an elliptical cross-section by calendering tend to resume a more circular cross-section and the fibres within the yarns become less tightly consolidated after wetting and subsequent drying (see Figures 4.1(i) and (ii)).

In knitted cotton fabrics the shape and orientation of the loops changes as the fibres swell and relax towards their minimum energy conformation. This often occurs without significant changes in knitted loop (stitch) length. The loops usually become rounder in shape (Figure 4.2) causing shrinkage in the wale (length) direction, especially when the fabric has been previously dried under tension. In order to release the stresses imposed by bending twisted yarns into loops, the loops themselves tend to twist out of the plane

(i)

(ii)

(iii)

(iv)

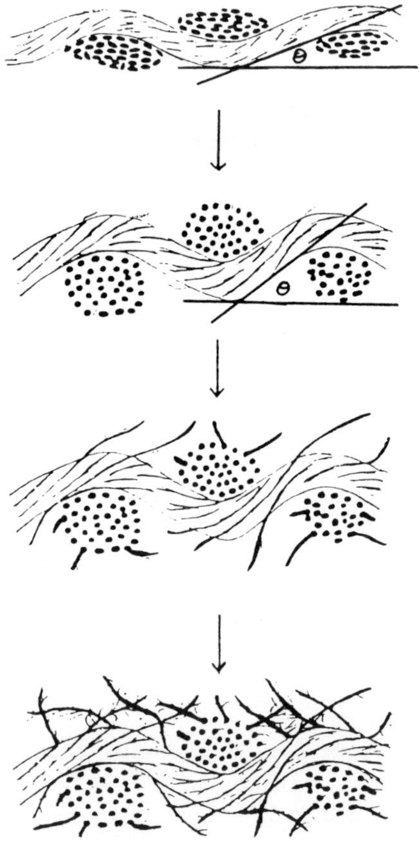

Figure 4.1 Schematic representation of progressive changes caused by repeated washing in a woven cotton fabric: (i) new fabric; (ii) relaxed fabric after one wash/tumble dry cycle; (iii) fibre migration; 'fuzz fibres' appearing after a few washes; (iv) fibre migration and fibre fibrillation damage after many washes.

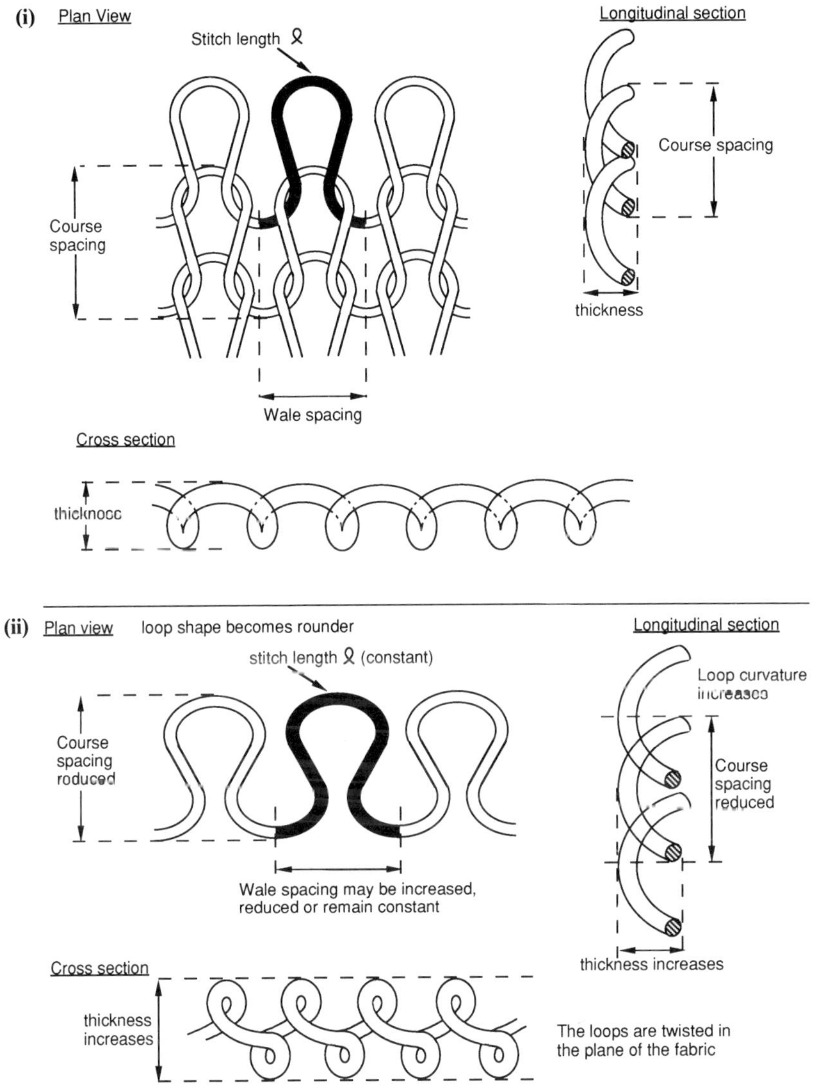

Figure 4.2 Schematic representation of relaxation shrinkage processes in a knitted fabric: (i) new fabric; (ii) relaxed fabric.

of the fabric (Figure 4.2(ii)). This causes shrinkage in the course (width) direction, and often produces significant differences in twist level in the two legs of the knitted loops (Anand *et al.*, 1990).

Fibre swelling causes many cellulosic textiles to become 'jammed' in the wet state. This may limit the relaxation that can occur in a single wet treatment, even when considerable mechanical agitation is applied. In woven fabrics, in which the wet swollen yarns cannot move readily, fabric buckling or wrinkling may develop as a means of relaxing the 'swelling stresses' in the yarns. The end results of wet treatments are evaluated on dry fabrics, but the final fabric properties also depend on the nature of the drying process.

Under static drying conditions, drainage and evaporation of water from the fabric causes increasing capillary attraction between fibres and yarns. This tends to hold the fabric in the wet conformation, and as the fibres de-swell and the moisture content falls, strong inter- and intra-fibre adhesions develop as hydrogen bonds reform in the cellulosic matrix. Any wrinkles, folds or creases then become hydrogen-bonded into the dry fabric structure. These cannot easily be removed unless T_g is exceeded by an appropriate combination of moisture and heat. Thus under hot-pressing, calendering, or domestic ironing conditions, wrinkles and creases are removed, and the fabric is re-set in a smooth conformation, by breaking and reforming inter- and intra-fibre hydrogen bonds.

In tumble drying, the constant agitation of the fabric structure prevents capillary attraction between yarns and fibres from forming inter-fibre adhesions in the structure. Consequently as the fibres de-swell, there is sufficient mobility in the structure for further relaxation to occur. At tumble-drying temperatures, intra-fibre hydrogen bonds will not reform until the fabric is almost 'bone-dry'. Consequently, as long as cotton fabrics are removed from the drier before they are completely dry, and before the drier cools down, much of the wrinkling and creasing associated with 100% cotton fabrics can be avoided.

(b) Initial changes in fabric properties. It will be apparent from the preceding discussion that the first few washes, and the choice of drying method, can cause a variety of changes in the geometry of new cellulosic fabrics. These changes must be expected significantly to affect both fabric mechanical properties, and the perception of fabric handle. In extreme cases, such as gross shrinkage of knitted cotton garments, there may also be visible changes.

Figures 4.3–4.5 illustrate some of the changes in fabric mechanical properties that occurred when a new, plain woven cotton sheeting fabric was washed and tumble dried ten times. It can be appreciated that the measured changes were consistent with the woven fabric geometry changes described earlier, and shown schematically in Figures 4.1(i) to (iii). Increased yarn crimp in the washed fabric led to increased tensile extensibility (Figure 4.3) and

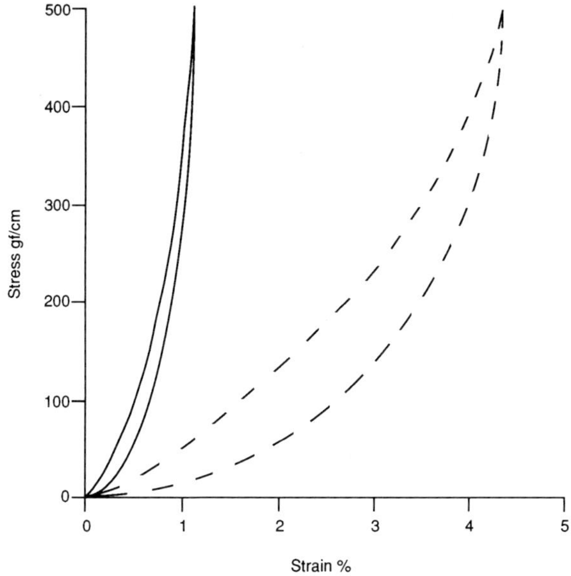

Figure 4.3 Changes in the tensile properties of a cotton sheeting fabric after ten machine wash/tumble dry cycles; ———, new fabric; — — —, fabric after ten wash cycles.

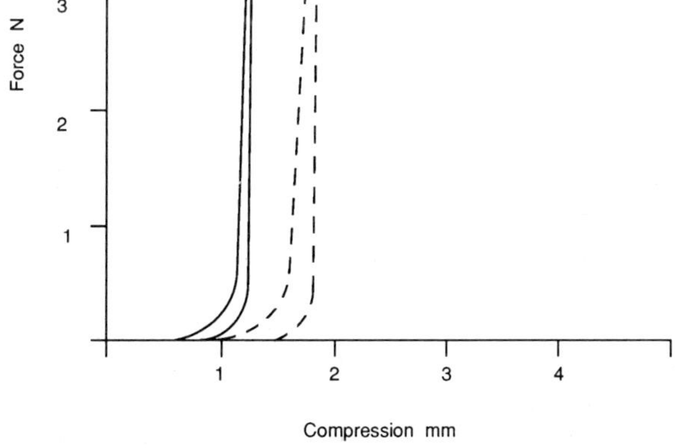

Figure 4.4 Changes in the compressibility of a cotton sheeting fabric after ten machine wash/tumble dry cycles; ———, new fabric; — — —, fabric after ten wash cycles.

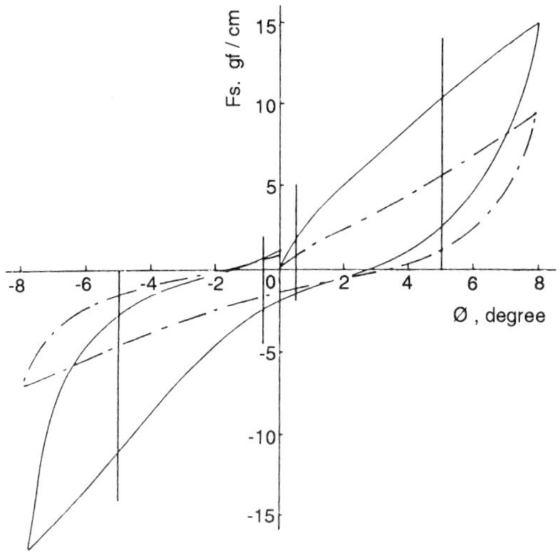

Figure 4.5 Changes in the shear stiffness and shear hysteresis of a cotton sheeting fabric after ten machine wash/tumble dry cycles; ———, new fabric; — - — - —, fabric after ten wash cycles.

increased fabric thickness (resulting from both increased yarn crimp and yarn thickening) led to increased compressibility (Figure 4.4). The reductions in shear stiffness and shear hysteresis in the washed fabric (Figure 4.5) resulted from the reduced number of fibre–fibre contacts at yarn crossover points, and from the generally reduced inter-yarn pressure in the fully relaxed, washed and tumble-dried fabric.

The effects of wash temperature and drying method on the shrinkage of a cotton interlock fabric are illustrated in Figure 4.6. Since fibre swelling, and hence the yarn retraction forces, increase with increasing temperature, it might be expected that wash temperature should determine the final dimensions of knitted cotton fabrics. For the present example, this trend is apparent only after the first wash (line dried). The results of further washing suggest that the wet, swollen structure became 'jammed' at about 25% area shrinkage, and that these dimensions became 'frozen-in' under static, line-drying conditions. Since mechanical agitation during tumble drying prevented inter-fibre adhesion, the structure approached a more relaxed conformation as the fibres began to de-swell (but still remained above T_g) and this led to further shrinkage.

It is notable that between five and ten wash/tumble dry cycles were required to attain the fully relaxed fabric dimensions, and this is believed to be typical for knitted cotton fabrics. The final percentage shrinkage of any fabric is of course a function of its original 'new' fabric dimensions. In this particular

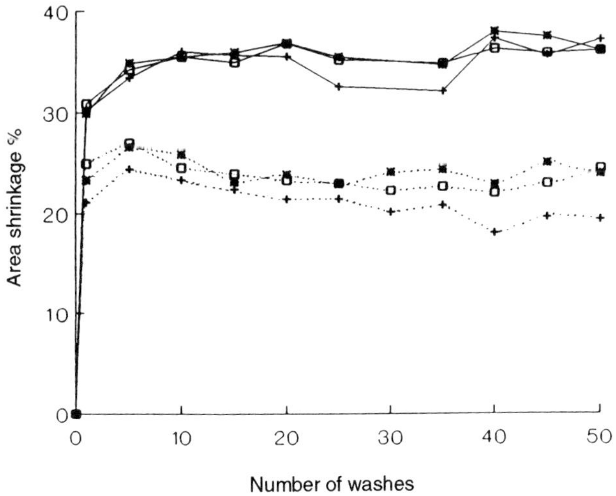

Figure 4.6 Shrinkage behaviour of a cotton interlock fabric; effects of wash temperature and drying method. Solid line, tumble dried; dotted line, line dried; +, 15°C; ×, 40°C; □, 60°C.

case the new dimensions resulted from stenter-drying after scouring on a winch. Clearly the stenter-drying conditions set the fabric in a stretched conformation which then led to very high (35%) area shrinkge on washing and tumble drying. This is a typical example of previous processing history determining a fabric's in-wash performance.

(*c*) *Longer-term changes in cellulosic fabrics.* When water-swollen cellulosic fibres are subjected to vigorous mechanical action—such as beating in papermaking—they disintegrate into their constituent fibrillar structures. The mechanical action used in fabric washing is very gentle by comparison, but nevertheless its repeated application soon begins to have similar effects. From Figures 4.7(i) and (ii) it can be seen that cotton fibres have been found to show clear signs of damage after only six machine wash cycles. After 50 washes (Figure 4.7(iv)) serious fibre fibrillation damage is invariably present at levels which may be expected to change significantly the mechanical and sensory properties of the fabric.

Figure 4.1(iv) suggests schematically how a woven structure may be affected by fibre damage. Certainly surface fibre in fabrics with a raised finish, and fibres in the loop pile of terry weave fabrics, are more prone to damage than those in the underlying fabric structure. It can be expected that the greatly increased number of hydrogen-bonded contacts, which must result on static drying a fibrillated mass, will produce structures with increased shear (and

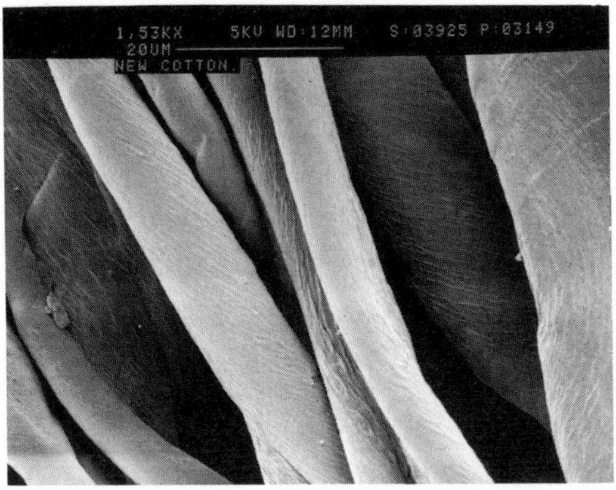

(i)

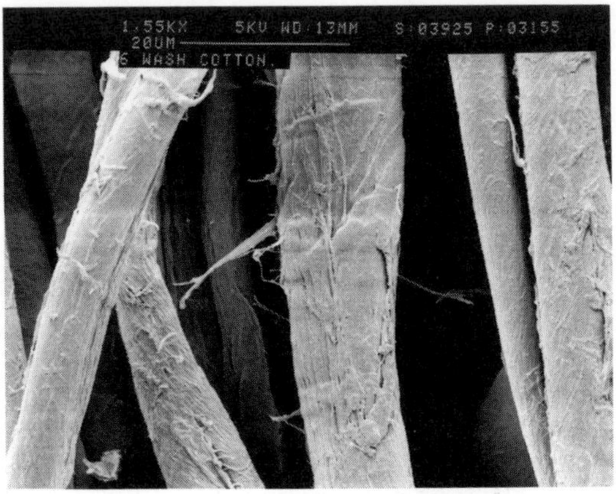

(ii)

Figure 4.7 Cotton fibres in new and washed fabric: (i) new; (ii) after six machine wash cycles; (iii) after eleven machine wash cycles; (iv) after fifty machine wash cycles.

bending) stiffness and hysteresis characteristics. This has been confirmed by experimental results such as those shown in Table 4.2 (Bishop, 1991). These demonstrate the progressive effect of multi-washing followed by static drying, on the shear stiffness (G) and shear hysteresis ($2HG5$) of a cotton terry towel fabric.

It is also shown that when the same towelling fabric was tumble dried after every wash, the G and $2HG5$ values were, as expected, significantly lower than those of the line-dried fabric, even after a single wash/dry cycle.

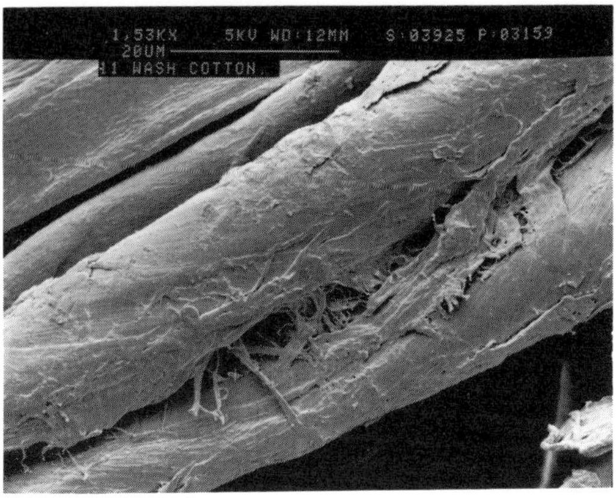

(iii)

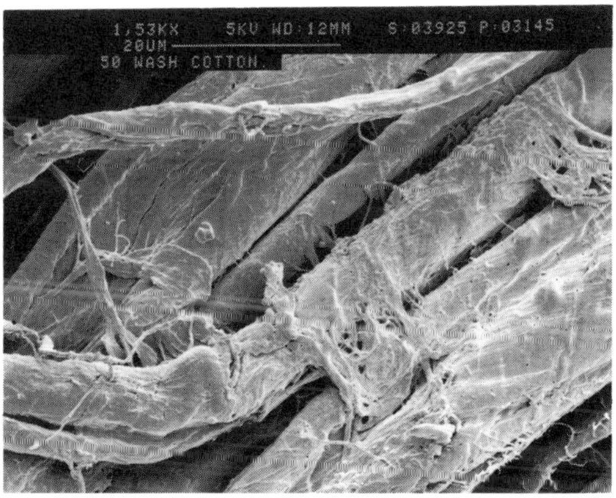

(iv)

Figure 4.7 (Continued)

The fully relaxed structure, in which adhesions between fibres are largely prevented by tumble drying, would be expected to achieve constant *G* and *2HG5* values after a few wash cycles (*cf.* the cotton interlock shrinkage shown in Figure 4.6). This seems to be the case from the constant values recorded at ten and 25 wash/dry cycles. After a further 25 cycles, however, the *G* and *2HG5* values were further reduced. It has been suggested that this was due to the continual shedding of small quantities of fibre during the tumble drying

Table 4.2 Effects of repeated washing (followed by line drying and tumble drying) on the shear stiffness (*G*) and shear hysteresis (*2HG5*) of cotton terry towelling

	Line-dried		Tumble-dried	
No. of washes	*G* ($mN\,cm^{-1}\,deg^{-1}$)	*2HG5* ($mN\,cm^{-1}$)	*G* ($mN\,cm^{-1}\,deg^{-1}$)	*2HG5* ($mN\,cm^{-1}$)
1	1.35	6.7	1.25	6.4
10	1.50	6.9	1.20	5.8
25	2.15	9.7	1.20	5.8
50	2.60	10.4	1.10	5.1
	Re-wetted and tumble-dried		Re-wetted and line-dried	
50	1.25	6.35	2.70	10.4

cycles. Certainly the contents of the lint traps of regularly used tumble driers do suggest that this could be the case.

Since it is the beating, during washing and rinsing operations, that causes cellulosic fibres to fibrillate, it should be expected that the two towelling samples washed 50 times (line-dried and tumble-dried) would exhibit similar overall levels of fibre fibrillation damage. This was in fact confirmed by remeasuring the *G* and *2HG5* values of both samples after re-wetting them, and then tumble drying the previously line-dried sample and line drying the previously tumble-dried sample. The results are also shown in Table 4.2. It is interesting to note that while the previously tumble-dried sample acquired similar *G* and *2HG5* values to the 50 times wash/line-dried material, the previously line-dried sample acquired *G* and *2HG5* values similar to the once wash/tumble-dried sample. This is, of course, exactly in line with expectation, because the 50 times wash/line-dried sample had only undergone the relaxation achievable in one tumble-drying cycle. This result may also be considered to provide indirect evidence that repeated tumble drying causes further reduction in shear stiffness and shear hysteresis via fibre shedding in the drying process.

4.2.3.2 *Resin-finished cellulosic fabrics.* The chemical finishing processes applied to cellulosic fibres that are variously described as resin finishing, crease-resist, easy-care, durable-press or permanent-press finishing encompass reactions with a variety of organic molecules capable of forming covalent crosslinks in cellulosic networks. The application conditions for different products include acid- and alkali-catalysed reactions, and high- and low-temperature reaction conditions on wet or dry fabrics. The performance of different finishes in terms of wet or dry crease recovery, abrasion resistance, whiteness retention, formaldehyde release, and durability to repeated laundering may vary within wide limits, but in principle they all confer important improvements in washing performance.

The introduction of permanent, intra-fibre crosslinks, which are not broken when internal hydrogen bonding is disrupted by imbibed water, restricts fibre swelling, and largely preserves the fibre and fabric conformation that pertained when the crosslinks were formed. From the earlier discussion of swelling, shrinking and wrinkling (section 4.2.3.1) it will be apparent that fibres treated in this way will have greatly enhanced dimensional stability, as well as reduced wet wrinkling and improved wet and dry crease shedding properties.

It may also be anticipated that fibres that are less swollen, and are strengthened by internal cross-links, may be less susceptible to fibrillation damage during washing. For many regenerated cellulose fibres this is certainly the case, and historically this type of fibre has often had unacceptably low wet strength if not resin-finished.

The fibrillation of cotton fibres may also sometimes be retarded by crosslinking, but any benefit is usually confined to the first few washes. Experience had shown (Bishop, 1990) that after about 25 wash cycles, resin-finished cotton fabrics are equally or even more severely damaged in comparison with non-resinated controls. Scanning electron microscope studies have shown that in cross-linked cotton the secondary wall begins to tear away from the fibre in 'sheets', whereas individual fibrillar strands are detached from untreated fibres. At this stage, the crease recovery performance of resin-finished fabrics also begins to deteriorate. While this may be partially attributable to gradual removal of cross-links by alkaline hydrolysis (see section 4.3.2.2), it is suggested that a contributing cause must be the generation of large areas of damaged fibre surface, which can hydrogen-bond together during drying, thus locking-in wrinkles and creases.

4.2.3.3 *Celulose acetate fabrics.* The importance of cellulose triacetate and diacetate (secondary cellulose acetate) in washable end-uses has declined over the years, as easier-care, bright-polyester filament yarns have replaced them in many 'silk-like' fabric constructions. Nevertheless, it is of some interest to compare the properties of cellulose acetates with those of cellulosic fibres.

It will be seen from Table 4.1 that both triacetate and partially hydrolysed diacetate have a dry T_g at around 180°C. This probably corresponds to the temperature at which increasing main-chain vibrations overcome the dipole attractions between the ester carbonyl groups in adjacent polymer molecules.

Cellulose acetates swell by 5–10% in water, and hydration of the semipolar carbonyl sites depresses the wet T_g to about 90°C in triacetate, and about 60°C in diacetate. The lower value for diacetate is attributable to its more hydrophilic character resulting from its hydroxyl group content.

It is essential to wash all cellulose acetate fabrics well below their wet T_g in order to prevent wrinkling, creasing and shrinking from occurring in the rubbery state. If this does occur, the result is often irretrievable, because any distortions become set in by rinsing in cold water. Many acetate garments have 'permanent' pleating or shaping which has been set by cooling after steam pressing. This is particularly susceptible to damage by washing in hot water, and such articles are often labelled 'hand wash only', or sometimes 'dry clean only'.

4.2.3.4 *Polyamides.* The polyamide fibres used in washable domestic textiles are mainly of two types which are still widely referred to as nylon 6 and nylon 6.6 types, although other polyamide fibre names are well established. The nylon 6 type is made by ring-opening and self-condensation of caprolactam, and nylon 6.6 is made by condensation of adipic acid with hexamethylene diamine. Thus both are aliphatic polyamides, and the fibres spun from them have sufficiently similar properties, from the point of view of domestic washing, for them to be discussed together.

Nylon 6 and nylon 6.6 have dry T_g values of 56°C and 62°C respectively, which are thought to be associated with the thermal breakdown of hydrogen bonding between the N—H···O=C groups in adjacent polymer chains. In water, both nylon types swell about 10% by volume, and the hydration effects of imbibed water reduce their wet T_g values to close to 0°C. Thus polyamide fibres are always washed above T_g and revert to the glassy state on drying. It might therefore be anticipated that polyamide fabrics would have poor dimensional stability, and creasing problems similar to those described for cellulosic fabrics.

In fact, polyamide fabrics are remarkably stable and smooth drying. This is widely attributed to heat setting at temperatures well above the dry T_g range, but the moisture sensitivity of the relaxation process should nevertheless be expected to cause loss of set under washing conditions. A possible explanation (Thompson, 1963) for their dimensional stability in washing, and their smooth drying properties, is that the spectrum of their relaxation times and temperatures is sufficiently broad that although the centre of the transition range is depressed to around 0°C by moisture, there remain some relaxation times long enough to retain set even in wet processes.

By comparison with cellulosic fibres, polyamides are mechanically tough. They are not generally susceptible to mechanical damage under domestic wash conditions, except where they are used in fine knitted articles such as stockings and tights in which 'snagging' can easily lead to holes or 'ladders'.

Polyamides have a high affinity for many polar and nonpolar species that may be present in typical wash liquors. Since the fibres are washed at above T_g these species readily diffuse into the fibres, often causing permanent and unacceptable discolouration of whites and pastel shades. This problem is discussed further in section 4.4 in the context of dye desorption and cross-staining in washing product solutions.

4.2.3.5 *Polyester fabrics and polyester blends.* Polyester fibres used in washable textiles are invariably poly(ethylene terephthalate)s and although these are marketed in many physical modifications and some minor chemical modifications (to confer differential dyeing properties), they are all sufficiently similar to be discussed together. Polyester fibres are highly crystalline, mechanically tough and hydrophobic. They do not swell significantly in water, and the thermal transition that occurs at around 100°C in the dry state is therefore relatively moisture-insensitive, and occurs at around 85°C in water. Since polyester fabrics are normally washed at 60°C or below, they can be heat-set to give excellent dimensional stability, permanent pleats, and smooth drying properties which last throughout their wash/use lifetimes.

The inherent toughness and stability of polyester fibres is extremely valuable in blends with other fibres. Fabrics constructed from optimum polyester–cotton, polyester–viscose, and polyester–wool blend yarns have dramatically improved easy-care characteristics, and their resistance to wet and dry abrasion is also increased. These benefits are obtained while maintaining to a large extent the more desirable aesthetic and comfort-in-wear properties of the natural fibres.

A disadvantage of fibre blends can be that differential friction effects between fibre types, and the possible mismatch of fibre diameters and cross-section shapes, may lead to increased fibre mobility, and hence to increased pilling. Although pilling is widely associated with in-use abrasion, especially on knitwear, there is undoubtedly a problem of in-wash pilling which appears all over certain types of fabrics after about ten wash cycles. The most affected articles tend to be knitted from blended yarns, although some 100% cotton articles such as T-shirts and sweatshirts are also affected. Where one of the fibres in a blend is polyester there is little or no tendency for the pills to wear off after further washing or use, because the strength of polyester fibres anchors the pill to the fabric structure. The introduction of deliberately weakened 'low-pill' polyester has not eliminated this problem.

The hydrophobic nature of polyester fibres makes them particulary prone to oily soiling, and to the redeposition of oily or fatty soils during washing. Such soiling is often difficult to remove, but since polyester is washed below its wet T_g there is little penetration of fatty soils into the fibre. As a consequence they can be stripped by washing in strongly alkaline solutions (section 4.3.2). Furthermore, the fibre surface can be rendered somewhat more hydrophilic (and less prone to oily soiling problems) by minor chemical modification of the fibre surface. This is discussed further in sections 4.3.2 and 4.3.6.1(c).

4.2.3.6 *Acrylic fabrics.* The acrylic fibres used in washable end-uses are mainly copolymers containing at least 85% acrylonitrile and up to 15% of other vinyl monomers, which are used to introduce anionic (carboxyl, sulphate or sulphonate) sites into the polymer in order to confer dyeability with basic (cationic) dyes. Dry T_g values for these fibres are in the region of 70–90°C; the transition is made relatively moisture-sensitive by the presence of the hydrophilic anionic groups and occurs at about 50–70°C in water.

Unlike polyamides, acrylics do not retain heat-set dimensions when T_g is exceeded. This property is utilised to produce soft, highly bulked yarns which compare favourably with woollen knitting yarns. Bulking is achieved by blending a small percentage of fibres that have been hot-stretched and cooled (to give temporary set) with normally oriented fibres produced by wet spinning and drawing. When the blended yarn is subsequently heated, without tension, to above T_g, the hot-stretched fibres return to their original dimensions, causing yarn shrinkage of 20–40% and thus buckling the normal fibres to produce a highly bulked yarn.

Provided that acrylic articles knitted from these yarns are washed below their wet T_g they retain their shape and bulk. Care labelling on acrylics always recommends washing at 40°C or below, and usually indicates a gentle or 'wool wash' cycle, and drying 'flat' or 'supported'. Unfortunately acrylics are often washed too hot because some washing machines either do not have adequate temperature control or do not clearly indicate the temperature associated with particular wash cycles or programmes. There is a tendency for stress applied during washing (or drying) above the wet T_g to restretch the bulking fibres (which remain under tension) and subsequently to re-set them on cooling. Consequently consumers often complain that acrylic knitwear "stretches when you wash it" and that it "goes flat and lifeless" after a few washes.

Apart from their dimensional instability, acrylic fibres are not susceptible to mechanical damage during washing. They also have good chemical resistance, but the polymer is prone to yellowing by light and heat and there is some evidence that this is exacerbated by washing in strongly alkaline products.

4.2.3.7 *Wool.* Wool fibres have no T_g in dry air below their thermal decomposition temperature, but they swell (about 40% by volume) and become rubbery in water at all temperatures. The fibres have good recovery properties under both wet and dry conditions which are at least partly attributable to covalent cystine crosslinkages in the keratin structure. 'Dried-in' creases are, however, set by temporary salt linkages and hydrogen bonding.

The fibre is protected from abrasion damage by a tough cuticle, but the scale-like structure of this layer gives rise to the differential friction between the fibres which is responsible for the felting of wool fibre assemblies. Felting is promoted by mechanical agitation; it is greatly increased in water, and is further exacerbated by the presence of surfactants.

Although shrink-resist treatments, which can render woollen knitwear fully machine-washable, are now being widely used, felting shrinkage still remains an area of concern in domestic washing. Furthermore, wool keratin is particularly sensitive to chemical damage by several ingredients that have become commonplace in modern fabric washing products (see sections 4.3.2.2, 4.3.3.1, 4.3.4.2, and 4.3.5.1). Fortunately, this is now leading to the marketing of specialist wool wash products which should allow the domestic consumer to wash, and machine-wash, wool with more confidence.

4.2.3.8 *Silk.* Traditionally, silk has been treated with extreme care, many articles being labelled 'dry clean only' and relatively few 'hand wash with care'. In recent years there has been an increase in silk imports, particularly from China, which has led to relatively inexpensive ranges of underwear, blouses and shirts, and even outerwear such as anoraks, being sold in high street stores. Many of these articles are labelled as machine-washable at 30°C or 40°C or 'wash as wool'.

While some experts are confident that silk can be washed without difficulty, others recommend only dry cleaning, and suggest that silk cannot be regarded as a washable fibre (Payot, 1991). Certainly, fabric washing products and processes do have the capacity to cause rapid deterioration in the desirable aesthetic qualities for which silk fabrics are so highly prized.

Like cotton and wool, silk swells in water and is above its wet T_g at all wash temperatures. In this state, the fibre is particularly prone to fibrillation, and even low levels of damage can give rise to a floury appearance (chafing or 'farinage') on the surface of lustrous satin and taffeta fabrics. This is less of a problem on the 'sand-washed' or 'peach-skin' finished silks that are currently popular, but washing and drying these causes a marked increase in fabric stiffness which impairs their soft, luxurious handle.

Creases in the wet fabric are set by intermolecular hydrogen bonding as it dries, and some skill is needed to achieve a satisfactory crease-free finish after washing.

Silk fibroin is sensitive to chemical degradation by several ingredients commonly used in modern fabric washing products (see sections 4.3.2.2, 4.3.4.2, and 4.3.5.1). The fineness of silk filaments and the lightweight construction of many silk fabrics makes them less able than other fibres to sustain even low levels of chemical attack without impairing their most desirable qualities. In particular, the lustre and lively handle of many silk fabrics is easily destroyed by washing in anything but the mildest products. Specialist wool wash liquids and hair shampoos are however suitable for handwashing 'washable' silk articles.

4.3 Chemical effects of fabric washing product ingredients on textile fibres

4.3.1 *Water softeners or 'builders'*

It has long been recognised that calcium and magnesium hardness ions, present in supply waters, are deleterious to textile washing processes. Not only are soaps and anionic detergents wasted by the precipitation of their insoluble calcium and magnesium salts, but these salts are often also deposited on and bound by textile fibres, where they build up to cause greying or yellowing, and the development of rancid, fatty odours on the fabric.

Hardness ions are invariably bound to fibres containing carboxylic acid groups, or other anionic sites provided by dyes and fluorescers, because rinsing at the end of the wash is carried out in hard water. All fibres also carry dried-on (rather than ion-exchanged) hardness ions from the last rinse into the next use, wear and subsequent wash cycle. The hardness ions present in washed textiles provide potential sites for the attachment of anionic soils during use. The fatty acid components of sebum are often tenaciously held, especially where they are able to diffuse into the interior of fibres such as cotton, during warm, moist use conditions.

Clearly it is important to effect the removal of hardness ions both from the water and from textile fibres, if efficient washing is to be achieved. Although the term 'water softening' is widely used to describe the various means of removing hardness ions from water, the manufacturers of fabric washing products generally use the term 'building' to describe the process, and hence describe water softeners as 'builders'. The reason for this is that when water softeners such as sodium carbonate or sodium triphosphate (STP) are used in excess of the stoichiometric amounts needed for removing hardness ions, they provide alkalinity, and in the case of STP, soil dispersing and suspending properties, essential to effective fabric washing. That is, they 'build' on the wetting and detergent functions of surfactants.

In order to prevent precipitation of the anionic surfactants used in fabric washing products, and of the soaps generated from fatty soils on the wash load, it is necessary to reduce free hardness ion concentrations to less than 10^{-4} M (1° French hardness) and preferably to about 10^{-5} M (0.1° French hardness). This may be achieved by sequestration (i.e. forming stable complexes) by precipitation, or by ion exchange (e.g. for sodium ions in zeolites).

Whenever precipitant builders such as sodium carbonate or soap are used, or if sequestrant builders are used in 'underbuilt' or precipitating conditions, there is a risk that insoluble calcium and magnesium salts will precipitate, and build up inside or on the surfaces of textile fibres. Such deposits, depending on their level, location, and chemical constitution can cause several undesirable effects such as:

apparent colour fading
increased rate of wear
loss of flame retardancy
adverse changes in fabric mechanical properties and fabric handle.

Calcium phosphate deposits. STP ($Na_5P_3O_{10}$) forms a soluble 1:1 calcium complex ($CaNa_3P_3O_{10}$) which is responsible for its excellent sequestrant building performance. When incorporated into spray-dried powders, however, STP invariably suffers some breakdown into other phosphate species which have insoluble calcium salts. Where there is a molar excess of STP over hardness ions, no insoluble phosphates are precipitated, but in marginally built conditions insoluble calcium ortho- and pyro-phosphates occur. In underbuilt washes the 2:1 calcium STP complex is also precipitated.

When cotton fabrics are washed repeatedly in underbuilt conditions, significant quantities of calcium phosphates build up on the fabric. Examples have been found in which up to 25% of the fabric weight is accounted for by phosphate 'ash'. (Inorganic deposits in textiles are often referred to as 'ash' because their levels are readily determined by burning off the fibre and weighing the residual ash.)

Surprisingly, when cotton fabrics having high phosphate ash contents are examined microscopically, there is little or no evidence of inorganic deposits on fibre surfaces. Examination of the ash, however, reveals a complete inorganic skeleton of the fibre structure, and it may be concluded that insoluble phosphates are precipitated inside the fibre, rather than being precipitated from solution and deposited, or 'seeded' and grown at the fibre surface.

The process of repeatedly rinsing in hard water, and re-wetting in the next wash with a solution containing phosphate ions, is not unlike that used for weighting silk, and it is perhaps not surprising that such high levels of insoluble phosphates can accumulate in cotton fibres. Even at high levels, these deposits have surprisingly small adverse effects on fabric handle, although some stiffening is detectable, especially in knitted fabrics (Mackay, 1992).

Increased light-scattering by inorganic deposits can cause apparent fading of dyed fibres, but this is small compared with other effects of repeated washing, such as dye desorption, and dye damage by bleaching.

Calcium carbonate. Unlike calcium phosphates, calcium carbonate deposits tend to grow on fibre surfaces. Even relatively low levels ($<5\%$) can be readily seen under the microscope as crystalline encrustations of calcite, which give the fabric a harsh and sticky handle, and cause 'frosting' of coloured fabrics. Furthermore, since calcite is mildly abrasive, the wear rate of textiles regularly washed in carbonate-built products is significantly increased compared with that of similar articles washed in phosphate-built detergents (Mohamed, 1982).

The growth of calcite at fibre surfaces, rather than inside the fibre, may be explained by the fact that calcite crystal growth is relatively slow and is easily 'poisoned' by other ionic species present in the wash liquor. As a consequence, the initially high concentration of calcium ions in the dry fibre is greatly reduced by dilution, and by ion-exchange with sodium ions, before calcite precipitation occurs.

The slow growth of calcite, and hence the slow rate of lowering of free calcium ion concentrations in the wash liquor, makes sodium carbonate a poor builder for fabric washing products. Soaps and other anionic surfactants are irreversibly precipitated before the builder does its job.

In the days before 'self-acting' detergent products, washing soda (sodium carbonate decahydrate) was used effectively by adding the soda to the water and boiling the wash load before adding grated soap, or rubbing individual articles with bar soap. This reflected the normal practice in commercial laundries (and in textile processing) of softening the process-water before use.

Calcium soaps. The deposition of calcium and magnesium soaps in poorly-built washing conditions has already been mentioned. The problems can be much more severe if fabrics are washed regularly in soap-based products, which rely on soap as both (precipitant) builder and surfactant. Although cellulosic fabrics in particular acquire a pleasantly soft handle, they also become increasingly hydrophobic, and prone to yellowing and the development of rancid odours. More seriously, the substantial soap deposits increase the flammability of the fabrics, and this has been shown to impair the effectiveness of flame-retardant finishes (Carfagno and Pacheco, 1973). A full discussion of the effects of washing on flame-retardant finishes is beyond the scope of this chapter, but the subject has recently been reviewed (Horrocks *et al.*, 1993); see also chapter 3.

4.3.2 *Alkalinity*

4.3.2.1 *The need for alkalinity in fabric washing products.* Alkali is essential in fabric washing operations that call for the saponification of fatty soils. A substantial part of the soil on domestic laundry is sebum secreted by the sebaceous glands in the skin. Since it is found at levels of 0.5–1.0% on bed linen and clothing worn next to the skin, it can be estimated that around 20 g of sebum is present in a typical 3 kg wash load.

Although it contains some hydrocarbons such as squalene, sebum consists largely of triglycerides and fatty acids which can be rendered effectively self-dispersing by appropriate quantities of alkali. In principle, fats can be dispersed by surfactants alone, but enormously increased levels would be required to achieve this from neutral fabric washing products.

In STP-built detergents, additional alkalinity is required, over and above that provided by the phosphate itself. This is because calcium ion binding by STP is pH-dependent, and deteriorates markedly below about pH 9.5. Sodium silicate is normally used to provide this additional alkalinity because it confers improved structure to spray-dried powders, as well as providing increased alkaline buffer capacity. In zeolite-built products, however, the addition of silicates promotes unacceptable powder 'caking', and sodium carbonate is used instead.

Alkalinity is also required for the efficient functioning of the hydrogen peroxide-based bleaching systems that are used in many fabric washing products. Since sodium perborate or sodium percarbonate are the sources of hydrogen peroxide in these systems, they bring their own additional alkalinity to the wash solution.

4.3.2.2 *Effects of detergent alkalis on fibres.*

Cellulosic fibres. These are very slowly degraded by hot alkaline solutions, hydrolysis proceeding stepwise from the aldehydic end group of the cellulose chain (Golova and Nosova, 1973). Oxidative treatments that produce increased numbers of aldehydic or ketonic groups in cellulose greatly increase its sensitivity to alkaline hydrolysis. This is discussed further in the context of detergent bleaching systems in section 4.3.4.2. In general, however, cellulosic fibres are resistant to damage by alkaline detergents, and indeed to the strongly alkaline processes used in scouring and mercerising.

The covalent cross-links introduced into cellulosic fibres to impart crease-resistance and crease-recovery properties are susceptible to slow hydrolysis by repeated washing in alkaline detergents. As a consequence the performance of crease-resist finished fabrics soon deteriorates if they are washed frequently.

Cellulose acetate fibres. These are susceptible to deacetylation by alkaline hydrolysis, but since acetate fabrics require very careful wash treatment to avoid creasing (see section 4.2.2.3) they are not normally subjected to potentially damaging wash conditions.

Wool. Wool absorbs alkali initially by 'tritation' of carboxyl groups, but before an end-point is reached, alkali begins to be consumed by reaction with cystine cross-links. These are converted to lanthionine, with the partial elimination of sulphur as sodium sulphide. Ultimately, alkaline hydrolysis of peptide bonds leads to gradual dissolution of the fibre. Clearly, wool is subject to progressive deterioration by repeated washing in conventional alkaline detergents. In the past this was not a serious problem because wool articles were, in the main, washed carefully by hand in mild 'hand wash' products to avoid felting shrinkage. Nowadays many wool articles are fully machine-washable. As a result, they are washed more frequently in conventional machine-wash products. Such products are not only highly alkaline, but they often contain both proteolytic and lipolytic enzymes and bleaching systems, which are also particularly damaging to wool (see sections 4.3.4.2, 4.3.5.1 and 4.3.5.2). A marked deterioration in the handle of wool, often described as 'impoverishment', becomes clearly noticeable after a few washes in these products.

As a result of the serious effects that conventional domestic detergent products can have on wool, the International Wool Secretariat has introduced a new product endorsement scheme for products that meet appropriately low wool-damage criteria. In response, there has been an increase in the availability and advertising of specialist products which are suitable for washing wool in the machine, as well as by hand.

Silk. Silk fibroin, like wool keratin, is damaged by alkaline hydrolysis on washing in conventional detergent solutions, particularly when they also contain proteolytic enzymes. The recent increase in the number of washable silk articles in European markets has, however, led major fabric washing product manufacturers to identify products suitable for washing silk, and even to draw particular attention to their performance on silk in television advertising.

Polyester, polyamide and acrylic fibres. When washed at recommended wash temperatures, the relatively hydrophobic synthetic fibres are not noticeably affected by alkaline detergent solutions. Polyesters are, however, subject to slow surface hydrolysis which renders the fibre surface more hydrophilic and less prone to the retention and redeposition of oily and fatty soils. In bright polyesters, this may eventually lead to dulling of the fabric lustre, as the fibre surfaces become roughened and pitted. This is not, however, regarded to be a problem within the normal lifetimes of bright polyester garments.

The polyester-based soil-release polymers that are often applied to polyester fabrics are gradually stripped from the fibre by repeated alkaline wash treatments. It is, however, possible to replenish soil-release performance by incorporating soil-release polymers in fabric washing products (see section 4.3.6.1(c)).

The discolouration of white and pastel shades is the most characteristic form of deterioration in polyamide fibres. Whilst alkalinity may play a role in the long-term yellowing and degradation of polyamides, the principal cause of discolouration, within the lifetime of washable textiles, is undoubtedly the adsorption of fugitive dyes, and other coloured species, from the wash liquor. This is discussed more fully in section 4.4.3.

The deterioration of elastication, which is a feature of many polyamide garments, is often due to the hydrolytic degradation of elastomeric polyurethane yarns. In other cases, however, oxidative attack on hydrocarbon elastomers is the cause of failure. In either case, washing in low-pH, bleach-free detergents should be recommended for garments of this type.

It is unlikely that washing acrylic fibres in conventional alkaline detergents at recommended temperatures ($\leqslant 40°C$) contributes to their long-term yellowing. The problems that consumers identify with acrylics are confined to their lack of dimensional stability when washed too hot or tumble-dried (see section 4.2.3.6) and to the harsh handle which develops, as softeners are removed by repeated washing (see section 4.3.3).

4.3.3 *Anionic, nonionic and cationic surfactants*

4.3.3.1 *Adsorption of anionic and nonionic surfactants.* All surfactants, by definition, adsorb at interfaces—they are 'surface-active agents'. Although the anionic and nonionic surfactants used in fabric washing products adsorb on and diffuse into textile fibres to promote wetting and detergency, they are largely desorbed and washed away during the rinsing process. Thus they do not generally cause modification of fibre and fabric properties. Wool is, however, an exception. Anionic surfactants are bound at protonated amino sites, the concentration of which clearly depends on wash pH. Unfortunately, at near neutral pH, which is necessary to protect the wool from hydrolytic attack, all amino end-groups are protonated, and the potential to bind large amounts of anionic surfactants is at a maximum. Apart from displacing acid dyes (see section 4.4.3), the adsorption of anionic surfactants has also been shown to have deleterious effects on fabric handle, and on the fibre modulus and tensile strength (Nolte, 1994).

Much lower concentrations of nonionic surfactants are adsorbed by wool, and the mechanical properties of the fibre are relatively unaffected (Nolte, 1994). Washing with nonionics does, however, leave the fibre with a particularly harsh handle, and there is clearly a need for effective wool softening agents to be incorporated in specialist wool wash products. Many of the marketed products do contain softeners, and some make specific softening claims, but their performance still leaves room for improvement.

The continual improvement in fatty soil detergency (which is being achieved by careful selection of nonionic surfactant species) also increases the efficiency with which detergents remove finishing agents from textiles. In particular, semi-permanent lubricants, which give the luxurious handle currently demanded by customers, are more efficiently removed. This gives rise to frequent complaints that cotton and acrylic knitwear becomes harsh, even after only one or two washes.

4.3.3.2 *Adsorption of cationic surfactants—rinse conditioners.* The use of rinse conditioners (fabric softeners) goes some way towards alleviating the harsh handle caused by (a) the removal of softeners applied during textile finishing and (b) the fibrillation damage caused by the repeated washing of cotton textiles.

For many years the cationic softeners used in rinse conditioners were either di-tallowyl dimethyl ammonium salts or di-tallowyl imidazolinium salts. They adsorb readily on negatively charged fibre surfaces, where they form close-packed monolayers or multilayers which act as excellent boundary lubricants between yarns and fibres (Chugg, 1990). In recent years these actives have come under environmental pressure as 'unnecessary non-biodegradable chemicals'. Although, increasingly, they are being replaced by biodegradable alternatives, they continue to come under pressure, and the market for them, at least in Europe, seems to be in decline.

Other criticisms levelled at rinse conditioners are that (i) they cause hydrophilic fibres to become hydrophobic and (ii) they increase 'snagging', thread-pulling and pilling in some fabric constructions. While all of these phenomena can certainly be demonstrated in controlled experiments, none of them appears to be a problem perceived by consumers. One reason for this is that the strike rate of cationics is so fast that patchy delivery to the fabric is almost inevitable, especially in automatic dosing washing machines. As a consequence, although localised regions of large articles may be rendered hydrophobic, their overall absorbancy is not impaired. Similar arguments may apply to the overall snagging and pilling propensities of rinse-conditioned wash loads.

For many consumers, the other benefits obtainable from rinse conditioners far outweigh these possible negatives. For some, the delivery of perfume to the fabrics is an important attribute, while for others (especially those who use tumble driers) the reduction of electrostatic charging and 'fabric cling' may be even more important than improved softness. Other less well advertised conditioner benefits include improved wrinkle shedding and crease recovery, and greater comfort in wear of resin-finished shirts. The elasticity of knitted cottons is also improved, and although this is closely related to the perception of softness, it can manifest itself as better 'fit' and shape retention during wear. Thus, on balance, the adsorption of cationic surfactants from rinse conditioners provides more positive than negative fabric attributes, and their continued use should not be discouraged.

4.3.4 *Bleaching systems*

4.3.4.1 *The development of bleaching systems for fabric washing products.* For many years, bleaching from fabric washing products was achieved by using sodium perborate as a source of hydrogen peroxide. At about 70°C and above, hydrogen peroxide is effective on a broad spectrum of common stains, but its performance in fabric washing products is extremely variable unless some care is taken to control its decomposition. There are two main causes of hydrogen peroxide decomposition in fabric washing systems: the enzyme catalase; and transition-metal ions such as copper and iron.

Catalase is a haemoprotein, containing four Fe^{3+} ions per molecule, which is produced by many organisms as a defence against the hydrogen peroxide generated by various oxidases. It therefore occurs very widely, and is present on naturally soiled wash loads at concentrations which depend on the level of soiling, the composition of the load, and the storage time of the soiled clothes before washing. No safe method of effectively controlling catalase activity in domestic washing systems has yet been found, but some alternative bleaching agents (discussed below) are less sensitive than hydrogen peroxide to decomposition by catalase.

Catalytic quantities of transition metals, especially copper and iron, are brought into washing systems by domestic supply waters, soils on textiles, the textiles themselves, and as impurities in detergent products. These metals not only catalyse the wasteful decomposition of hydrogen peroxide:

$$2H_2O_2 \xrightarrow{M^{n+}} 2H_2O + O_2$$

but they also catalyse oxidative attack at alcoholic functions, converting them to aldehydes, ketones or carboxylic acids. Such reactions can be particularly damaging to cellulosic fibres (see section 4.3.4.2) and for this reason, peroxide-based bleaching systems have usually contained powerful metal ion sequestrants, such as ethylene diamine tetra-acetic acid (EDTA) and magnesium silicate, to complex free copper and iron.

Throughout the 1970s average wash temperatures in Europe were falling. This was partly because of the increasing incidence of brightly coloured textiles, which require washing at 60°C or below, and partly also because of the concern over the increasing cost of energy. In order to maintain bleaching performance at these lower wash temperatures, products based on peracid precursors such as tetra-acetyl ethylene diamine (TAED) were introduced. The principle of this system is to use hydrogen peroxide (from sodium perborate) to perhydrolyse TAED in the wash liquor, thus generating peracetic acid, a species which is known to give significant bleaching effects at 50°C to 60°C (equation 4.1).

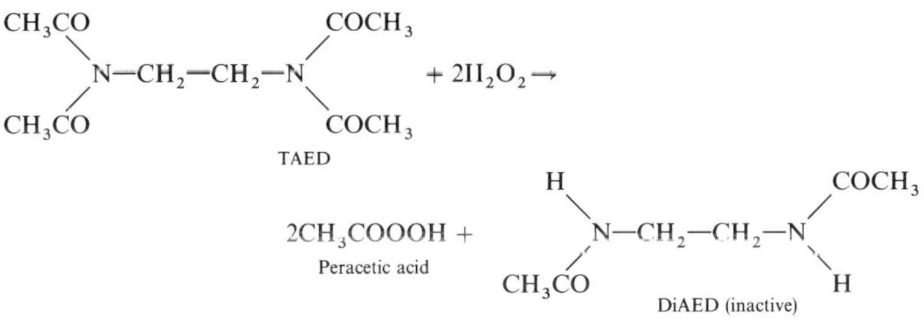

(4.1)

At first, the TAED level was kept down by high initial cost and low availability. The early products therefore used a high stoichiometric excess of hydrogen peroxide (as sodium perborate) over TAED. While this excess was partly to allow for the uncertain losses of H_2O_2 through catalase activity, it was also essential to boost high-temperature bleaching performance so that the new products showed no deficiency in 90°C washes when compared with the traditional perborated products.

Unfortunately, excess hydrogen peroxide drives a mutual decomposition reaction with peracetic acid, which is strongly catalysed by transition metals:

$$CH_3COOOH + H_2O_2 \xrightarrow{M^{n+}} CH_3COOH + H_2O + O_2$$

This led to a need for more powerful transition metal ion sequestrants, which was met by the introduction of low levels of the sodium or calcium salts of ethylene diamine tetramethylphosphonate (EDTMP) (**1**) and other similar molecules.

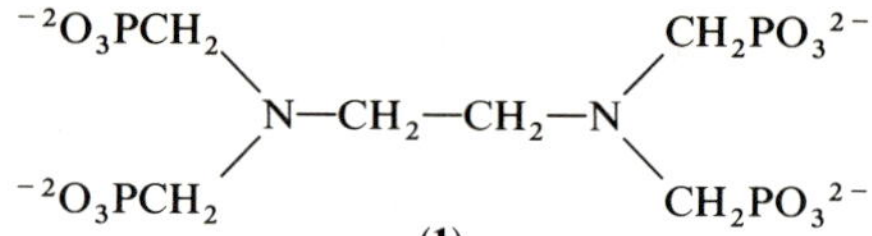

(1)

As TAED became more readily available and relatively less costly, the need to maintain boil wash performance was also declining because average wash temperatures continued to fall. Higher levels of TAED with lower levels of perborate were therefore gradually introduced. Since these higher TAED/perborate ratios suffered less from mutual decomposition, it became possible to reduce the level of, and eventually to omit, the expensive sequestrants such as EDTMP.

Apart from providing stain bleaching at lower wash temperatures, TAED/perborate, and other peracid bleaching systems, such as sodium nonanoyloxybenzene sulphonate (SNOBS)/perborate (see Figure 4.8) have three further advantages over hydrogen peroxide:

(i) they are less likely to cause oxidative damage to cellulosic fibres
(ii) once generated, the peracids are not decomposed by catalase, and
(iii) they kill bacteria that are resistant to hydrogen peroxide, thus giving a more hygienic wash, even at low temperatures.

On the other hand, the greater specificity of peracids in oxidising disulphide bonds leads to increased damage to wool, and to sulphur dyes. These aspects of bleaching are discussed more fully in sections 4.3.4.2. and 4.4.2.

By the early 1990s the most widely used domestic wash programmes, in the majority of European countries, were being carried out at 40°C. This fact, together with the ever-present need to improve stain removal, especially under the ambient-temperature wash conditions that pertain in most other parts of the world, has led one of the major detergent manufacturers to develop a manganese-catalysed perborate bleach system. This system has not however proved to be completely free of fibre and colour safety problems. Understandably this has led to adverse comment by representatives of the coloration industry (Smith, 1994).

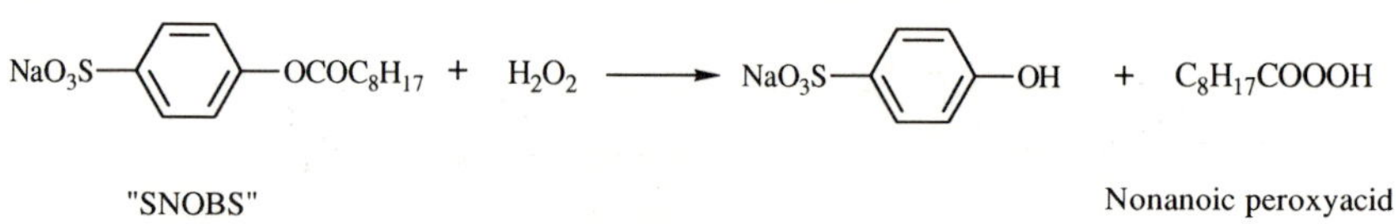

Figure 4.8 The generation of nonanionic peroxyacid from SNOBS/perborate, a more hydrophobic alternative to peracetic acid.

4.3.4.2 *Effects of bleaching systems on fibres*

(*a*) *Cellulosic fibres.* It was noted in section 4.3.4.1 that transition metals catalyse the oxidation of alcohols by hydrogen peroxide. In cellulosic polymers this may result in oxidation of the primary alcohol function at the C6 position of the anhydroglucose ring to an aldehyde and thence to a carboxylic acid group. Similarly, oxidation of the secondary alcohol functions at the C2 and C3 positions leads to the formation of ketone or aldehyde and thence carboxylic groups (Figure 4.9).

When aldehyde groups occur at the C2 and C3 positions, the polymer is particularly sensitive to alkaline hydrolysis (Figure 4.10). Consequently, under alkaline washing conditions, any metal-catalysed oxidation is likely to lead to cellulose degradation, and hence to a loss in the tensile and tear strengths of cotton and viscose fabrics.

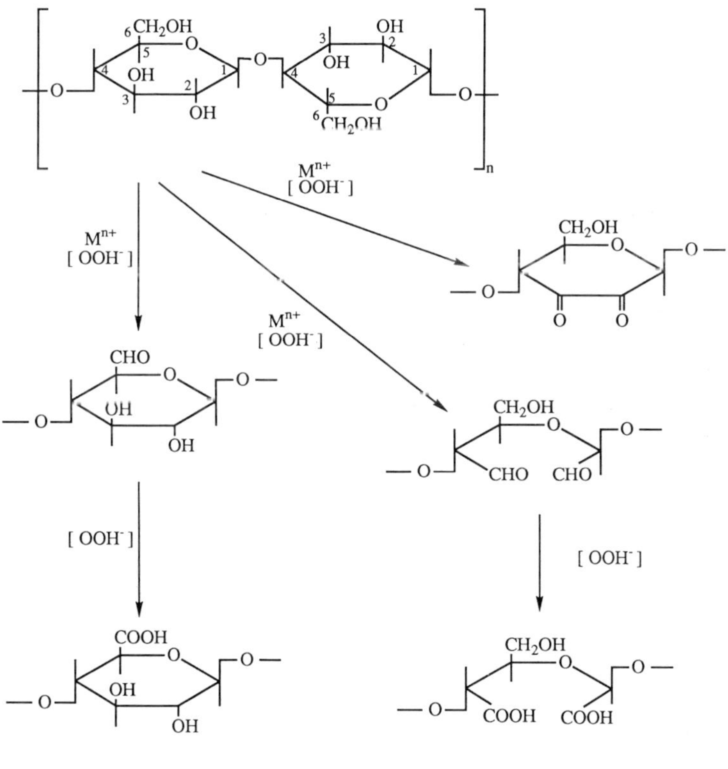

Figure 4.9 Possible mechanisms for the metal catalysed oxidation of cellulose.

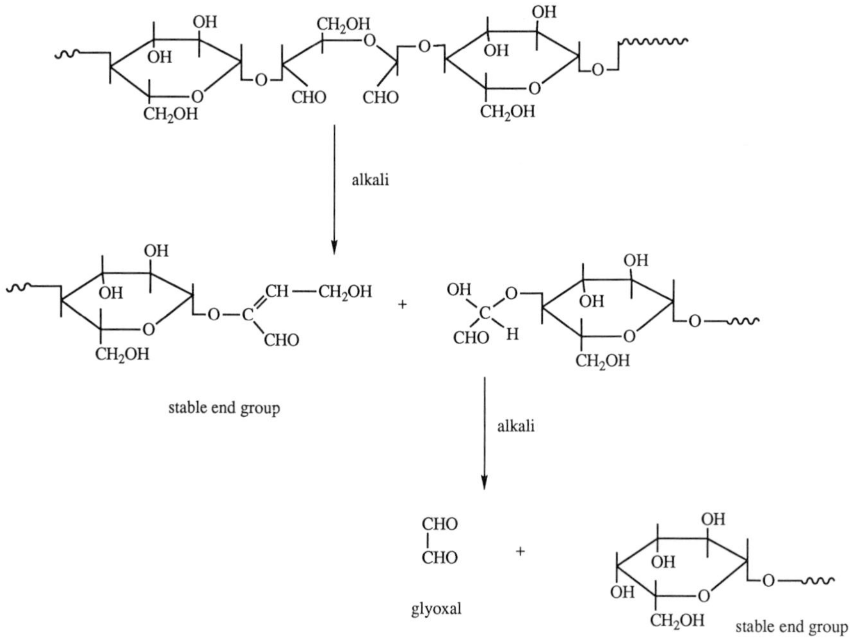

Figure 4.10 Alkaline hydrolysis of oxycellulose.

(*b*) *Wool and silk.* The damaging effects that alkaline fabric washing products have on wool and silk were discussed in section 4.3.2.2. The combination of alkalinity with conventional hydrogen peroxide-based bleaching systems has the potential to cause serious oxidative damage at the peptide bonds in wool and silk (and cystine links in wool), especially if free transition-metal ions are present. In practice, however, provided that the recommended short, low-temperature (30–40°C) wash programmes are used, no significant oxidative damage occurs.

Although organic peracids are more reactive than hydrogen peroxide at lower wash temperatures, they do not cause general oxidation at peptide bonds. They are, however, highly specific in their attack on disulphide bonds. Consequently, the cystine cross links in wool keratin, which are so important to the physical properties of wool fibres and fabrics, are particularly vulnerable.

Under mildly acidic conditions, peracetic acid quantitatively oxidises cystine to cysteic acid (Figure 4.11(i)). Under alkaline conditions, however, the expected increase in the concentration of sulphonate groups is not detected. Nevertheless, there is marked deterioration in the mechanical properties of fibres, and in the handle of woollen garments, washed under these conditions. It is believed that this results from a different, but nonetheless damaging, mechanism of cystine oxidation and hydrolysis (Figure 4.11(ii)).

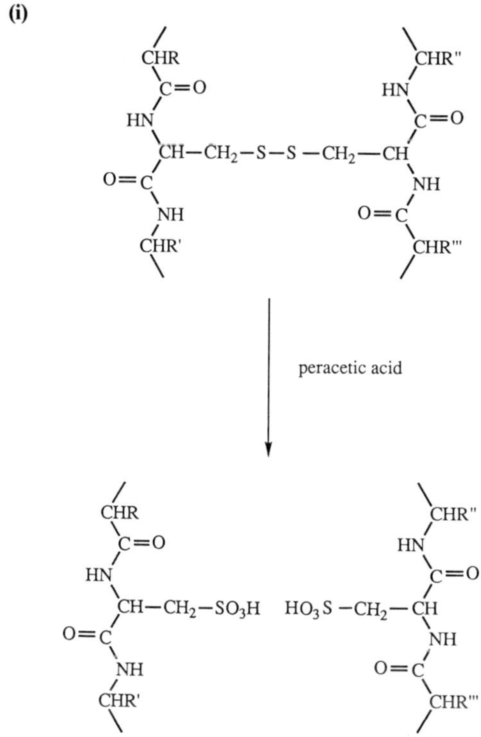

Figure 4.11(i) Oxidation of cystine to cysteic acid under acidic conditions.

4.3.5 *Enzymes and their effects on textile fibres*

4.3.5.1 *Proteases.* Proteolytic enzymes (proteases) have been included in many fabric washing products since the late 1960s. Such products are often described as 'biological' because of the natural origin and function of the enzymes, and the emphasis that this has been given in certain advertising campaigns. The function of these enzymes is to assist the removal of tenaciously bound proteinaceous soils, by breaking them down into smaller fragments, which are then more easily washed off by detergent action. Their optimum pH and temperature ranges are, of course, selected to fall within those of normal washing conditions.

Since a wide variety of protein types may constitute soils on textiles, non-specific proteases are used. These hydrolyse peptide bonds, irrespective of the structure of the adjacent side chains. As a consequence, protein fibres such as wool and silk are also subject to proteolytic attack. Although

(ii)

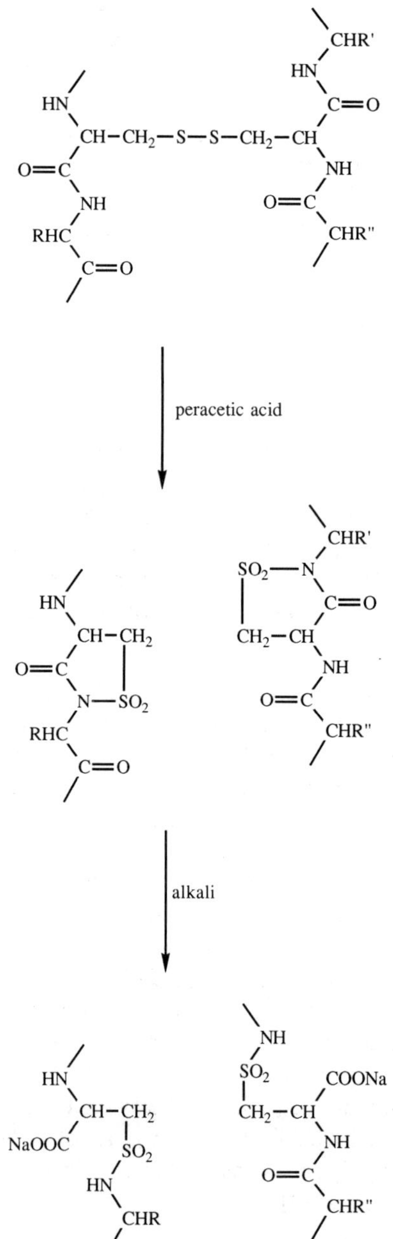

Figure 4.11(ii) Oxidation of cystine by peracetic acid under alkaline conditions.

detergents manufacturers normally warn consumers not to soak wool or silk in these products, they do not label products as 'unsuitable for wool and silk', probably because under recommended use conditions, little or no damage is apparent until articles have been washed many times.

Products which both generate peracid and contain proteolytic enzymes are particularly damaging to wool because of the synergy between peracid attack at the cystine cross links (section 4.3.4.2) and proteolytic attack at the peptide bonds. This can lead to the removal of substantial amounts of protein from the fibre even under recommended wash conditions. It has been reported, for example (Smith, 1994, personal communication), that a machine-washable wool sweater lost in excess of 1% by weight per wash in such a product, and that holes appeared in white areas of the garment after only about ten washes. The white areas were knitted from bleached wool, which is known to be extremely vulnerable to subsequent attack by proteases and peracids.

Since the shrink-resistance of machine-washable wools is usually achieved by oxidation followed by application of a polymer, their susceptibility to further oxidative and enzymatic attack, compared with that of untreated wools, is also of some interest. Recent work (Nolte, 1994) has shown that although the applied polymer layer appears to provide a temporary barrier against enzymatic attack, once this has been breached, proteolytic attack then follows much more rapidly in the shrink-resist treated fibre.

4.3.5.2 *Lipases.* During the past few years, detergent products containing lipases, as well as proteases, have been introduced. Lipases are enzymes which specifically hydrolyse the ester linkage in fats, such as triglycerides and other fatty esters, but do not have general esterase activity. Clearly, their function in washing products is to facilitate the hydrolysis and so improve the detergency of fatty soils. In practice, however, it is difficult to show convincingly that lipases are effective under washing conditions, because their activities are greatly reduced by the absorption of surfactants at their active centres (Benzonana and Desnuelle, 1968; Nolte, 1994). Suppliers of lipases optimised for use in detergents have suggested, however, that lipases absorbed on to fatty soil in one wash are not easily rinsed away, and that they continue to degrade the fatty stains 'between washes' and so give more complete removal in the next wash (Novo,1994).

There is no evidence to suggest that lipases have adverse effects on fibre properties when they are used in detergent products. They do, however, have the potential to modify lipoproteins on the cuticle and in the cell membrane complex of wool (Nolte, 1994).

4.3.5.3 *Cellulases.* Cellulases of bacterial and fungal origin are responsible for breaking down the cellulosic components of fallen leaves and dead plants. They are complex mixtures of enzymes which exhibit three types of cellulolytic activity, each of which is nevertheless specific to 1,4-β-D-glycosidic bonds. Thus total, crude cellulases always contain:

(a) Endoglucanases (EGs) which cause random chain scission at accessible points of the polymer chain, producing one new reducing end-group and one new non-reducing end-group at each cleavage point.
(b) Cellobiohydrolases (CBHs) which exhibit so called 'exoglucanase' activity—stepwise cleavage of cellobiose units from the ends of cellulose chains.
(c) Cellobiase or β-D-glucosidase, which hydrolyses cellobiose to glucose.

Natural cellulases are always found to contain several (possibly as many as six) different EG components. Their individual roles and the interactions between them are not yet fully understood (Emari and Niku-Paavola, 1987), but clearly there is synergy between the EGs which continually produce new end groups and the CBHs which attack them. At least two CBH components (CBH I and CBH II) always occur in natural cellulases. These components have differing affinities for cotton cellulose and cause differing rates of hydrolysis. There is recent evidence that CBH I preferentially attacks reducing end-groups while CBH II attacks non-reducing end-groups (Biely *et al.*, 1993). Only one cellobiase component is known; its role is essential to the complete degradation of cellulose, since hydrolysis is retarded by increasing concentrations of cellobiose.

Advances in biotechnology have made it possible to manipulate the composition of crude cellulases so that mixtures can be produced in which the activities of particular components have been deleted. Furthermore, pure, single cellulase components can also be produced, so that in principle it is now possible to tailor cellulase mixtures for specific textile finishing applications.

The use of cellulases in domestic detergents was first proposed over 20 years ago, as a means of alleviating the harshness that develops in cotton textiles as a result of fibrillation damage (section 4.2.3.1(c)). At the time, however, there were few commercially available cellulases, and none that was sufficiently effective, at alkaline wash pH, for successful exploitation in detergents. More recently, as a wider variety of cellulases has become commercially available, there has been renewed interest in their exploitation in detergents, not only to give a softening benefit, but also to remove pills from cotton articles, and to improve colour brightness and stitch definition by removing fuzz fibres.

Since cellulases degrade cotton cellulose, a compromise between delivering the required benefit within a reasonable number of washes and causing excessive cotton damage has to be achieved. By using a cellulase mixture with the EGs deleted, considerable weight loss can be tolerated without causing serious loss in tensile strength. On the other hand, mixtures with CBHs deleted, or pure EGs, can be used to cause drastic modification in fibre mechanical properties with very little weight loss. The few cellulase-containing detergent products, that are currently being marketed, do very little damage to cellulosic fibres, even over 10–20 wash cycles. It is questionable, however, whether they contain sufficient cellulase activity to deliver any perceptible benefit.

4.3.5.4 *Amylases.* A few domestic detergent products contain amylase to break down starch-based soils and so aid their removal. Since amylases are specific to the hydrolysis of 1,4-α-D-glucosidic bonds, they cause no damage to textile fibres.

4.3.6 *Other ingredients*

4.3.6.1 *Antiredeposition agents.* When soils have been removed from textiles in washing processes, it is important to keep them dispersed in the aqueous phase, and to prevent their redeposition on to the textiles. This is not a problem when a sufficiently high product dosage is used to deal with the total water hardness and the soil level present in the system. In domestic washing, however, the variability of wash load size, soil level, water hardness, and dosage selected by the consumer can often lead to marginally under-dosed conditions. In these conditions, a variety of polymers can help to reduce soil redeposition by mechanisms which are specific to particular types of soils and fabrics.

(*a*) *Sodium carboxymethylcellulose* (*SCMC*). Sodium carboxymethylcelluloses (SCMC) with a degree of substitution of about 0.6 are water-soluble, but are strongly adsorbed on to cellulosic fibres under washing conditions. A monolayer of SCMC adsorbed on cellulosic fibres significantly increases their negative surface potential by virtue of the increased density of $—COO^-$ sites. This has the beneficial effect of increasing the repulsion of negatively charged soil particles, especially clay soils, and precipitating calcium phosphates.

SCMC is therefore incorporated in most fabric washing products, at levels which will maintain monolayer coverage of all the cellulosic fibres in typical wash loads, and assuming typical levels of product dosage by the consumer. This provides protection against the redeposition of clay soils on to cellulosic fibres in marginally 'built' conditions (see section 4.3.1), but once significant levels of calcium phosphates, calcium soaps, or calcium salts of anionic surfactants begin to precipitate, the $—COO^-$ sites provided by the SCMC become occupied by free calcium ions and the benefits of electrostatic repulsion are lost. Under these circumstances the binding of the redeposited soils on fibre surfaces (via calcium salt 'bridges') becomes a problem.

(*b*) *Polyacrylates.* A variety of polyacrylates, and other polycarboxylic acids such as ethylene/maleic acid co-polymers, has been used in fabric washing products to help delay the onset of precipitation and deposition of insoluble calcium salts. Unlike SCMC, these polymers do not function by adsorption on the fibre, and although they provide some additional calcium 'building' capacity, their main role is to become incorporated into growing precipitates thereby poisoning crystal growth, and inhibiting flocculation and deposition.

(*c*) *Soil release polymers and cellulose ethers.* Since polyester fibres are relatively hydrophobic, they strongly adsorb hydrophobic, fatty and oily soils during use or wear, and tend to retain them even under ideal wash conditions. Under poor wash conditions polyester fabrics readily readsorb any precipitating or non-solubilised fatty soils. Then they become grey, and develop fatty odours after only a few washes. This problem can be alleviated by making the fibre surface more hydrophilic. Textile finishers can achieve this by the application of soil-release polymers, which are often block copolymers of polyethylene glycols and polyesters.

The polyester blocks can be adsorbed on, and, under heat-setting conditions, co-crystallised with the fibre surface. The polyethylene glycol blocks provide some hydrophilic character, and on washing they become heavily hydrated and so aid the release of oily soils, and prevent their redeposition. Such polymers can also be applied during fabric washing or rinsing processes to provide similar benefits, albeit at lower efficacy. The major manufacturers of detergents have, however, been slow to adopt them, and they are perhaps not yet convinced of their cost-effectiveness.

An alternative route to providing similar benefits in the wash was applied in a few detergent brands during the 1970s. Cellulose ethers, such as methyl cellulose, and certain ethylhydroxyethyl celluloses are water-soluble, but have the correct hydrophobic–hydrophilic balance to adsorb on to polyester under washing conditions, thereby rendering it somewhat more hydrophilic. The use of cellulose ethers, as oily soil anti-redeposition polymers for polyester, appears to have been discontinued. Again, the detergents manufacturers are, perhaps, unconvinced of their cost-effectiveness.

(*d*) *Polyvinylpyrrolidone* (*PVP*). It has been known for many years that PVP interacts strongly with some dyes (Scholtan, 1954), especially those with large planar chromophores such as dis- and tris-azo direct dyes. From time to time this property has been utilised in fabric washing products in order to reduce cross-staining by fugitive dyes. Unfortunately, powerful dye scavengers not only reduce the activity of dyes in solution (thereby reducing dye staining) but also encourage the desorption of more dye from the source fabrics. Indeed, PVP solutions have been used in the dyeing industry as an aid to stripping dyed fabrics which require shade correction.

There are certainly combinations of dye, fabric, and wash process conditions, under which PVP reduces cross-staining with a barely measurable increase in the amount of dye desorbed from the source fabric. Such combinations are not, however, so widespread that PVP could be claimed to have generally beneficial effects in fabric washing. Furthermore, the interaction of dyes with PVP is strongly influenced by surfactant type and level.

In general, the interaction between anionic dyes and PVP is reduced in the presence of other anionic surface-active species. Thus, potentially beneficial effects of PVP are reduced, or eliminated, not only by the presence of anionic surfactants in the detergent product, but by the fatty acids which are invariably present in natural soiling on domestic wash loads.

4.3.6.2 *Fluorescers (optical brightening agents).* Fluorescers, fluorescent whitening agents (FWAs) and optical brightening agents (OBAs) are all terms frequently used to describe essentially colourless dyes which have strong absorption maxima in the near UV between 300 and 400 nm (normally around 350 nm) and which re-emit the absorbed energy as violet to blue visible light at about 400–440 nm. Such dyes have been widely used in fabric washing products since the 1950s (as well as in the textiles and paper industries) to improve the perceived whiteness of bleached textiles. They were, and still are, the basis of the many detergents' advertising claims for washing 'whiter than white' and 'adds brightness to whiteness' and so on.

Although some 'nylon fluorescers' were used in the past (when white nylon was more widely used in washable end-uses), the majority of fluorescers are direct cotton dyes. Many of these are based on diaminostilbene, with the dianilino, dimorpholino, cyanuric chloride derivatives (Figure 4.12) probably being the most popular, for their good all-round performance in terms of solubility in a variety of product types, and reasonable stability to light when adsorbed on cotton. As for direct dyes in general, the substantivity of cotton fluorescers is inversely correlated with their water solubility. Similarly, their rates of adsorption and their equilibrium concentrations in the fibre are strongly influenced by temperature and electrolyte concentration. In countries where washing is carried out in cold water, alternative fluorescers, with

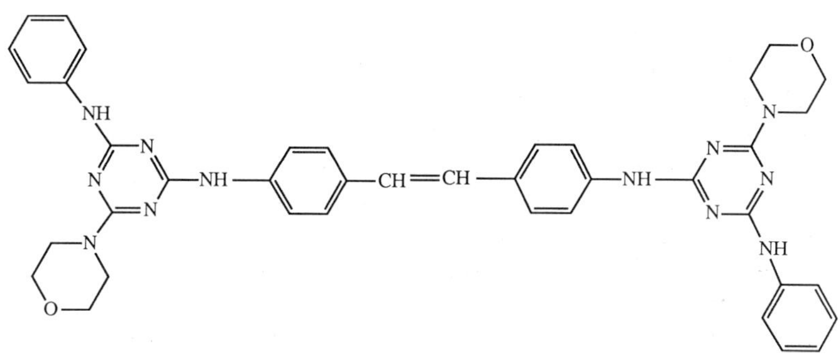

Figure 4.12 A typical cotton fluorescer derived from diaminostilbene.

increased cold-water solubility, may be required. Usually, however, these are variants of the same diaminostilbene derivatives, which have been micronised, or produced in modified, more readily soluble crystalline forms.

In tropical climates, where the lightfastness of conventional fluorescers is inadequate, 'superbrighteners' with greatly improved light (and chlorine) stability may be preferred. They are, however, expensive, and while they have good cold-water solubility, they have relatively low substantivity to cotton.

Fluorescers give white cotton a bluer and brighter whiteness, which is almost universally preferred by consumers, but this has the disadvantage of making pale yellow and beige shades look 'washed-out', and of making pink and red shades look noticeably bluer. For this reason they are omitted from 'coloureds wash' detergent brands. This is only a partial solution to the problem, however, since fluorescers, like other direct dyes, have poor wash-fastness and tend to desorb and transfer from one fabric to another during the wash process.

4.3.6.3 *Antifoams.* It has been mentioned (section 4.2.2.4) that front-loading automatic, or other forms of horizontal-axis, rotating-drum machines can generate excessive amounts of foam. Consequently, products intended for use in these machines contain antifoam systems. Historically, these were based on finely divided hydrophobic particles formed by the precipitation of the calcium salts of long-chain C_{18-22} fatty acids. During the 1980s these were replaced by more efficient and cost-effective systems, based on the combination of much lower levels of hydrophobic particles (such as calcium stearyl phosphate) with hydrocarbon or silicone oils.

The danger with any antifoam system based on hydrophobic oils and/or particles is that in poor wash conditions these will be deposited on hydrophobic fibres such as polyester (see section 4.3.6.1(c)). It is certainly the case that the early calcium soap-based antifoams could be found deposited on polyester fabrics. The much lower levels that are required for particle-in-oil antifoams have reduced this problem, and this may be the reason why the use of cellulose ethers to control fatty soil redeposition on polyesters (section 4.3.6.1) has been discontinued.

4.3.6.4 *Perfume.* For many years the role of perfume in fabric washing products was to mask unpleasant odours of product ingredients, both in the pack, and when the product was dissolved in water. Since the introduction of rinse conditioners, the consumer has also come to expect the washing product to leave a pleasant odour on the clean washing—not only while it remains wet, but even after drying and in susequent use.

This has necessitated the development of fragrances containing components which partition between textile fibres and surfactant solutions, so that perfume is slowly released from fibres at their normal regain moisture contents.

Attempts have also been made to utilise so-called 'deodorant–perfumes' (which are used successfully in toilet soaps and other personal products) to reduce the perception of body odour while wearing clothes that have been freshly washed.

An alternative approach to providing long-lasting fragrance on textiles is to deposit micro-encapsulated perfumes on the fabric, which are slowly released during use or wear as capsules are continually being ruptured. This route has been used by some fashion houses and stores to give their goods a distinctive 'house fragrance' which is intended to influence the future buying habits of their customers!

4.4 Effects of wash process and product formulation variables on dyed fabrics

4.4.1 *Introduction*

It will be apparent from the preceding sections that there are several factors in the domestic washing process which can contribute to colour changes in dyed textiles. Some of these factors, such as the effects of cotton fibrillation damage (section 4.2.3.1(c)), the encrustation of fibres with insoluble calcium salts (section 4.3.1), and the adsorption of fluorescers (section 4.3.6.2), have already been described. These will not be discussed further here, except to note that they are of relatively minor importance by comparison with the colour changes that may result from the effects of bleaching agents or the desorption of fugitive dyes.

4.4.2 *Effects of bleaching systems*

The tendency for any detergent bleaching system to damage dyes is strongly dependent on the wash temperature and product concentration selected by the consumer. At the recommended wash temperature for a particular dyed textile, and recommended detergent product dosages, the potential for bleaching systems to cause damage increases with their increasing reactivity: hydrogen peroxide < organic peracids < catalysed hydrogen peroxide. The sensitivity of dyes to oxidative damage depends on the structure of their individual chromophores, rather than on dye application class. The characteristics of dyes in some application classes do, however, make them more or less sensitive, as a class, to oxidative damage. This is particularly true for vat dyes and sulphur dyes. The anthraquinonoid, indigoid and other polynuclear aromatic ketonic chromophores that are characteristic of vat dyes are developed in the fibre by mild oxidation, and are not sensitive to damage by further oxidation. Sulphur dyes are also applied in their soluble, reduced form and are developed by oxidation. In this case, however, the polysulphide structures, which constitute many of their insoluble chromophores, are particularly sensitive to further oxidation by organic peracids. This results in cleavage of —S—S— bonds by analogous mechanisms to the oxidation of cystine in wool (section 4.3.4.2(b)). Thus, the chromophore is broken down into smaller, more soluble fragments, which can be washed off by detergent action. Consequently, sulphur dyes which have good wash-fastness, and adequate fastness to perborated products, are likely to have very poor fastness to products containing TAED/perborate or other peracid bleach systems.

The sensitivity to peroxide oxidation of dyes based on azo chromophores can be increased by chelation of transition metal ions, and the associated damage to cellulosic fibres dyed with such metallised chromophores may also be increased (Anon, 1985). For this reason some metal complex dyes have poor fastness to washing in perborated detergent products. Similarly azo chromophores with the potential to complex metal ions may also have poor fastness in perborated detergent products which do not contain stabilisers to sequester transition metal ions.

4.4.3 *Dye desorption and cross-staining*

4.4.3.1 *Effects of washing process variables: time, temperature, liquor/cloth ratio and agitation.* Dyes that are applied by exhaustion from solution, and which are not subsequently either covalently bonded to the fibre or rendered insoluble, will tend to re-equilibrate with washing solutions. This applies in principle to direct, acid, disperse and basic dyes, and to any hydrolysed reactive dyes that have not been thoroughly washed off after dyeing.

In practice, however, basic dyes do not cause cross-staining problems in washing because they are used almost exclusively on acrylic fabrics that must be washed below T_g (section 4.2.2.2). The diffusion of dyes in fibres below T_g is so slow that significant desorption does not occur within the duration of typical washing cycles. This also applies to disperse dyes on polyester and on cellulose acetate fabrics, but since disperse dyes diffuse readily in nylon (which is above T_g at all wash temperatures) they can give rise to significant cross-staining problems. Disperse dyes on polyester microfibre fabrics may also be expected increasingly to contribute to dye staining problems in domestic washing. The problem here is that the high surface area of the fibres, and the increased dye-in-fibre concentrations required to achieve a given depth of shade, combine greatly to increase the amount of dye that desorbs from the fibre surface, especially in the first few washes.

Acid dyes on nylon and wool, and direct dyes on cellulosic fibres (especially if not after-treated) re-equilibrate readily with washing solutions. Hydrolysed reactive dyes behave like direct or acid dyes having low affinity for the fibre.

The desorption and cross-staining behaviour of dyes under washing conditions clearly has much in common with their dyeing characteristics. Consequently, the effects of the main wash process variables are predictable, at least in general terms, from the relationships used to predict dyeing equilibria and dyeing kinetics. For example, the quantity $\Delta\mu_{sf}^0$ in the equation:

$$-\Delta\mu_{sf}^0 = RT \ln \frac{[D]_f}{[D]_s}$$

represents the partial molar free energy change associated with the transfer of one mol of dye from its standard state in solution to its standard state in the fibre. This is the 'standard affinity' of the dye for the fibre, and once determined (from a series of equilibrium dyeings) it can be used to predict the equilibrium concentrations of an undissociated dye (such as a disperse dye) in the fibre $[D]_f$ and in solution $[D]_s$ at any temperature T, or for any liquor/cloth ratio. Since this applies equally to dyeing and dye desorption, it is predictable that washing at long liquor/cloth ratios drives dye desorption in order to achieve the equilibrium dye activity in the large volume of water. Similarly, the presence of a large amount of undyed fibre must drive desorption and cross-staining in order to achieve the equilibrium dye activity in both the originally dyed and undyed fibres.

The effect of temperature on dye desorption is also predictable; high wash temperatures cause greater equilibrium dye desorption. In any given wash, however, the effect of temperature on the rate of dye desorption is more important than its effect on equilibrium dye distribution.

The rate of dyeing — or dye desorption — is determined by (a) the driving force for dye transport and (b) the resistance to dye transport:

$$\text{Rate} = \frac{\text{Driving force}}{\text{Resistance}}$$

Clearly the driving force at any instant in time is determined by how far the system is from equilibrium. In dyeing this can be represented by the 'instantaneous' chemical potential gradient between the fibre and solution phases. At short times this approximates to the dye concentration gradient (grad C).

Resistance to dye transport is provided primarily by the slow diffusion of relatively large molecules through the fibre bulk and is therefore inversely proportional to the diffusion coefficient of dye in the fibre (D). Thus the mass (M) of dye diffusing through the fibre surface at any instant can be represented by the general equation:

$$M = -D\,(\text{grad } C)$$

It has been mentioned (section 4.2.2.2) that diffusion coefficients increase exponentially with increasing absolute temperature according to the Arrhenius equation. Consequently, the rate of dye transport is approximately doubled for every 10°C rise in wash temperature. Clearly the choice that the consumer makes between a 40°C and a 60°C wash for a coloured fabric can influence the amount of dye transported by a factor of four.

In practice, washing machines go through heat-up cycles and the real effect of programme choice is determined by the combination of heat-up time and the length of time for which maximum temperature is maintained. The longer the total cycle time, however, the greater the dye transport, and in general higher temperature wash programmes have longer total cycle times.

In poorly stirred systems, dye transport to the fibre surface from the bulk solution (or vice versa in dye desorption) may also become a rate-determining factor. This has not been found to be the case in washing machines, since they provide relatively high agitation levels to assist the detergency process.

4.4.3.2 *Effects of product ingredients*

(*a*) *Alkalinity and electrolyte concentration*

Direct dyes. Increasing alkalinity increases the negative potential of cellulosic fibres by increasing the ionisation of carboxyl groups and any other anionic groups that may be present as a result of processing and finishing operations. This would be expected to reduce the equilibrium concentration of direct dyes in the fibre according to well-established relationships (Peters, 1975a). In fact, the effects in washing processes are very small because anionic groups in the fibre are almost completely ionised under (neutral) dyeing conditions. The increase to pH 9–10 in washing therefore has very little additional effect on fibre potential.

By contrast, the effect of electrolyte concentration on the equilibrium distribution of direct dyes is extremely large. The salt concentration used in direct dyeing may be of the order of 20–30 g/l. The purpose of this is to increase the activity of dye in solution (by increasing the sodium ion concentration), thereby forcing the equilibrium in favour of the fibre:

$$-\Delta\mu_{\mathrm{sf}}^{0} = RT\ln\frac{[\mathrm{Na}]_{\mathrm{f}}^{z}\,[\mathrm{D}]_{\mathrm{f}}}{[\mathrm{Na}]_{\mathrm{s}}^{z}\,[\mathrm{D}]_{\mathrm{s}}}$$

where z = the number of sulphonate groups in the dye molecule. Washing at sodium ion concentrations that are equivalent to only about 2 g/l of sodium chloride leads to equilibrium desorption of about 80% for typical tetrasulphonate direct dyes (Marshall and Peters, 1947). Clearly, this does not happen in the course of a single wash, because at moderate wash temperatures the dye diffusion coefficient is greatly reduced. The driving force for desorption, however, remains essentially the same for successive washes.

Although washing products which provide relatively high ionic strengths help to reduce direct dye desorption, they also drive the re-adsorption (cross-staining) of desorbed dye on to other undyed (or pale shade) cellulosic fibres in the wash. Since cross-staining is perceived to be a more serious problem than the colour fading that may result from dye desorption, it may be considered preferable to formulate 'coloureds wash' products to a very low ionic strength, in order to minimise the cross-staining problem.

Acid dyes. Acid dyes are adsorbed at protonated amino end-groups and side-chains during the dyeing of nylon and wool. The concentration of such sites is maximised during dyeing by setting the dye bath at about pH 4.0 or lower. The concentration of protonated sites is progressively reduced by increasing pH, as the pK_b values of the various amino groups in the fibre are exceeded. It can be shown for a monobasic acid HA (such as monosulphonated acid dye) that:

$$\frac{-\Delta\mu^0_{HA}}{2.303} = 2RT\left\{\log\left(\frac{\theta}{1-\theta}\right) + pH\right\}$$

where θ is the fractional occupation of positive and negative sites by A^- and H^+ ions respectively, and it is assumed that the concentrations of positive and negative sites are always equal.

From this equation the equilibrium acid dye distribution can be predicted for washing product solutions of varying pH. Clearly it is essential to minimise the wash pH in order to minimise dye desorption. On the other hand, cross-staining by desorbed dye increases with decreasing pH. In practice, however, neutral or very weakly alkaline products are most appropriate for wool washing because of the damaging effect of alkali on the fibre (section 4.3.2.2).

It will be apparent from the preceding discussion that increasing the concentration of anions other than A^- dye anions must increase the competition for protonated dye sites. Thus, high electrolyte concentrations promote the desorption of acid dyes, but retard their cross-staining—the opposite dilemma to that posed by direct dyes. It is, however, usual to formulate specialist wool wash products as low ionic strength liquids. Relatively low concentrations of anionic surfactants (see below) have similar effects to high concentrations of inorganic anions and act as effective retarding agents against cross-staining by any small amount of acid dyes desorbed under neutral washing conditions.

Reactive dyes. The covalent bonds formed between the triazinyl and pyrimidyl reactive dyes and cellulose have more 'ester-like' than 'ether-like' character. They are therefore prone to slow hydrolysis under strongly alkaline washing conditions, and full shades tend to 'wash down' over many wash cycles. Hydrolysed dye washed from the fibre behaves towards cellulosic fibres in the same way as direct dyes of low affinity, and towards wool and nylon in the same way as acid dyes.

(*b*) *Surfactant effects.* Water-soluble dyes have some features in common with surfactants. They are relatively large organic molecules with one or more solubilising groups, and indeed many dyes show marked surface activity. Like surfactants, dyes aggregate in solution above 'critical' concentrations. Typical aggregation numbers are in the range 2–20 and some dyes (with alkyl side chains) form true micellar solutions (Peters, 1975b). Mixtures of dyes and surfactants interact strongly in solution in a variety of ways depending on the charge of the individual dyes and surfactants. Consequently, surfactant solutions have a marked influence on both the desorption of dyes from dyed fabrics, and the uptake of dyes from solution.

Disperse dyes. Since disperse dyes have no charge, or other solubilising groups, they are only sparingly soluble in water. Typical solubilities are in the range 0.2–50 mg/l, but their solubility increases dramatically at the critical micelle concentration of surfactant solutions. The dye may be considered to be solubilised in the hydrocarbon-rich interior of the surfactant micelles, and since disperse dyes are nonionic, the effect is independent of surfactant charge.

When fabrics dyed with disperse dyes are washed in surfactant solutions, dye is desorbed until the equilibrium partitioning of dye between the fibre and micellar surfactant phases is achieved. Clearly, the higher the concentration of micellar surfactant, the greater the equilibrium dye desorption. Thus, high product or surfactant concentrations tend to increase dye desorption, but also tend to retard cross-staining, at least until the micellar phase is saturated with dye.

Direct and acid dyes. Interactions between surfactants and anionic dyes, such as direct and acid dyes, are dominated by the ionic character of the surfactant:

Cationic	> Zwitterionic > Amphoteric	≫ Nonionic	> Anionic
Insoluble complexes formed	Dyes taken into soluble complexes in the micelle	Many dyes strongly solubilised	Some dyes weakly solubilised

The formation of insoluble complexes between cationic surfactants and direct dyes may be used to insolubilise dyes on cellulosic fibres in order to improve their wet-fastness. These complexes can, however, be dispersed and solubilised in excess nonionic surfactant. Such systems have been used to control the strike rate and levelling of anionic dyes (Hughes *et al*., 1971) and to reduce cross-staining by anionic dyes in fabric washing products (Bishop and Nelson, 1978).

The anionic dye restraining effects achievable with zwitterionic and amphoteric surfactants are generally more powerful than those obtained with optimised nonionic–cationic mixtures (Bishop, 1978). Their exploitation in domestic detergents has until now been limited by the high cost and low availability of the appropriate surfactants.

Interactions of anionic dyes with nonionic surfactants have been widely studied (Peters, 1975b) and there is good evidence from spectral shifts, the raising of the cloud points of the nonionics, and lowering of the critical micellar concentration, that mixed dye/surfactant micelles are formed. In general, the effect of nonionics is to increase dye solubility and to retard dyeing by lowering the dye monomer concentration in solution. In washing processes, however, where dye cross-staining results from the combination of the dye desorbing and restraining properties of the surfactant system, the net effect is not always beneficial compared with that obtained with predominantly anionic active systems.

Anionic surfactants do not interact strongly with anionic dyes in solution, but they are used to retard dyeing with acid dyes on wool and nylon where they compete with the dye for cationic sites in the fibre. In washing systems, this has the beneficial effect of reducing cross-staining by acid dyes on these fibres, but high concentrations of anionic surfactants also increase the driving force for the displacement of dyes from the dyed fabrics.

Acknowledgement

The author wishes to thank Unilever Research Laboratory for data included in this chapter.

References

Anand, S., Bishop, D.P. and Mackay, C. (1990) Unpublished work.
Anon (1985) *Textile Chemist and Colourist*, **17**(12), 239.
Benzonana, C. and Desnuelle, P. (1968) *Biochim. Biophys. Acta*, **164**, 47.
Biely, P., Vrsanska, M. and Claeyssens, M. (1993) Mode of action of *Trichoderma reesci* β-1,4-glucanases on cellooligosaccharides. In *Proceedings of the Second TRICEL Symposium*, Espoo, 1993 (eds P. Suominen and T. Reinnikainen) Foundation for Biotechnical and Industrial Fermentation Research, vol. 8, pp. 99–108.
Bishop, D.P. (1990) Unpublished work.
Bishop, D.P. (1991) In *Proceedings, UMIST Annual Symposium, Fibre Fracture and Product Performance Industrial Research Group*, 5–6 June, p. 7.
Bishop, D.P. and Nelson, R.T. (1978) UK Patent GB 2002022A (assigned to Unilever).
Brussels (1987) *The Quest for Cleanliness in our Environment*, Brussels: Association Internationale de la Savonnerie et de la Detergence.
Carfagno, P.P. and Pacheco, J.F. (1973) Laundering variables and their effect on flame retardant fabrics, *University of Utah Polymer Conference Series*, June.
Chugg, K.J. (1990) *Mechanisms of Fabric Softening*, PhD Thesis, University of Cambridge.
Cutler, W.G. and Davies, R.C. (eds) (1972) *Detergency Theory and Test Methods, Part I*. Vol. 5 of the Surfactant Science Series, Dekker, New York.
Cutler, W.G. and Davies, R.C. (eds) (1975) *Detergency Theory and Test Methods, Part II*. Vol. 5 of the Surfactant Science Series, Dekker, New York.
Cutler, W.G. and Davies, R.C. (eds) (1981) *Detergency Theory and Test Methods, Part III*, Vol. 5 of the Surfactant Science Series, Dekker, New York.
Cutler, W.G. and Kissa, E. (eds) (1987) *Detergency: Theory and Technology*. Vol. 20 of the Surfactant Science Series, Dekker, New York.
Emari, T.M. and Niku-Paavola, M.L. (1987) Enzymatic hydrolysis of cellulose: Is the current theory of the mechanisms of hydrolysis valid? *Crit. Rev. Biochem.*, **5**, 67–87.
Flory, P.J. (1953) *Principles of Polymer Chemistry*, Cornell University Press. Chapter XIII, equations (38) and (42), pp. 578 and 580.
Golova, O.P. and Nosova, N.I. (1973) Degradation of cellulose by alkaline oxidation, *Russian Chemical Reviews*, **42**(4), 327–335.

Horrocks, A., Hall, M.E. and Roberts, D.L. (1993) Environmental implications of using flame retardant textiles, *Meilland Textilber*, **74**(5) E190–E193, 417–418, 425–428.

Hughes, J.A., Sumner, H.H. and Taylor, B. (1971) *J. Soc. Dyers and Colourists*, **87**, 463.

Karsa, D.R. (1990) *Review of Progress in Colouration*, **20**, 70.

Mackay, C. (1992) *Effects of Laundering on the Sensory and Mechanical Properties of 1 × 1 Rib Knitted Fabrics*. MPhil Thesis, Bolton Institute.

Marshall, W.J. and Peters, R.H. (1947) *J. Soc. Dyers Colourists*, **63**, 446.

Mohamed, S.S. (1982) *Textile Chemist and Colourist*, 14 March, 65/37.

Nolte, H. (1994) PhD Thesis, Deutsches Wollforschungsinstitut, An der T.H. Aachen, e.V., Aachen.

Novo Nordisk AIS (1994) Personal communication.

Payot, F. (1991) Objective approach to measurement of changes of hand properties on treated silk fabrics. In *Proceedings of Seminar: Practical Developments in Objective Measurements for Textiles and Clothing*. Organised by BTTG as part of EC (DG XIII) SPRINT project RA115, December.

Peters, R.H. (1975a) *Textile Chemistry III, The Physical Chemistry of Dyeing*. Amsterdam, Oxford and New York: Elsevier, Chapter 12, pp. 412 and 427.

Peters, R.H. (1975b) *Textile Chemistry III, The Physical Chemistry of Dyeing*. Amsterdam, Oxford and New York, Elsevier, Chapter 27, p. 851.

SDIA (1989a) *Is it Biodegradable?—the Facts about Detergents in the UK*. Soap and Detergent Industry Association.

SDIA (1989b) Summary of SDIA Seminar on *Detergents and Care for the Environment—the UK Position*, 9 October. Soap and Detergency Industry Association.

Scholtan, W. (1954) The absorptive capacity of water-soluble polymer compounds, particularly of polyvinylpyrrolidone. *Makromol. Chem.*, **11**, 131–230.

Smith, P.J. (1994) *J. Soc. Dyers and Colourists*, **110**, 289.

Thompson, A.B. (1963) Fibres from condensation polymers. In *Fibre Structure* (eds J.W.S. Hearle and R.H. Peters) The Textile Institute/Butterworths, Chapter 4, pp. 513–516.

5 Easy care

D.W. HEYWOOD

5.1 Introduction

Amino resin finishing commenced in the early 1920s when Tootal Broadhurst Lee took patents out on the manufacture of simple urea–formaldehyde resins for the production of crease-resistant fabrics (Foulds *et al.*, 1926). Those original patents also included phenol formaldehyde as well as urea; the phenol formaldehyde could be discounted due to its colour. But Tootal Broadhurst Lee commercialised on their patent and built up a world wide licensing organisation where companies licensed the process, and when they produced the required standards of crease resistance and tensile strength were granted the Tebilized (Registered Trade Mark) label.

The general formulation for production of the resin was as follows. The product itself would have a limited shelf life dependent on local conditions, pH conditions and temperature.

$$\mathrm{CO}\left\langle\begin{matrix}\mathrm{NH_2}\\ \mathrm{NH_2}\end{matrix}\right. + \mathrm{HCHO} \rightarrow \mathrm{CO}\left\langle\begin{matrix}\mathrm{NHCH_2OH}\\ \mathrm{NH_2}\end{matrix}\right. \qquad \text{Monomethylol urea}$$

$$\mathrm{CO}\left\langle\begin{matrix}\mathrm{NHCH_2OH}\\ \mathrm{NH_2}\end{matrix}\right. + \mathrm{HCHO} \rightarrow \mathrm{CO}\left\langle\begin{matrix}\mathrm{NHCH_2OH}\\ \mathrm{NHCH_2OH}\end{matrix}\right. \qquad \text{Dimethylol urea}$$

The process required an acid catalyst together with a drying and curing schedule which is still practised in modern resin finishing. The manufacture of own-made resin was still prevalent in the UK until 1993, when the last two companies producing their own-made precondensate ceased production and went over to a manufactured precondensate.

After urea a full range of products was developed, each one containing formaldehyde, such as the following (Figure 5.1).

1. Urea–formaldehyde
2. Highly condensed urea–formaldehyde
3. Methylated urea–formaldehyde
4. Melamine–formaldehyde
5. Methylated melamine–formaldehyde

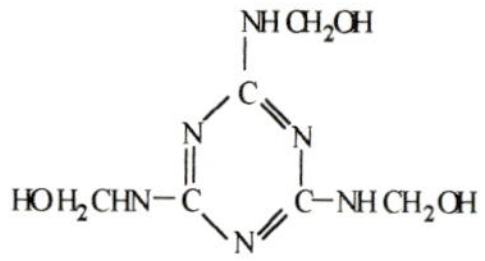

Trimethylol melamine

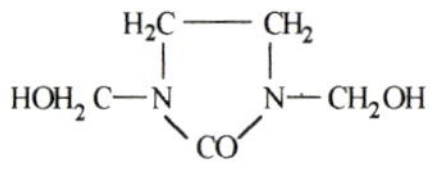

Dimethylolethylene urea

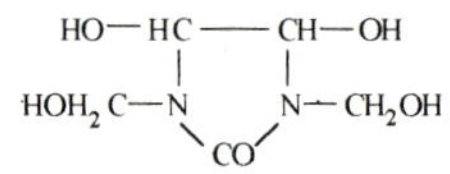

Dimethyloldihydroxyethylene urea (DMDHEU)

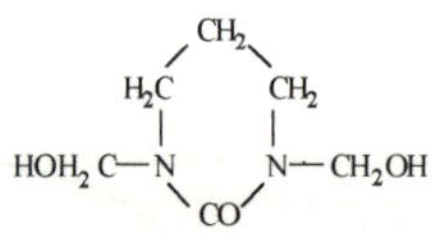

Dimethylolpropylene urea

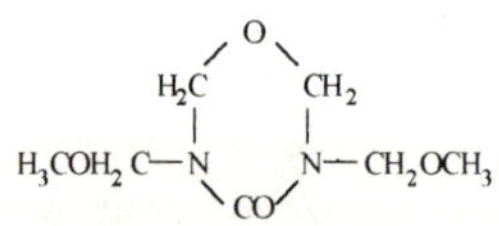

Dimethoxymethyl uron

Figure 5.1 Examples of major resins used in the industry and their chemical structures.

6. Ethylene urea–formaldehyde
7. Heterocyclic crosslinking agents based on melamine–formaldehyde
8. Glycol based reactants and derivatives thereof.

Each of these resins has its own particular part in the evolution of easy care finishes to produce present day standards.

The properties required by an easy care finish are:

1. Dry and wet crease resistance
2. Good dimensional stability
3. Little loss in physical properties (i.e. tensile and abrasion)
4. Compatibility with other finishing agents (e.g. water-repellent, softeners, optical brightening agent)
5. Minimum effect on handle
6. Little effect on dye or print shade or the light-fastness of the dye
7. Little effect on rub fastness of dye or print
8. No effect on the whiteness of the fabric
9. As low as possible formaldehyde release
10. Good environmental condition both in the application of the resin and the final making up of the fabric when finished
11. Ease of application and low cost add-on factor in both chemical and application cost.

5.2 Urea–formaldehyde resins

These products were initially produced in-house by various organisations and licencees of the holders of original patents (mainly Tootals and Courtaulds). However, due to the high free formaldehyde content of the resin, which ranged around 5% (present products have less than 0.5% on comparative solid levels), washing-off of the fabric was required after finishing. During the period of development of urea–formaldehyde resins, the advent of viscose rayon came about, and without resin finishing, this type of fabric would never have gained the popularity it did in the years prior and subsequent to the Second World War.

5.3 Methylated urea–formaldehyde resins

Coming onto the market in the early 1950s, these products (such as Kaurit W, BASF, BT 322, BIP) produced product stability well in advance of six months, and therefore opened up the field of resin finishing to companies who had previously been unable to carry out chemical finishing due to lack of knowledge of resin manufacture. Also at the same time a fashion trend for nylon petticoat finishing came about, and it was found that methylated urea–formaldehyde resins were able to produce durable stiff finishes on nylon. By the addition of thiourea, flame resistance of the nylon fabric was also obtained.

Comparison of the finishing of viscose rayon or cotton with that of nylon with a methylated urea–formaldehyde resin may be illustrated as follows. With the cotton fibre, the resin enters into the fibre and, when catalysed, produces a polymeric network structure in the fibre that prevents the fibre from creasing and shrinking on washing. With nylon, the resin forms a complete bond and produces a sheath-like coating to the fabric which is durable and stiff (Figure 5.2).

Methylated urea–formaldehyde resins also had the advantage over home-made precondensates that, when applied to nylon, they did not produce 'chalking or writing on the fabric' as did simple urea–formaldehyde resins.

Other advantages, methylated urea–formaldehyde resins have greater durability to washing and also produce better shrink-stability of the treated fabric.

The molecular sizes of the methylated urea–formaldehyde resins are very small, which enables them to enter the fibre to produce the crease resistance and stability, the size of the molecule remaining stable. However, in the case of straight urea–formaldehyde resins, the molecular size is not stable and is increasing all the time. Subsequently the size of the molecule becomes so large that it falls out and the resin becomes unusable. This is not the case with the methylated urea resins until they are catalysed, when the stability of the bath may be in the order of 24–48 h.

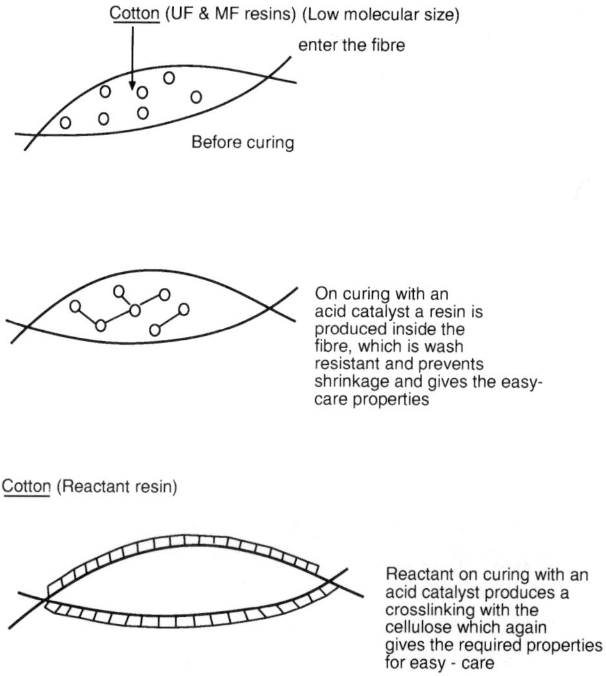

Figure 5.2 Comparison of reaction of resin former and a reactant with cellulose fibre.

Using the above knowledge, chemical manufacturers have been able to produce high molecular size urea–formaldehyde resins which may be used to stiffen or bond fibres together. These resins simply stay on the surface of the fibre and produce a stiffened effect rather than a crease-resistant effect. These resins are very rarely used on their own for finishing fabrics but are generally incorporated with crease-resistant resin to produce the ‘handle’ required. Typical formulations for a normal crease-resistant finish on viscose rayon are as follows.

Urea–formaldehyde resin (40% solids)	300 g
Cationic softener	20 g
Monoammonium dihydrogen phosphate	3 g
Water to 1 litre	
Mangle expression	75%

To increase the ‘spring’ and ‘handle’ the formulation could be

Urea–formaldehyde resin (40% solids)	260 g
Highly condensed urea–formaldehyde (50% solids)	40 g
Cationic softener	20 g

Monoammonium dihydrogen phosphate	3 g
Water to 1 litre	
Mangle expression	75%

5.4 Melamine resins

These resins were the next resin group to appear, and initially two major products were predominant, dimethylol melamine and trimethylol melamine. Dimethylol melamine was unstable and was mainly produced in a spray-dried (powder) form.

Trimethylol melamines were produced by A.C.C. in the USA (Aerotex M3), became widely used and are still used. The melamine resins were found to give a fuller handle on cotton and viscose than the urea–formaldehyde resins, but commercially they were more expensive. Four major niche markets were found for melamine resins and these still exist:

1. For durable mechanical finishes on cellulose
2. As a crosslinker for polymers such as acrylonitrile, acrylics, polyvinyl acetate, polyurethane, etc.
3. As a stiffener for polyester or nylon
4. As a handle builder in conjunction with other resins.

5.5 Mechanical finishing

Durable mechanical finishing came into vogue in the early 1950s and Joseph Bancroft produced the initial Everglaze (Registered Trade Mark) patent.

The principle of Everglaze is to produce a mechanical finish on the fabric that is wash-fast. This is carried out by impregnating the fabric with a solution of melamine resin plus catalyst, and drying the fabric to a moisture content of approximately 6–8% before carrying out the mechanical effect prior to a final curing. Typical formulations for this would be:

Trimethylol melamine resin (80% cured solids)	60 g
Magnesium chloride (40% solution)	20 g
Silicone softener	5 g
Water to 1 litre	

Based on a 65% pick-up, this would give approximately 3.1% solids on the fabric. The required amount of melamine solids in the fabric should be in the region of 2.5–3.2%; higher solids may produce a better mechanical finish but would impair the physical properties of the fabric.

After impregnation, the operation of drying the fabric is a crucial part of the process. The moisture content must be fully controlled and a final moisture content of 6–8% should be obtained before mechanical finishing. Too much moisture may produce an initial better mechanical finish but the durability will be poor, whilst too little moisture will give poor mechanical effects. It should also be remembered that if the fabric is overdried precure is brought about, and re-wetting cannot reverse the reaction. When precure takes place poor durability of the mechanical effect can take place. Types of mechanical finishes that are possible are as follows.

1. Glaze
2. Schriener
3. Sculptured or embossed
4. Permanent pleating

Once the mechanical effect has been produced on the fabric, the effect has to be made permanent by a curing process.

Figure 5.3 illustrates the type of equipment normally used for the process.

Glazing and schrienering of fabrics have kept in vogue of the last 25 years, but very little call has been made for permanent pleating, except for a brief period when Koratron (Registered Trade Mark) pleating became popular, this being overtaken by the advent of polyester–cotton, which itself was capable of producing its own permanent crease effects.

The use of melamine and urea resins for finishing produced a number of unwanted elements, the major one being ‘fishy odour’ or amine smell, and the next section deals in depth with this.

5.6 The problem of fishy odour in resin-treated fabrics

One of the problems which sometimes arises out of the application of crease-resist finishes and related processes is the spasmodic and frequently

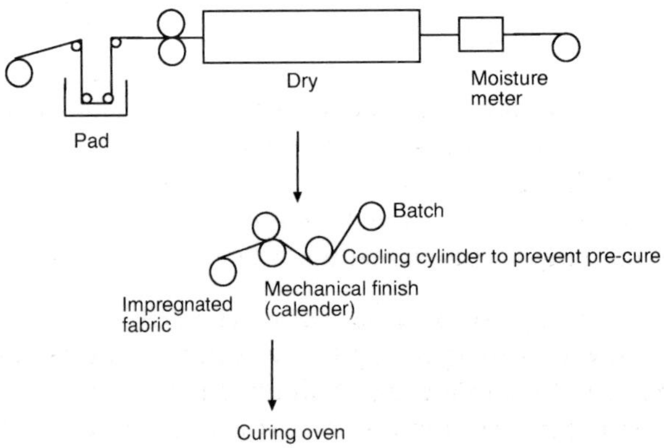

Figure 5.3 Typical layout for mechanical finishing.

unpredictable development in the fabric of a fish odour. This can occur in the baking operation but is more often associated with storage of the fabric, or with subsequent treatments involving heat and moisture such as Hofmann pressing.

In the past, the standard recommendation for preventing odour development was to wash-off the fabric after the resination process. This procedure is expensive, however, and is now regarded as unnecessary, provided that the causes of fishy odour are appreciated and adequate precautions are taken.

The fishy odour has long been attributed to the presence of methylamines in the finished fabric, and to trimethylamine in particular. The trimethylamine was said to be produced from the reaction of ammonia and free formaldehyde in the fabric, and the possibility of sodium formate being involved as well was suggested by Fluck (1951). The subject was subsequently examined in detail by Nuessle *et al.* (1959) and the significance of free formaldehyde, ammonia and formate was confirmed. Although their findings are not absolutely conclusive, they have thrown considerable light on the chemical reaction involved and have suggested that those shown below are some of the more likely (see equations 5.1–5.6).

$$2NH_3 + 3CH_2O \rightarrow 2CH_3NH_2 + CO_2 + H_2O \quad (5.1)$$

$$2CH_3NH_2 + 3CH_2O \rightarrow 2(CH_3)_2NH + CO_2 + H_2O \quad (5.2)$$

$$2(CH_3)_2NH + 3CH_2O \rightarrow 2(CH_3)_3N + CO_2 + H_2O \quad (5.3)$$

$$NH_3 + CH_2O + HCOOH \rightarrow CH_3NH_2 + CO_2 + H_2O \quad (5.4)$$

$$CH_3NH_2 + CH_2O + HCOOH \rightarrow (CH_3)_2NH + CO_2 + H_2O \quad (5.5)$$

$$(CH_3)_2NH + CH_2O + HCOOH \rightarrow (CH_3)_2N + CO_2 + H_2O \quad (5.6)$$

The problem of fishy odour in fabrics may be prevented by the following.

(1) Use of low-formaldehyde resin (e.g. the use of DMDHEU resin, which is fully illustrated in section 5.7).
(2) Use of non-ammoniacal catalysts, although with urea–formaldehyde resins this is difficult, as the other catalysts may not have the efficiency that ammonia-based catalysts have, and are certainly not as economical.
(3) Use of formaldehyde acceptors, such as urea or dicyandiamide.
(4) Washing-off of the fabrics after finishing.
(5) So-called masking agents, which one would only use as a last resort.

Other effects of easy-care finishes on fabric properties may be summarised in Table 5.1. The disadvantages may be overcome by the use of auxiliaries in the resin bath, and the following products may be employed.

For tensile strength, the addition of softeners, acrylics and preferably water-based polyurethane resin can compensate. However, the too liberal use of softeners may eventually reduce the tensile strength, and slippage of

Table 5.1 Advantages and disadvantages of easy-care finishes

Advantages	Disadvantages
	Loss of tensile strength
Crease resistance, wet and dry	Loss of abrasion
Improvement of shrink resistance	Loss of tear strength
Easy care	Slight change in shade of the dyestuff which may be due to the resin or the catalyst employed in the process

the fibres during the test will occur. Acrylics will affect the handle and also may produce back soiling from dirty wash liquors. Water-based polyurethane resins are probably the most effective; polyurethane resins are available with glass transition temperatures as low as −60°C, thus giving very soft handles without back soiling occurring.

With regard to abrasion, softeners are recommended, but it is debatable whether the improvement of abrasion occurs due to the transference of the softener from the fabric to the abrasive material, giving a false result. In general, again water-based polyurethanes tend to give the better results, and on cotton fabrics it is possible to improve the original abrasion above that of the fabric before resin finishing. For example, with cotton rep an original abrasion figure of 20 000 Martindale falls to 16 000 with a resin finish but rises to 24 000 with polyurethane added to the resin finish.

The amounts of polyurethane solids used are relatively small, approximately 0.75–1.0% on fabric, and increased amounts give no increase in abrasion. In fact, a drop-off is noted as the amount of polyurethane is increased above the 1.5% level.

Silicone polymers and elastomers may be incorporated with resin formulations to improve the physical properties. Tests have shown that especially on polyester–cotton blends, the use of silicone-based products does give excellent results, and the amount of resin may be reduced by the inclusion of these products.

5.7 Reactant-type products

These have been on the market since the early 1960s, and the following illustration shows that the popularity of these resins has increased in proportion to the use of synthetic–cellulosic mixture fabrics. Also, it can be noted that a decrease in viscose rayon is also in proportion (Figure 5.4). The main reasons for this are as follows:

(a) Urea–formaldehyde resins produce a stiff handle on synthetic fibres.
(b) Urea–formaldehyde resins produce excellent crease-resistant finishes on viscose rayon with good stability and are cheap; they are still used for specific end-uses, i.e. tie linings.
(c) DMDHEU/reactant resins do not affect the handle of synthetic fabrics and only react with the cellulosic component, thus giving a soft handle.

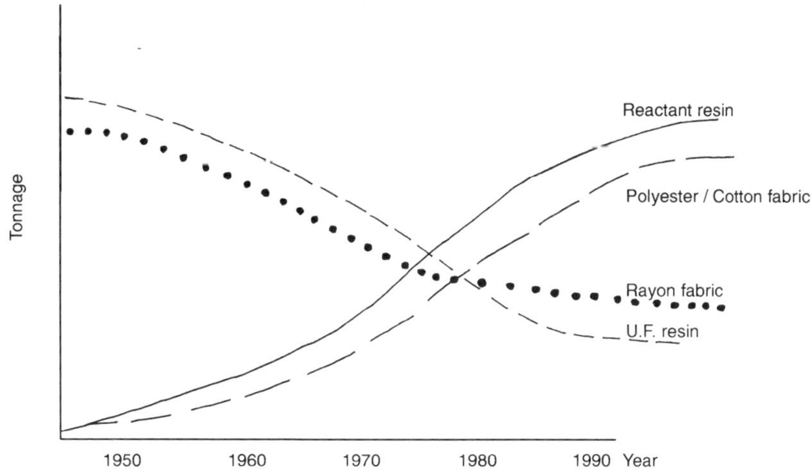

Figure 5.4 Changes in amounts of reactant resins and urea–formaldehyde resins used since 1950, together with changes in amounts of polyester–cotton and rayon used.

(d) Urea–formaldehyde resins, in the main, give high-formaldehyde contents.
(e) Reactant resins have a lower formaldehyde content, and they require smaller amounts of resin on to the fabric to produce results comparable with resin formers such as urea– and melamine–formaldehyde resins, again helping to reduce the formaldehyde level. Unlike older resins which form a polymeric network within the fibre, reactant products actually react with the fibre.

Reactant resins may be listed as the following:

Ethylene urea	(a)	Ethylene urea–formaldehyde
Propylene urea	(b)	Propylene urea–formaldehyde
Methylated uron	(c)	Methylated uron–formaldehyde
Glyoxal	(d)	Dihydroxyethylene urea–formaldehyde (DMDHEU)

But the most widely used product is the DMDHEU type and modifications of this. DMDHEU was introduced to replace cyclic ethylene formaldehyde resin, which was based on ethylene urea. The DMDHEU resin is based on a reaction of glyoxal, urea and formaldehyde, and reacts with cellulose by forming ether linkages under the influence of the acid catalyst.

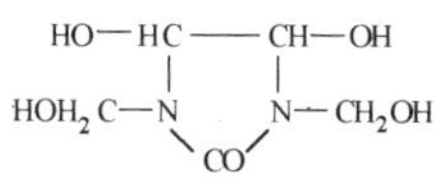

Dimethyloldihydroxyethylene urea
(DMDHEU)

Cyclic ethylene urea–formaldehyde resins (Fixapret EU; Rhonite R1 BT324) were produced by a simple reaction of ethylene urea with formaldehyde, and came on to the market for the following reasons:

(1) Soft handle on cotton and polyester–cotton
(2) Good chlorine resistance on white fabrics if fully cured
(3) Soft mechanical finishes.

However, the disadvantages of ethylene urea–formaldehyde resins were high price, and the fact that they required a full cure case to close the ring structure, in order to obtain full resistance to chlorine and prevent HCl being formed on impregnation with chlorine. These two disadvantages made the way open for new products.

Propylene urea–formaldehyde (Fixapret PHS BASF) produced excellent chlorine-resistant finishes, and the amount of cure had no effect on the finish result (i.e. non-yellowing of the finish when subjected to chlorine-containing wash liquors). The product, although expensive, filled the 'niche' market for the type of finish. The majority of these products are used in the production of white collar interlinings, and fabrics that would be used in the work-wear market and will possibly be subject to chlorination and laundering in the 'hot-box system'.

5.8 Formulations for use of propylene urea–formaldehyde resin

White collar interlining. Addition of a stiffener alongside the reactant resin is normally required to produce the correct handle; the one most used is polyvinyl alcohol. This product produces the correct handle but does not affect the crease recovery of the fabric; it also does not affect the chlorine resistance of the material. Other products may be used to produce a more plasticised effect, e.g. durable softeners, but again care must be taken, as the product may give a release surface to the interlining. As most interlinings go forward for coating with a heat-seal product, this obviously will produce an undesirable effect. One product used successfully as a plasticiser is a melamine–stearamide softener (BT 323 (BIP) PERMFEL B (ACC)). This tends to give the softer handle required, but also gives a 'Velcro' effect to the coating. (Key Coating.)

1. Interlining finish stiff

Cotton pick-up	75%
Propylene urea–formaldehyde (50% solids product)	80 g
Low-viscosity high-solids polyvinyl alcohol	30 g
Magnesium chloride hexahydrate	10 g
Water to 1 litre	

To give a fuller softener effect, 10–20 g/l of melamine–stearamide may be added to the formulation.

2. Work-wear — 50/50 polyester–cotton

Pick-up	65%
Propylene urea–formaldehyde (50% solution)	50 g
Silicone micro-emulsion	10 g
Magnesium chloride hexahydrate	6 g
Water to 1 litre	

Alongside the propylene urea–formaldehyde resin came the heterocyclic crosslinking agents based on melamine. These are made by a reaction between ethylene urea and melamine. Examples are Aerotex 23 special ACC; BT 328 — BIP; and Lyofix NC — Ciba-Geigy. Together with the propylene urea–formaldehyde, these products held the white fabric market for a number of years.

However, the swing away from these types of resin came about with the advent of the DMDHEU, which produced similar results (with the exception of chlorine resistance) at a lower cost.

DMDHEU resins are produced by all the major chemical suppliers and are produced by a reaction of glyoxal, urea and formaldehyde. The proportion of glyoxal is high in the formulation of the resin, and therefore the price of the resin is controlled by the market price of glyoxal. Initially, unmodified DMDHEU resins were produced at 40–42% dry solids, the best-known product being Fixapret CP (CPN) BASF. However, higher-solids resins were requested from the market, and the resin solids of these products were increased to 62–65% dry solids. Various modifications of these resins are now produced, and the complexity of these are seen in Petersen (1984).

Probably the most widely used DMDHEU resins are now the low-formaldehyde versions of DMDHEU. These have been modified to produce a low-formaldehyde resin by using diethylene glycol as a formaldehyde acceptor. The formaldehyde figure can be reduced by approximately 50%, compared with the original DHDHEU resin.

Typical formulations for finishing with DMDHEU are as follows:

Polyester–cotton 50/50

Pick up	55–60%
40% DMDHEU resin	60 g
Silicone softener	10 g
40% magnesium chloride solution	25 g
Water to 1 litre	

When using low-formaldehyde resins, care should be taken with regard to the solids of the product, owing to the fact that they may have been modified with diethylene glycol. As this is 100% solids, the active solids of the product may be lower than estimated, i.e. the cured resin solids could be as much as 25% lower than the figure obtained on evaporation of resin.

The use of formaldehyde has for some time come under closer scrutiny leading to zero formaldehyde finishes, e.g. polycarboxylic acids, being developed. However, the performance of such finishes is generally either poorer and/or more expensive to achieve.

5.9 Solids of resins

All resins have two solids values. Examples are given in Table 5.2.

(*a*) *Calculated solids.* This value is the solids of resin that is available for reaction in the finishing process. This may take the form of formaldehyde or free methanol, but calculating the amount of resin to treat the fabrics is difficult, owing to the fact that the amount of reactable product varies so much from product to product.

(*b*) *Cured solids.* This value is much preferred, and is found by heating the resin for 3 h at 120°C. The test is usually carried out on a 20 g sample of resin in a metal dish. The total solids value is the difference between before and after 3 h at 120°C (BS 2782).

There are a number of exceptions to the rule; these are normally where resins are sophisticated types of reactant such as a Fixapret AC (BASF) and BT350 (BIP).

Where the resins contain reaction products which couple with the cellulose, these products will react at low temperatures and sublime, and therefore it is necessary to carry out a solids determination at lower temperatures, e.g. 70°C for 2 h, to obtain a correct figure.

Table 5.2 Examples of solids values

	Calculated	Cured
Melamine resins (BT336)	80%	72%
Urea resins (BT322)	50%	42%
DMDHEU (BT733)	48%	40%

Table 5.3 Recommended amounts of solids to apply to fabrics

	Urea	Melamine	Reactant
Cotton	6%	4.5%	2.5–3.0%
Polyester–cotton	Not recommended	2.5%	1.5–1.75%
* Viscose rayon	8–15%	6–9%	6–8%
Polyester–viscose	Not recommended	5%	5–6%

* On certain types of viscose, e.g. tie linings, an exceptional amount of resin is required.

The approximate amounts of solids to apply to the textile fabric are illustrated in Table 5.3. These amounts will give a good easy-care finish, and a stable finish to the fabric when fully cured. To obtain a good cure it is essential to take into account the following considerations:

1. That all resins require acid conditions to carry out a cure.
2. That the fabric should be fully dried before applying the resin.
3. Correct drying should be carried out.
4. The correct catalyst and amount should be used.

1. Preparation of the fabric. Before resin finishing, the fabric should be completely desized, and after bleaching, printing or dyeing, the fabric should be of acid or neutral pH. It is suggested that in the last rinse-off of the fabric prior to resin finishing, acetic acid be added in order to ensure an acid or neutral pH.

2. Padding of the fabric. It is very difficult to access the amount of resin applied to the fabric if the fabric is wet before padding. Therefore the fabric should be fully dried before padding.

3. Drying of the fabric after application of the resin. Thc amount of liquor applied to the fabric should be as low as possible. Therefore the mangle expression should be low, i.e. pressure on the mangle should be high.

If the fabric containing the resin is dried at an initial high temperature, the resin will migrate to the surface and give a harsh handle. For example, with a five-bay stenter the first compartment should be at 100–120°C, then the second compartment at 130–140°C, and the third, fourth and fifth may be increased to 170–180°C. This will prevent migration, and the resin will dry in the correct part of the fabric.

The use of drying cylinders for complete drying of resin-treated fabrics is not recommended; they may be used to take off initial moisture but not for full drying, as this again will give surface migration and a harsh handle will be obtained.

Table 5.4 Possible catalysts for various resin systems

	Urea	Melamine	Reactant
Ammonium chloride	Yes	Yes	Yes
Monoammonium dihydrogen phosphate	Yes	Yes	No
Ammonium sulphate	Yes	Yes	No
Zinc nitrate	Yes	Yes	Yes
Magnesium chloride	No	Yes	Yes
Amine hydrochloride	No	Yes	Yes
Modified $MgCl_2$ catalyst (flash catalysts)	No	Yes	Yes
Ammonium thiocyanate	Yes	Yes	No

Table 5.5 Amounts of catalysts required, as % of total solids

	Urea	Melamine	Reactant
Ammonium chloride	2%	2%	—
Monoammonium dihydrogen phosphate	2%	2%	—
Ammonium sulphate	2%	2%	—
Magnesium chloride (anhydrous)	—	10%	10%
Amine hydrochloride	—	15%	15%
Modified $MgCl_2$ catalyst (flash catalyst)	—	12%	12%

4. Catalyst. Table 5.4 illustrates the possible catalysts that may be used for various resin systems. It is suggested that, even though zinc nitrate is an excellent catalyst, it is ignored as it produces adverse effects on optical brightening agents.

With the advent of flash catalyst systems, such as Condensal F.B (BASF), on the market, these have been able to increase the speed of cure. They have been used mainly on cotton–polyester blends of fabric. The flash catalyst systems should be avoided on mechanical finishes, as these tend to give 'pre-cure', giving a poor mechanical effect.

Ammonium-based catalysts should be avoided when used alongside high-formaldehyde resin, because of the possibility of amine-type odours.

The amount of catalyst applied to the resin varies with each type of resin, and Table 5.5 illustrates the amount of catalyst calculated on the total resin solids. Note that $MgCl_2$ is normally sold in crystal form as $MgCl_2 \cdot 6H_2O$, and therefore approximately 18% of $MgCl_2 \cdot 6H_2O$ should be used on 100% solids.

Registered trademarks

Fixapret — BASF
BT (BEETLE)—BIP Speciality Resins Ltd.
Aerotex — American Cyanide Co.
Rhonite — Rohn and Haas Co.

References

Bancroft, J. (1952) US Patents 2,103,243 and 2,148,316.
Foulds, R.P., Marsh, J.T. and Wood, F.C. (1926) British Patents 291,473 and 291,474.
Fluck, L.A. (1951) *American Dyestuff Reporter*, **40**, 769.
Nuessle, A.C. *et al.* (1959) *Textile Research Journal*, **29**, 41.
Peterson, H.A. (1984) Crosslinking with formaldehyde coating reactants. In *Handbook of Fibre Sciences, Functional Finishes*, Part A (eds Lewin, M. and Sello, S.B.), pp. 217–235.

6 Machine-washable knitwear — Production routes

K.M. BYRNE

6.1 Introduction

The inherent ability of wool to shrink, or felt, during wet finishing has been used to good advantage for generations to produce a huge variety of products, ranging from woollen spun knitted and woven goods to felts for use either in hat manufacture or as coverings for snooker tables. Whilst this property divides for endless property development potential, it does constitute a major disadvantage in terms of end-product performance, and impacts considerably on the care labelling associated with wool products. Consequently, products made from wool which has not been subjected to chemical treatment may only be labelled as being suitable for either dry cleaning or hand washing.

Techniques for the production of machine-washable wool products have, of course, been a commercial reality for many years, but the importance of this type of finishing has recently assumed greater importance for a number of reasons, the most important of which is the progressive shift by the modern consumer to a more casual lifestyle, which, in turn, has emphasised the need for easy-care properties such as machine-washability.

Environmental issues and concerns have also now assumed greater importance in terms of product marketing, as evidenced by the high profile generic promotion of eco-labelled products, and the inherent machine-washability of a garment often warrants such a label, because of the legislation to which the dry-cleaning industry is progressively being subjected.

The end product which is the primary subject of this article is wool knitwear, and recent years have seen a number of developments in chemical finishing techniques which have been of particular relevance to the production of machine-washable garments. Although the options available to the knitwear finisher for producing machine-washable products might appear somewhat diverse and confusing (Figure 6.1), many of these processes have been developed and introduced with a particular end-product or market in mind; in most cases, therefore, only one or two options realistically need to be considered.

When deciding upon the finishing route to be used, the primary consideration is the nature of the product, since, whereas the production of a washable worsted spun product will feature techniques designed to prevent facing-up, woollen spun products will require a route which includes a milling cycle in order to produce a typical woollen finish.

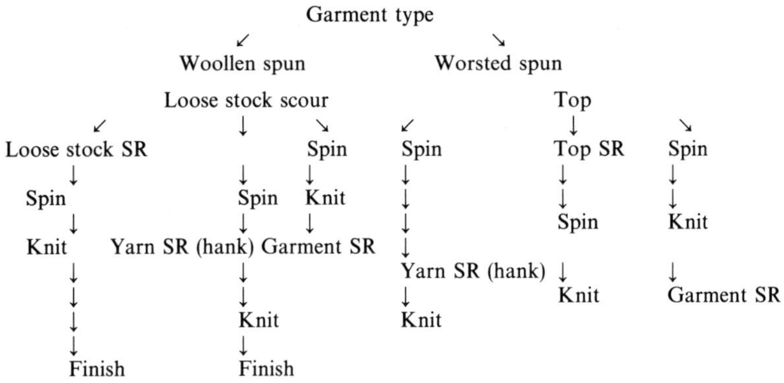

Figure 6.1 Options available for producing machine-washable products; SR = shrink-resist treatment.

Milling, the process by which yarns are burst to produce a raised finish, is a controlled form of felting, and clearly this procedure must be carried out before the shrink-resist treatment is applied; consequently, such products are most effectively produced using garment treatment techniques. Worsted spun products may be produced either from continuously treated top or via garment treatment routes, although yarn treatments in hank form may also be considered.

Woolmark specifications in terms of washability vary according to the nature of the product, with garments worn next to the skin, such as underwear, hosiery and shirts, requiring a higher performance standard than, for example, trousers or knitwear (Table 6.1). The specifications for knitted sweaters cover both felting and relaxation shrinkage, the primary distinction between them being that only felting shrinkage is irreversible.

Table 6.1 Woolmark shrinkage specifications for machine washable knitwear

	IWS test method	Sweaters and cardigans
Relaxation shrinkage		
% Extension — width maximum		5
% Shrinkage — length maximum	31	10
% shrinkage — width maximum		8
No. and type of wash cycle		1 × 7A*
Felting shrinkage		
% Area shrinkage — based on dimensions after 1 × 7A* wash		8
No. and type of wash cycle		2 × 5A**

* 1 × 7A Cycle is equivalent to 1 Woolmark Machine Wash Cycle. ** 1 × 5A Cycle is equivalent to 10 × 7A Cycles.
IWS = International Wool Secretariat.

The maximum area felting shrinkage permitted is 8% after a total of 20 Woolmark machine wash cycles (40°C), whereas the dimensional changes occurring during relaxation shrinkage encompass the possibilities of length and width shrinkage as well as width extension.

Relaxation shrinkage is normally introduced during finishing by oversize framing, i.e. stretching, immediately prior to steam pressing, and is effectively removed during the first wash cycle. Although, from a production point of view, it is important to be able to differentiate between, and measure, both components of shrinkage, it is of course of little consequence to the customer, whose only concerns in terms of garment performance are those of after-wash appearance and toal shrinkage.

6.2 Garment treatments

6.2.1 *Equipment*

Although the treatment route, or process, chosen depends upon a number of different factors related to the aesthetics of the product, e.g. handle required, colour combination/design, etc., an equally important consideration is that of the equipment available in which to treat the garments. However, before any form of shrink-resist finish is carried out, the garments must first be scoured and, for woollen spun products, milled to give the characteristic woollen aesthetics.

Numerous types of equipment are used for the scouring and milling of woollen spun knitwear, and many of the earlier models were based upon industrial washing machines featuring central rotatable drums. Such machines have a relatively severe mechanical action, and later models have incorporated a number of useful features such as high-speed hydro-extraction and a high degree of automation over heating and drum rotation speeds.

Another machine type which has been successfully modified for the scouring and milling of wool knitwear is the Pegg dyeing machine, in which the liquor is circulated by an impellor situated under a perforated false bottom. The garments are circulated in a spiral (toroid) fashion and, because it operates so slowly, it allows garments to be easily inspected without either draining or stopping the machine. However, because the action of the machine is so gentle, the milling times are somewhat longer than for horizontal, rotary drum-type machines.

Worsted-type products are generally processed in machines with a gentle action such as a side-paddle or overhead-paddle machine and, without doubt, the most widely used type of equipment for the shrink-resist finishing of wool knitwear is the side-paddle. The degree of sophistication and control associated with these machines varies considerably from those which feature a single-speed paddle to models, such as those from Flainox, which feature

Figure 6.2 Flainox side paddle.

programming facilities, variable-speed paddle drive, chemical addition facilities, temperature control and automatic emptying (Figure 6.2).

Side-paddle machines operate at a liquor ratio of approximately 25–30:1, and because they exhibit a characteristically gentle mechanical action, are ideal for obtaining even and consistent results during both shrink-resist finishing and dyeing. However, the major disadvantage of this equipment lies in the relatively large volumes of water consumed. Overhead-paddles, which are usually used for garment dyeing, may also be used for the finishing of knitwear but, because the mechanical action of these machines is even less severe than the side-paddle, the process times are extended still further. Furthermore, the degree of liquor exchange operating is also substantially less than that operating in a side-paddle which, for critical processes such as shrink-proofing, may lead to unsatisfactory results.

Most of the major developments in knitwear dyeing and finishing equipment in recent years have been associated with front-loading machines, which offer the advantage of lower liquor ratios (10–15:1) and a high degree of automation and, furthermore, integral hydro-extraction facilities provide for much more rapid and efficient rinsing facilities. However, it is precisely because of the low liquor ratios which are a feature of these machines that they are not entirely suitable for shrink-resist finishing, and those companies which do operate this type of equipment may utilise its higher productivity only for certain parts of the production route, e.g. scouring and milling of garment dyeing, following piece chlorination in a side paddle.

6.2.2 *Scouring and milling procedures for woollen spun knitwear*

The scouring and milling of woollen spun products is an extremely important part of the finishing routine, and many companies have developed their own in-house techniques to produce the aesthetics typically associated with their products. Woollen spun products contain relatively large amounts of oil, which must be removed either before or during milling; this is normally achieved by using anionic or nonionic detergents under mildly alkaline conditions, e.g. sodium bicarbonate or carbonate, and a typical procedure would be as follows:

Scour	Set bath at 40°C
	Add 3–6% detergent
	Run 3–15 min
	Drain bath
Mill	Set bath at 40°C
	Add 1–3% detergent
	Run 3–40 min
	Drain bath
	Warm rinse (3–5 min)
	Drain bath
	Cold rinse (3–5 min)

In preparation for shrinkproofing it is essential that a maximum oil level of 0.8% on weight of wool (o.w.w.) is achieved (DCM Extract). The scouring procedure adopted to achieve this level will depend upon the nature of the lubricant used in yarn production and, although most modern woollen spinners use low levels of synthetic, water-soluble lubricants which are easily removed during scouring, higher lubricant levels which contain vegetable or mineral oils may still be encountered. Products of this type may only be removed efficiently by scouring under alkaline conditions using sodium carbonate, but, wherever possible, this procedure should be avoided because it can result in yellowing and also impair the handle of the wool. Such conditions may also result in the removal of dyestuffs leading to shade change or, when processing high colour-contrast designs, backstaining on to pale shades or undyed areas. However, several auxiliaries are now available which, provided that the amount of dyestuff removed is not excessive, can effectively prevent this from occurring.

A temperature of between 35 and 40°C is recommended for both scouring and milling because it is within this temperature range that the detergents are most effective and, because most of the residual oil is removed during the scouring cycle, the detergent concentration can be normally reduced by half in the milling cycle. Some companies do in fact scour and mill their garments in a single operation and, although the use of modern spinning lubricants does make this option more viable, its use as a preparation for

Table 6.2 Typical scouring and milling times for different types of equipment

		Process time (min)	
Machine type	Liquor ratio	Scouring	Milling
Rotary	12:1–20:1	3–10	3–10
Paddle	30:1–40:1	10–15	20–40
Toroid	10:1–20:1	5–10	10–20

shrink-proofing should first be carefully assessed before committing garments to production. As a general recommendation, the minimum scouring–milling time used should not be less than 3 min, to ensure efficient cleaning of the garment. Efficient scouring is a particular concern for the commision shrink-proofer, especially for woollen spun products where the level and type of lubricant used in yarn manufacture may vary significantly depending on the yarn source. Under these circumstances, solvent scouring techniques have been found to be an extremely effective option. Solvent milling may also be carried out following scouring by the selective introduction of a mixture of water and emulsifier into the machine, but in practice a superior handle is often obtained using aqueous milling techniques and it is not uncommon for an aqueous mill and shrink-resist finish to be applied following solvent scouring.

The extent to which woollen spun products such as lambs' wool and Shetland garments are milled can vary considerably, and this is particularly the case with lambs' wool, since certain markets now require a clean finish which is more reminiscent of a worsted spun product. For machine-washable products the degree of finish obtained during scouring and milling must be reduced to compensate for the additional milling which may occur during subsequent shrink-resist finishing, and the reduction in the length of the milling cycle will of course vary, depending on a number of different structural factors and the type of equipment being used. Typical scouring and milling times for different types of machine are shown in Table 6.2, but for a fixed milling time the degree of finish achieved will increase with increasing fibre fineness, increasing yarn count, reduction in yarn twist, reduction in cover factor, and even the colour of the wool, since dark, heavy shades can often require milling times of up to three times as long to achieve the same finish as that obtained on a pastel or pale shade. However, once these parameters have been established for a particular product and machine combination, reproducible results in terms of finish and handle may be readily obtained.

6.2.3 *Scouring and anti-cockle procedures for worsted spun knitwear*

The criteria which must be satisfied in the production of worsted spun products differ considerably from those of woollen spun products, since a clean finish in which the knit structure is clearly visible is required and every effort is therefore made to prevent milling or facing-up of the garments during processing. However, worsted spun products can often exhibit a phenomenon termed as cockling, which is described as 'an irregular surface effect caused by loop distortion'. This effect, which occurs most often in plain knit structures, is due to the relaxation of mechanical strains introduced during spinning or knitting, and almost invariably occurs when the garment is relaxed in water leading to a three-dimensional distortion of the knit structure. This effect occurs either at the rib–panel interface, near panel edges or, in severe cases, randomly distributed over the whole garment (Figure 6.3).

Provided that these effects are not too severe they can be eliminated, or at least significantly reduced, by carrying out an anti-cockle process. This is designed to set the yarn in the knitted configuration, and the normal wet-finishing routine adopted for processing worsted spun garments is therefore as follows:

Anti-cockle → Scour (optional) → Shrink-resist treatment

Again, because the processing aids used in the production of worsted spun yarns are applied at such low levels and are also often water-soluble, the scouring requirements are minimal, and in fact the anti-cockle procedure itself is normally sufficient to reduce the residual oil level to a value which is compatible with the application of a shrink-resist finish.

The conditions operating in the anti-cockle treatment will reflect the nature of the cockling which has been observed and, if it is not too severe, a mild treatment may be used as follows.

- Set and maintain the bath at the boil with the paddle operating at a speed which is just sufficient to prevent the garments from sinking.
- Add the garments individually to the bath and allow them to static-soak for 10 min, occasionally operating the paddle for 10–20-s periods.
- Cool the bath slowly to 40°C by adding cold water, again with the paddle operating at minimum speed. It should be noted that rapid cooling can cause permanent creases to be set into the garments.

This procedure often provides sufficient stabilisation to allow the garments to survive conventional shrink-resist finishing; however, if the cockling performance is particularly poor or if the garments are to be piece-dyed, a higher degree of set may be required. This is achieved by adding sodium metabisulphite or sodium sulphite (3–5% on weight of wool) to the anti-cockle bath immediately before adding the garments. This procedure will improve the performance considerably but, because some dyestuffs are sensitive to reducing agents, an effective compromise in terms of shade change and cockling may be obtained by treating at 80°C rather than at the boil. In both cases, the sensitivity of the dyestuffs being used must be tested before committing garments to bulk production.

The factors responsible for producing cockling in worsted spun knitwear are extremely varied and have been discussed at length in the literature

(a)

(b)

(c)

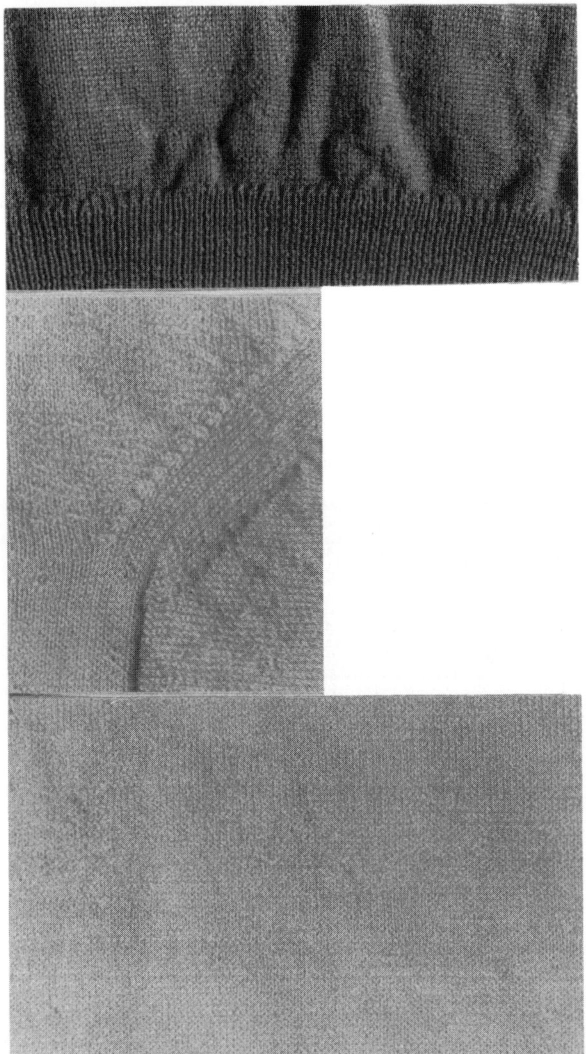

Figure 6.3 Examples of cockling: (a) cockling at the junction between the rib and plain knit area; (b) severe cockling near to the panel edge at fashioning points; (c) an example of severe random all-over loop distortion.

(Robinson *et al.*, 1965), but there is little doubt that its primary origin lies in the evenness and consistency of the yarn and ultimately those factors which can influence these properties. The quality of wool used in the production of a particular yarn is a major contributory factor and, as a general observation, the tendency of yarns to cockle reduces with decreasing fibre micron. The fibre micron range used can also influence cockling, and attempts to reduce manufacturing costs by diluting a blend with coarser fibre often leads to an increase in yarn rigidity, which in turn leads to yarn irregularity and ultimately cockling.

The degree of set imparted to yarns prior to knitting can also exert an influence since, if yarns cannot be reset in the knitted configuration, one might therefore expect an inferior result. The performance of package-dyed yarns in terms of cockling therefore tends to be inferior to an equivalent yarn which has been dyed either in hank form or produced from top dyed wool.

Where most of the garment is not affected, cockling may still occur at the rib–panel interface and at panel edges where loops are stretched during fully fashioned knitting. Both can be significantly improved by using conventional anti-cockling procedures, and further improvements at the rib–panel interface might also be expected by adjusting either the tightness of the rib or the cover factor of the plain knit area such that, on relaxation, the dimensions of both are similar; additional improvements may also be obtained by doubling when transferring the rib to the fully fashioned machine.

Panel edge cockling is more difficult to overcome but can be reduced by minimising the rate of width change during knitting, i.e. by reducing the angle between the line of the panel edge and the vertical wale lines.

Other properties such as yarn regain and tension control during knitting have also been implicated in the occurrence of cockling, albeit to a lesser extent than some of the other factors considered, but there is little doubt that the primary causes are those associated with yarn quality, and when these issues are addressed an anti-cockle procedure may not always be required, even when garment dyeing.

For garment finishers who do not routinely use, or have access to, yarns of this quality, the factors responsible for cockling are of little importance, unless action based on this knowledge can be taken to improve the situation. The primary concern of the finisher is to produce a garment which is of a commercially acceptable quality, and interest lies therefore in techniques and procedures which will allow this aim to be achieved. In most cases, an acceptable product can be obtained by carrying out an anti-cockle treatment and/or adressing some of the other points discussed in this context. However, when cockling is particularly bad, it may even occur in garments which have been stored prior to wet-finishing. Under these circumstances it is essential that the garments are subjected to an anti-cockle treatment as soon as possible after knitting, and if the necessary equipment is available, garments should also be steamed and framed immediately before treatment. Such a procedure would clearly only be used where no other option was available and is designed to deal with the symptoms rather than the cause of the problem. However, if an acceptable product may ultimately be produced by adopting these techniques, exposure to water should of course be avoided and the care label in the garments should therefore specify dry-cleaning only and not hand-washing.

6.2.4 *Garment shrink-resist treatments for knitwear*

Batch techniques for the treatment of knitted garments are by far the most versatile route for the production of machine-washable knitwear, since not only do they provide for the option of processing both woollen and worsted spun products, but they also allow the production of garments in a variety of shades and colour combinations and provide considerable scope in terms of handle modification.

Although there are many processes available to the modern knitwear finisher, in practice the options available are very much dictated by factors such as the nature of the equipment available, the type of garment (woollen or worsted), the shade or colour combination, the handle required and the inclusion, or omission, of dyeing in the finishing cycle.

Most batch processes for knitwear, in common with all other shrink-proofing treatments, feature an oxidative pretreatment in combination with the application of a suitable polymer, although two processes have been successfully used which feature polymer application only, namely Synthappret BAP (Bayer) and DC 109 (Dow Corning/PP(T)), the latter being applied from solvent in a modified dry-cleaning machine.

The relative merits of the batch routines currently available are considered in the following sections.

6.2.4.1 *Oxidation/polymer systems.* The oxidising agents most commonly used in the treatment of knitwear are either dichloroisocyanuric acid (DCCA) or potassium peroxymonosulphate (PMS). The former is available from BASF (Basolan DC) and PP(T) (Dylan Auxiliary), and the latter from BASF (Basolan 2448), PP(T) (Dylanise Salt), Interox (Curox), Dupont (Oxone) or Degussa (Caroat). Although an oxidative pretreatment alone will impart a degree of shrink-resistance, the extent of which will depend upon the application level used. Its primary function is to prepare the surface of the fibre for polymer application, and the combined effects of both lead to the production of a full machine-wash finish.

Both products are well-established in the field of knitwear finishing but each exhibits advantages and disadvantages which must be taken into consideration when deciding on which process is to be used. For example, DCCA treatment often results in a degree of yellowing, the extent of which will depend upon a number of factors including the treatment level used and the conditions under which it is applied; consequently it is unsuitable for the treatment of garments that feature pastel shades or high colour contrast designs. PMS treatment, however, imparts a mild bleaching effect and there are therefore fewer restrictions in terms of the colours or colour combinations that may be treated.

The potential for knitwear to shrink is influenced by a number of different factors and, as a result, some products are much more difficult to stabilise than others. For instance, worsted spun products are more difficult to stabilise than woollen spun products, because the finer quality wools which are often used in their production exhibit a higher tendency to shrink and also the tighter cover factors which are a feature in such products can prevent complete penetration of the oxidising agent into the yarn or knit structure. Of the two products, DCCA is far superior in terms of shrink-resist efficiency and, as such, is equally effective on a wide range of products. PMS is much less effective in this respect and its use is therefore restricted almost exclusively to the treatment of woollen spun products.

Another, and more fundamental, difference between these two oxidising agents is that, whereas DCCA treatment imparts shrink-resistance progressively as it reacts with the wool, this is not the case with PMS, which by itself imparts no resistance to felting whatsoever. Shrink-resistance with Caroat treatment is only subsequently achieved by after-treating (developing) the oxidised wool with sodium sulphite under alkaline conditions (Sweetman and McLaren, 1965). The degree of shrink-resistance imparted by this route increases progressively with increasing pH in the development stage, but in practice a maximum pH value of 8.0 is normally adopted. Such conditions can result in the removal of loosely bound dyestuffs and dyed soluble proteins which are generated as a result of the oxidative treatment, in which case their redeposition on to the garments can result in back-staining. The procedures adopted to minimise this effect will vary depending on the severity of the problem and the nature of the design.

Clearly, if large amounts of colour are removed during this stage of the process then a similar situation would have been observed during scouring and the dyeing procedure itself would ultimately have to be addressed. For solid shades, the consequences of treating such garments would then merely be a change in shade as a result of dye loss, but for high colour-contrast designs the situation following scouring may well be irretrievable. Where dyestuff loss is less apparent, back-staining can be prevented by including in the bath a proprietary auxiliary designed to hold dyestuff in solution, typical products being Milliclean IBA (PP(T)), Croscour SA25 (Crosfields) and Tinuvetin LB (Ciba). Further improvements may also be achieved by using a lower pH during the development stage following Caroat treatment, but this procedure realises only a proportion of the potential shrink-resistance imparted by the oxidative pretreatment, and therefore relies more on the shrink-resist efficiency of the polymer to achieve full machine washability.

Although a variety of polymers is used in combination with an oxidative pretreatment, not all products are compatible with both PMS and DCCA pretreatment. The polymers that may be used in combination with DCCA include Basolan SW and Basolan MW (BASF), Hercosett (Hercules) and Polymers RSM and GE (PP(T)), the latter being the polymeric component of the Dylan GRB process, by far the most widely used route for the treatment of knitted garments. All of these combinations are equally effective on both woollen and worsted spun garments, although the pretreatment level required varies depending upon the type of product and the polymer being used. Some polymers, namely Polymer GE and Hercosett, do require the application of a suitable softener to achieve an acceptable handle. Neither Polymer GE nor Hercosett is compatible with PMS treatment, but silicone-based products such as Basolan MW (BASF) and Ultrasoft (PP(T)) are used to good effect in the production of soft handling woollen spun lambs' wool. Further, non-silicone-based products, namely Polymer RSM and Polymer TM (PP(T)), are also suitable for use in the treatment of woollen spun products.

The reason for the difference in terms of shrink-resist efficiency between the two oxidative treatments is not immediately apparent, since both result in the production of a high concentration of sulphonic oxides at the fibre surface which, in turn, provide sites for the attraction of the cationic polymers during the subsequent exhaust treatments. It has been suggested that protein crosslink density, and ultimately the ability of such proteins to swell and prevent scale interaction, can influence the degree of shrink-resistance obtained with oxidation-only treatments, and this is consistent with the observation that DCCA leads to a lower protein crosslink density, and higher swelling capacity, than PMS treatment (Byrne *et al.*, 1990). However, of much more significance is the observation that, whereas chlorination produces a marked increase in wettability which ultimately leads to improved polymer spreading and increased wool/polymer adhesion (Bereck and Reincke, 1989), PMS has little or no effect in this respect (Byrne *et al.*, 1979). Consequently, only those polymer systems which exhibit a surface energy below that of PMS-treated wool, namely silicones, will spread and adhere to PMS-treated wool, whereas those polymers exhibiting a higher surface energy, such as Hercosett and Polymer GE, are only compatible with chlorine treatment.

However, there is one exception to this rule, namely Basolan SW, which is compatible with both PMS and DCCA treatment and which may be applied equally effectively to both woollen and worsted spun products. This product is perhaps the most efficient shrink-resist polymer available and as a result, the pretreatment conditions necessary to produce full machine-washability are not nearly as critical as with other polymer systems. In practice, this means that the development pH required following PMS treatment is much lower than that required for other polymer systems, and this observation has led to the development of a one-bath process in which PMS treatment is followed by simultaneous development and polymer application at pH 6.0 in the presence of sodium metabisulphite, features which all but eliminate any potential for back-staining, while also reducing water consumption and increasing productivity.

6.2.4.2 *Garment dyeing.* Shrink-resist technologies impact considerably on the dyeing process, not only because of the influence of the chemical treatment on the dyeing properties themselves, but also because of the higher dye fastness requirements. These specifications do impose some restrictions on dyestuff selection but, as a general observation, there are few shades which cannot be produced to a machine-washable standard. Dyestuff selection is beyond the scope of the article and for further information on this subject, either the International Wool Secretariat (IWS) or the dyestuff manufacturers themselves should be contacted.

The colour fastness of Woolmark products produced by garment, yarn, stock or top dyed routes must meet the IWS Machine Washable Colour Fastness Specifications. These are illustrated in Table 6.3.

Garment dyeing may be incorporated into a conventional shrink-resist finishing route, although the potential for facing-up or milling is much greater because of the extended processing time and high temperatures involved and, as with conventional shrink-resist finishing, the procedure adopted will depend very much upon the nature of the product and the type of equipment available.

The ideal route in terms of preventing excessive milling would be to dye immediately following a full oxidative/polymer shrink-resist process, but this option has not been widely adopted because any unevenness in terms of polymer distribution, which is cationic in nature, is invariably reflected in an uneven dyeing. The recommended procedure therefore is to rely upon the chlorination stage to stabilise the garments against felting during dyeing and, to this end, somewhat higher treatment levels are used than in conventional garment shrinkproofing (Ryder and Lavocah, 1984). However, chlorination also radically affects the dyeing properties of wool and, once again, any unevenness in treatment can lead to uneven dyeing, notably in terms of seam penetration. Particular care must therefore be taken to ensure that all areas of the garment are evenly treated.

Table 6.3 Colour fastness requirements for machine-washable knitwear

Colour fastness to	IWS test method	Minimum fastness rating		
Light	TM 5	3 at 1/12 standard depth and below; 4 above 1/12 standard depth		
		Change in shade	Staining of wool and nylon	Staining of other fibres
Washing	TM 193	3–4	4	3–4
Wet alkaline contact	TM 174	3–4	4	3–4
Rubbing wet and dry	TM 165	—	—	Wet/dry 3/3

6.2.4.3 *Equipment for garment dyeing.* Not surprisingly, the equipment recommended for garment dyeing is much the same as that used for shrinkproofing and, again, the side-paddle is the most widely used machine. The trend in all garment dyeing is towards the use of rotary machines because of the high degree of control and increased productivity that they provide, but when processing wool garments the low liquor ratios which operate in this type of equipment effectively increase the rate of chlorination, leading to an uneven and inconsistent treatment and ultimately poor quality dyeings. Consequently, in order to take advantage of the increase in productivity associated with such machines, one option is to carry out the chlorine pretreatment in a side-paddle and transfer the garments to the rotary machine for dyeing.

6.2.4.4 *Bright pastel shades.* Although the action of a side-paddle is gentle, milling and facing-up can be reduced still further by turning the garments inside out and operating at a paddle speed which is just sufficient to maintain constant movement of the garments through the liquor. The latter procedure may also be useful in controlling the rate and evenness of chlorination, but these parameters are more commonly controlled by adjusting the pH and temperature of the liquor. Where the water temperature varies considerably throughout the year, and cooling is not an option, consistency in terms of treatment time may be obtained by carrying out the chlorination stage at the maximum ambient temperature and heating the bath to this temperature when necessary. In addition, the rate of reaction may also be controlled by adjusting the liquor pH, with the rate of reaction decreasing with increasing pH. However, the disadvantage of this approach is that the degree of yellowing obtained also increases with increasing pH and, although this effect is of little consequence when dyeing to dark heavy shades, bright pastel colours cannot be produced unless this problem is addressed.

Because PMS treatment does not yellow wool, it might therefore be considered as a viable alternative to DCCA treatment; however, PMS does not impart sufficient shrink-resistance to enable the garments to survive the dyeing cycle. One solution therefore is to use wool which has been continuously chlorinated in top form; this route offers advantages in terms of evenness of treatment, and therefore dyeing quality, and furthermore the acidic conditions which operate during continuous top treatment minimise yellowing and therefore permit the production of pastel shades.

In the absence of treated top, such shades can be obtained via the batch route by carrying out a combined reductive bleach and antichlor immediately following chlorination, using a stabilised hydrosulphite product.

Additional improvements in terms of colour maintenance are obtained by dyeing at low temperatures (85°C) and by including in the dyeing a dyebath bleaching agent such as Lanalbin B (Sandoz), Erioclarite B (Ciba) or Lufibrol W (BASF). Further improvements in evenness of dyeing are obtained by using specific low-temperature dyeing auxiliaries such as Baylan NT (Bayer) and Lanasan LT (Sandoz), and improved control over dyestuff exhaustion is obtained by the use of acid donors. These products are esters which are progressively hydrolysed with increasing temperature, slowly releasing acid in the process. This produces a gradual shift in the pH from slightly alkaline or neutral pH to acid, and thus enables the dyeing to be started under conditions which provide for uniform dye uptake, i.e. pH 7.5–7.0.

6.2.5 *Polymer-only treatments*

Because most of the practical problems associated with the batch treatment of knitwear originate from the oxidative pretreatment, the ideal process is one which achieves full machine-washability via polymer application only. Two such processes are available for the treatment of knitwear, one of which is applied under aqueous conditions and the other from solvent in a modified dry-cleaning machine.

The aqueous-based route features the exhaust application of the product Synthappret BAP (Bayer), a polymer emulsion which is used extensively in certain markets for the shrink-resist treatment of woven wool fabrics. In this case, application is by padding and the treated fabric is then cured in a tenter at a temperature of not less than 140°C, but for the treatment of knitwear an exhaust application technique is used which relies upon destabilisation of the polymer emulsion using magnesium chloride; this leads to the polymer being taken up by the wool and complete curing is then achieved in the bath merely by raising the pH. The performance of garments processed by this route is excellent in terms of shrink-resistance, shape retention and pilling, but the conditions required to cure the polymer effectively restrict the shade range and, although considerable improvements can be made with microsilicone based softeners, the handle is still not consistent with current market requirements.

The second route features the application of a crosslinkable silicone polymer from solvent, the polymer being DC 109 from Dow Corning/PP(T). Although numerous application procedures can be adopted, the most common technique is to immerse the garments in a solution of the polymer, extract to the required pick-up, then to apply the catalyst by spraying on to the garments as they are tumbling. The process produces a product which, like the BAP route, justifies the label 'high performance' because of its excellent resistance to felting, good shape retention and low pilling performance; furthermore, being silicone-based, it also imparts a degree of enhanced water repellency, and this has led to this product becoming well established in the golf market.

Once again, however, the handle of treated garments is not consistent with the requirements of 'High Street' retailers, and this finish is only seen to its best advantage on woollen spun lambs' wool, with both worsted spun and

Shetland products exhibiting a synthetic, greasy handle. A further negative feature of this process is the legislation which is being introduced to restrict the use and discharge of solvents and, in the light of this, the long-term future of this finish must be in question.

6.3 Yarn treatments

Wool yarn can be effectively shrink-resist treated in hank form although, as with garment treatments, there are a number of factors which must first be considered. The production of piece-dyed machine-washable knitwear requires that the chlorination stage be carried out prior to dyeing to prevent milling or facing-up during dyeing, and although clearly this is not a requirement for hank dyeing, the sequence adopted is just as important and is influenced both by the shade being dyed and the type of machinery being used. Because of the potential for shade change and yellowing which accompanies chlorination, the ideal processing sequence would be to chlorinate prior to dyeing; however, on hank-dyeing machines which feature convential stick supports, the reduced circulation of the stick–wool interface is sufficient to reduce the overall level of chlorination achieved in this area. This in itself is not sufficient to impair shrink-resistance but, because the final shade produced is a function of the chlorination level achieved, this sequence can lead to the production of 'stick marks'. Under these circumstances, chlorination must be carried out after dyeing is complete, thereby leading to restrictions in terms of shade range. If, however, the dyeing machine features stick supports through which the dye liquor may be circulated, such marks can be prevented, and a full shade range may be produced by utilising the dye-bath bleaching techniques used for knitwear.

Because felting is not an issue for hank treatments, another option available to companies with conventional hank-dyeing equipment is to use a PMS-based shrink-resist process. This allows the treatment to be carried out following dyeing with minimal shade change, and therefore permits the production of a much wider shade range without the use of reductive bleaching techniques.

The use of PMS instead of DCCA once again restricts the options in terms of the polymers that may be used, but optimum results in terms of handle are obtained using either Basolan SW or MW from BASF. As with knitwear, the shrink-resist efficiency of PMS-based routes is somewhat lower than that of DCCA treatment, but even so, these processes are more than capable of attaining the machine-washable standards for knitwear. Most, if not all, of the production in this field has been on worsted spun yarns, but there is no reason why woollen spun yarns should not also be processed in this way. The main disadvantage is, of course, that products made from these yarns would have to be milled, in which case a treatment level would have to be determined which gave an effective compromise between milling propensity and shrink-resistance.

6.4 Continuous treatments

Approximately 75% of the wool treated to a machine-washable standard in top form using the IWS/CSIRO Chlorine Hercosett (Feldtman *et al.*, 1967) process, and at present about forty lines are operating throughout the world today. A detailed consideration of this technology is beyond the scope of this chapter, but the importance of this process and, in particular, the use of chlorine in top treatments, more than merit a review of such processes. Wool processed by this route meets the maximum Woolmark specifications for machine-washability and may therefore be used in the production of virtually any product, irrespective of structural considerations such as cover factor or yarn count. Machine-washable worsted spun knitwear may be produced in solid shades using either the top or garment dye routes, although the latter is not a widely adopted technique, because the cationic character of Hercosett resin alters the dyeing characteristics of the wool to the extent that a much higher degree of control over the dyeing cycle is required to ensure that evenness and consistency are obtained. The top dye route is therefore more widely used and obviously provides the option for the production of both solid shades and multicoloured designs. However, such a route does not lend itself to the quick-response philosophy which nowadays plays such an important part in retailing, and other factors, such as minimum lot sizes and high inventory costs, also restrict this route to the larger companies.

Those processors who offer the Chlorine Hercosett technology on a commission basis may often have little or no knowledge of the end-use to which the treated wool is put, in which case the wool is treated and tested to the maximum Woolmark specification. However, it is possible to achieve the Woolmark knitwear specification via a chlorine only treatment in top form, and this is a route which has been used to good effect by some knitwear producers, particularly in the production of garment dyed products. This is not an option which is widely adopted by garment processors using batch chlorination techniques, because the treatment levels required often lead to impaired product performance and excessive yellowing; the normal procedure therefore is to combine a low level of chlorination with polymer application to achieve full machine-washability. The acid conditions featured in continuous top chlorination produce very little discolouration and allow the machine-wash knitwear standards to be achieved by virtue of the improved levelness of chlorination when compared to batch routes, but there are some disadvantages associated with the use of wool treated in this way, notably the sinkage which occurs in the mechanical processing of chlorinated wool, the impairment in handle which is produced, and the additional weight loss which can occur both during dyeing and treatment. The increase in weight loss that occurs during treatment will be a function of the treatment level used, and this in turn will be dictated by the wool quality being processed, but the difference in weight loss observed between this route and Chlorine Hercosett treatment is partly compensated for by the application of 2% Hercosett solids following chlorination. The resin also provides a protective coating to the treated fibre which reduces weight loss during dyeing.

The handle associated with acid chlorination techniques is characteristically dry and is associated with a change in the frictional characteristics of the fibre surface (Makinson, 1979); consequently, the application of a suitable softener is required to provide an acceptable handle. However, the combination of acid chlorination and the application of a microsilicone softener does provide for a very significant improvement in handle, so much so that this combination, albeit using higher levels of chlorination, was used in the development of numerous handle modification treatments for wool top including the IWS Soft Lustre process (Rushforth, 1989).

Processes of this type were developed at a time when not only was the price of fine quality wools at its highest for some time, but they were also in short supply, and the demand was therefore for a chemical procedure that would allow coarser wool fibres to be used in certain end-products without compromising handle.

The forerunner of developments in this field was the Vantean process carried out at Nippon Hi Spinning Mill No. 21 (PCT, 1979). This was a continuous top process which was only carried out on commission, and details of the process are therefore limited to what can be gleaned from the patent literature. The effect of the process is to remove the scale structure completely and this is achieved by applying a transition-metal cation to the wool in a saturated salt solution and then passing it into a chlorination bowl. The former treatment ensures that the application is restricted to the surface of the fibre and this, in turn, leads to rapid surface-specific decomposition and reaction of chlorine, leading to scale removal. Although all efforts were made to restrict the chemical treatment to the fibre surface, this may not have been possible in all cases, and further evidence for this supposition is provided by the fact that a formalin treatment was also an integral part of the process; this presumably acted as a protein crosslinker.

The commercial introduction of this process never extended beyond the company that developed it, and indeed the large amount of equipment required to carry out the process would have precluded its use on conventional shrink-proofing lines. Developments in this field therefore concentrated upon utilising conventional shrink-proofing equipment and, although numerous companies operate processes under a variety of different names, they all feature acid chlorination, at a level approximately twice that applied in conventional top processing, followed by the application of a suitable microemulsion silicone softener. As a general observation the improvement in handle obtained is equivalent to approximately 2 μm, although the actual change in fibre diameter is less than 0.1 μm. As with the Vantean process, the scale structure is effectively removed, although in this case the restriction of chemical damage to fibre surface is a result of the mechanics of the chlorination units rather than the use of saturated salt solutions. The removal of the scale structure also leads to the production of full machine-washability and, as an additional bonus, the comfort properties of the treated wool are also significantly improved, thus leading to significant product development potential in the field of products worn against the skin (Garnsworthy *et al.*, 1988), e.g. sportswear, spring/summer knitwear and underwear.

The main advantage of acid chlorination techniques for the treatment of wool top is the low chemical cost involved; however, in certain markets, notably Italy, many companies continually treat wool top using DCCA via a pad application technique. This treatment is part of a production route that is designed to impart an improved handle to worsted spun yarns produced with fine quality wool, and also features hank dyeing followed by the application of a suitable softener. The DCCA route is somewhat more expensive than conventional top chlorination procedures that use either sodium hypochlorite in combination with sulphuric acid or chlorine gas dissolved in water, but the yarn bulk obtained is considerably higher than with acid chlorinated wool, a feature which is thought to be due to the additional protein crosslinking provided by the cyanuric acid liberated during treatment (Veldsman and Swanepoel, 1971). Many of the high quality Italian yarns that are sold on the basis of handle and colour feature this production route, and although machine-washability is rarely an issue, it can be achieved merely by increasing the DCCA application level. Such yarns supplied in ecru may therefore be used in the production of garment dyed products or, by applying a polymer after dyeing, fully machine-washable products.

Because the production of woollen spun knitwear invariably features a scouring and milling cycle, and because machine-washability and milling propensity are mutually exclusive, machine-washable woollen spun products have been traditionally produced only by batch techniques. These are extremely flexible in terms of the handle that can be produced and the colour combinations that can be featured; however, the design potential is limited by the fact that Jacquard knits cannot be processed because of the high felting propensity of the float stitches.

The solution to this problem is one that has vexed the collective minds of many workers in this field, since clearly the requirement is for a treatment which, while it is sufficient to produce the necessary level of shrink-resistance, can also allow the products to be milled. Such a combination, if it could be achieved, would allow any knitter with basic finishing equipment to produce machine-washable knitwear without recourse to conventional batch techniques, and provide for a dramatic increase in product design and development potential.

Some companies have pursued such a development via the batch treatment of loose stock in dyeing machines and, although some success has been achieved by this route, the evenness, consistency and reproducibility have in general proved to be an insuperable barrier to commercial success.

Clearly the key to success in this field lies in the precise control of the shrink-resist treatment, and this was ultimately achieved via the Kroy loose stock technology (Bourn *et al.*, 1985), which was originally introduced for the treatment of wool used as filling materials in bedding products.

The International Wool Secretariat is working with the two companies operating this technology, and also suitable woollen spinners, and each stage of the production pipeline has been optimised to produce a range of commercially acceptable yarns. At the time of writing, Chuwa of Taiwan are currently supplying yarns via two spinners, and yarn development with Jarmains, UK, is progressing along the same lines.

6.5 Environmental considerations

The environmental profile of all fibre production routes has in recent years attracted a great deal of interest, so much so that the concept of eco-labelling has become increasingly important in retailing. Although a discussion of the relative 'green' credentials of the different fibres is beyond the scope of this chapter, the facts that wool is a natural biodegradable fibre and is produced on land that can rarely sustain agriculture obviously offer advantages over many other fibres. It is the policy of the International Wool Secretariat to ensure that grower country wool is processed using production techniques that ensure minimal environmental impact, and the single most important issue as far as the production of machine-washable products is concerned is the use of chlorine.

Chlorine is not only cheap, it is unique in terms of the effect it has on the properties of the wool fibre, producing both efficient and rapid oxidation of the fibre surface and a high level of wettability. As a result, chlorination of one form or another features in a wide variety of chemical finishing procedures, including shrink-proofing techniques applied to wool in top, loose stock, yarn, woven fabric and knitted garments, as well as handle modification techniques, garment dyeing and printing.

An inevitable consequence of the chlorination of wool in aqueous solution is the release in the effluent of chlorinated organic products, mostly wool proteins, which are collectively termed as absorbable organic halides (AOX) (Waste Water Tax Law, 1987). Legislation has been introduced in some countries, notably Germany, to restrict the discharge of certain organohalides that are known to be, or suspected of being, injurious to health but, because the test method developed to analyse for these components does not differentiate between individual chemicals, the net effect is a blanket restriction of all such products, irrespective of their toxicological profile.

Whether or not AOX legislation will eventually spread to other countries is not yet clear, but even where this has not occurred, the number of restrictions being placed upon chlorine users is increasing progressively; the ultimate requirement is therefore for an environmentally acceptable alternative to chlorine for the chemical finishing of wool.

The most obvious replacement for chlorine in this respect is, of course, PMS, but its performance in terms of shrink-resist efficiency is such that commercial success has only been achieved in woven fabric, knitwear and yarn treatments, with little or no progress yet reported in garment dyeing, top and loose stock treatment, handle modification or printing.

Although much of the work being carried out in this field features the use of conventional wet finishing, other areas such as enzyme finishing and electrical discharge techniques are also being considered. Enzymes have been used in the past on wool to produce either shrink-resistance or handle modifications (Philips *et al.*, 1941; Philips and Middlebrook, 1948), but such techniques were never widely adopted because of the non-specific action of the enzyme, which resulted in an unacceptable degree of damage to the fibre. However, it is to be hoped that the advances that have been made in enzyme technology and manufacture will ultimately lead to the development of products which, either alone or in combination with conventional wet-finishing procedures, can duplicate the effects of chlorination and lead to an environmentally acceptable alternative.

The use of electrical discharge techniques has also attracted considerable attention in recent years, primarily because it is possible to achieve many of the effects of chlorination without either yellowing or chemical damage (Philips *et al.*, 1941; Philips and Middlebrook, 1948). Furthermore, it is a dry technology and, as such, its use is consistent with all of the criteria associated with an environmentally friendly process.

This route lends itself well to the treatment of wool in woven fabric form either under vacuum (glow discharge) or atmospheric conditions (corona discharge). A version of the former technology is used commercially in Japan by Unitika, and it is manufactured by Sando Iron Works of Wakayama, Japan, where it has been used to achieve a variety of different effects, including improved evenness of piece dyeing, to increase the depth of shade obtained with piece-dyed blacks, colloquially described as 'Real Black', and in the production of machine-washable fabrics.

A similar machine is also available from Niekmi of Ivanova in Russia; this is used commercially as a prepare for print on wool fabrics, and has recently been introduced into a mill in Italy (Mascioni). The cost of the Japanese equipment is considerable and, although information relating to the price of the Russian alternative is not readily available, it is likely to be such that its use will be restricted to large vertical companies who can fully utilise the multipurpose finishing capacity to good effect on substrates other than wool. Corona discharge equipment is a much cheaper alternative, because the process is carried out under atmospheric pressure, and results have shown that in terms of dyeing properties, printing and machine-washability, an equivalent effect may be obtained. Equipment of this type may be obtained from Softal of Hamburg.

The equipment referred to does not lend itself to the treatment of wool in any form other than fabric; however, a prototype glow discharge machine for the treatment of wool in top form has been built at The Textile Institute at Łódź, Poland, and a joint evaluation with IWS has again demonstrated the multipurpose finishing potential for this technology. Treated wool has been shown to exhibit significant processing advantages in terms of spinning performance and, when used in the production of woven fabrics, produces a marked improvement in dyeing properties and gives full machine-washability. Treated wool may also be used in the production of machine-washable yarns and knitted goods, although the application of a suitable polymer, namely Basolan SW (BASF), is invariably required.

6.6 Conclusion

Although many routes are now well established for the production of machine-washable knitwear and virtually any product can now be produced to this standard, the challenge to those working in the field is to eliminate the use of chlorine and thereby further improve the environmental profile of the wool production pipeline.

In order to achieve this end, techniques other than conventional wet finishing must also be considered, and the concepts of enzyme finishing and electrical discharge treatment, amongst others, must be firmly embraced by those wishing to progress in the field of easy-care finishing.

In addition, one must also ultimately accommodate a wider interpretation of the term 'easy-care' to include other properties such as stain-resist, stain-repellent, stain-release, water-repellent, anti-pill, tumble-dry, shape retention and so on, thereby ensuring that wool knitwear continues to meet the ever more stringent requirements of the modern consumer.

Shrink-resist polymer compositions

Basolan SW — aziridine terminated polyether
Basolan MW — polyamino siloxane
Hercosett — cationic polyamide epichlorohydrin resin
Polymer GE — proprietary mixture
Polymer RSM — proprietary mixture
Polymer TM — proprietary mixture
Synthappret BAP — carbamoyl sulphonate polyether
DC109 — silicone elastomer

References

Bereck, A and Reincke, K. (1989) *Melliand Text.*, **70**, 452.

Bourn, A., Inman, R., Jackson, J., Needham, P., Rushforth, M.A. and Smith, P. (1985) *Proc. 7th Int. Wool Text. Res. Conf.*, Tokyo, Japan, Vol. IV, p. 282.

Byrne, K.M., Roberts, M.W. and Ross, J.R.H. (1979) *Text. Res. J.*, **49**, 34.

Byrne, K.M., Bereck, K. and Rushforth, M.A. (1990) *Proc. 8th Int. Wool Text. Res. Conf.*, Christchurch, Vol. 4, p. 431.

Byrne, K.M., Barwick, A., Gerstenberg, D. and Bell, V. (1991) *Proceedings of Aachen Textile Conference.*, November.

Byrne, K.M., Ryder, A. and Rakowski, W. (1993) *Proceedings of Aachen Textile Conference*, November.

Feldtman, H.D., McPhee, J.R. and Morgan, W.V. (1967) *Textile Mfr.*, **93**, 122.

Garnsworthy, R.K. *et al.* (1988) *Australasian Textiles*, **8** (4), 26.

Makinson, R. (1979) *Shrinkproofing of Wool*, Marcel Dekker Inc., New York.

PCT (1979), Publ. Appl. No. WP 8102-752.

Philips, H. and Middlebrook, W.R. (1948) British Patent 513,919.

Philips, H., Middlebrook, W.R. and Higgins, A.E. (1941) British Patent 546,915.

Robinson, G.A., Cawood, M.P. and Dobson, D.A. (1965) *Proc. 6th Int. Wool Text. Res. Conf.*, Pretoria, Vol. IV, p. 169.

Rushforth, M.A. (1989) Annual World Textile Conference, Nottingham, UK.

Ryder, A. and Lavocah, W. (1984) IWS Internal Report CPD 483.

Sweetman, B.J. and McLaren, J.A. (1965) *Text. Res. J.*, **35**, 315.

Veldsman, D.P. and Swanepoel, O.A. (1971) *Applied Polym. Symp.* **18**, 691.

Waste Water Tax Law (1987). Federal Republic of Germany, Bonn, 5th March 1987 (Amendment of 1986 Law).

7 Coated and laminated fabrics

R.A. SCOTT

7.1 Introduction

Coated and laminated textiles usually consist of a textile substrate, which will typically be a woven, knitted or non-woven textile fabric, which has been combined with a thin, flexible film of a natural or synthetic polymeric substance.

A *coated* fabric (Figure 7.1a) usually consists of a textile substrate on to which the polymer is applied *directly as a viscous liquid*. The thickness of the film is controlled by applying it via a blade or similar aperture.

A *laminated* fabric (Figure 7.1b) usually consists of *one or more* textile substrates which are combined with a *pre-prepared polymer film or membrane* by adhesives or heat and pressure.

7.1.1 *History*

Textile coated fabrics have been used for over a thousand years. The earliest coated fabrics originated in Central and South America where the natives collected natural rubber latex which exuded from the trunks of certain trees.

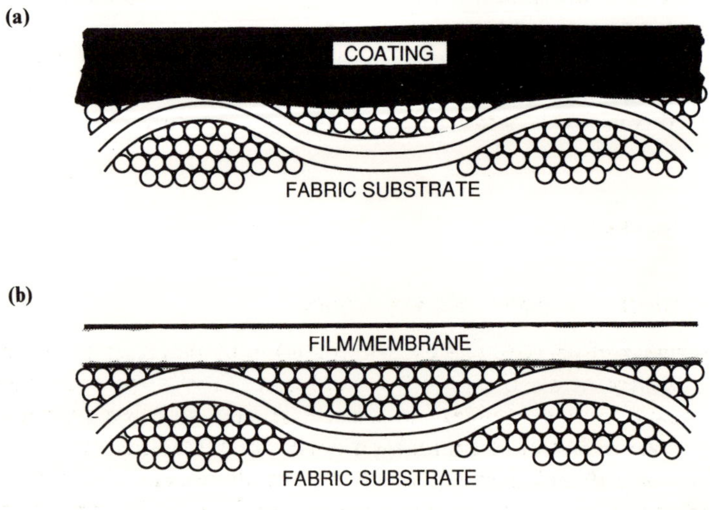

Figure 7.1 Cross-sections of: (a) coated fabric; (b) laminated fabric.

They could spread this milky latex on to fabrics which then coagulated in sunlight to form a flexible, elastic, liquid-proof fabric. This could be used for waterproof bags, covers and containers.

During the 16th and 18th centuries, the discovery of America made this knowledge widespread in Europe. By the late 18th century a rubber industry had been established. Natural rubber could be dissolved in benzene to produce a viscous spreading solution.

Other substances based upon waxes and setting oils, such as linseed oil and tars, were also used to impregnate textile fabrics to render them liquid-proof.

The vulcanisation process for crosslinking rubbers using sulphur was invented in 1840 by Goodyear but patented by Hancock. This led to a large period of growth in the uses of rubber, and plantations were started in Malaysia, Africa and Sri Lanka. Synthetic rubbers and plastic polymers were developed in the 1930s and 1940s. This work became particularly important during the Second World War, when supplies of natural rubber from the Far East were cut off by the Japanese invasion. This led to the abundant and versatile supply of products for a wide range of purposes.

7.1.2 *General usage*

Coated and laminated fabrics have certain advantages over unsupported films or sheets. They have much higher tensile and tear strengths, better puncture resistance and better handle. Coated and laminated fabrics are used for the following purposes.

(a) To confer liquid- and gas-resistant properties on flexible textile fabrics. The most common liquid is rainwater, and the most common gas is air.
(b) To provide a soil-resistant, easily cleaned surface.
(c) To provide materials having the appearance and wearing qualities of leather, or other animal skins, at a much lower cost, and to satisfy the ethical conscience of those who dislike the exploitation of animals.

7.2 Textile substrates

7.2.1 *Synthetic continuous filament fabrics*

Modern high-performance coated fabrics tend to be based upon continuous filament woven nylon 6 and 6.6 or polyester fabrics (polycaprolactam, poly(hexamethylene adipamide), and poly(ethylene terephthalate), respectively). These provide high tear strength, high dimensional stability and resistance to microbiological and chemical attack. Flat multifilament fabrics provide a thin hairless surface to which the continuous coatings can be applied. Where the filament fabric is used as the outer layer of the coated fabric, it prevents abrasion of the coating and sheds liquids and soil readily. In rainwear, the water-shedding facility is enhanced by applying a water-repellent treatment based upon silicones or fluorocarbon finishes.

Filament fabrics can be texturised (Taslan air jet process) to improve coating or film adhesion, to improve handle, and to reduce noise and gloss. This gives the aesthetic appearance of staple fibre fabrics.

Warp- or weft-knitted fabrics tend to have too low a cover factor to be used as the substrate for coated fabrics. They are used as the basis of lightweight laminates for linings or as the second substrate in three-layer sandwich-laminated fabrics.

Non-woven spun-bonded fabrics form the basis of laminates used for disposable items such as disposable clothing, covers and linens. Those utilising polyolefins (Tyvek®) or polyamide are widely used.

7.2.2 *Staple fibre fabrics*

Early coated and laminated fabrics were predominantly based upon woven or knitted cotton fabrics. Heavier coated fabrics may have used wool, flax, jute or hemp fabrics, although these were not so common. Some modern coated fabrics or laminates utilise blends of polyester and cotton or polyester and viscose.

Staple fibre substrates tend to preclude lightweight coatings due to the difficulty in maintaining a continuous film over a 'hairy' fabric. Thus, they tend to produce heavy coated fabrics for tarpaulins, covers, furniture, synthetic leathers, etc. Such fabrics tend to have a softer, quieter handle than their filament-fibre fabric analogues, although strength and puncture resistance tend to suffer.

7.2.3 *Special substrates*

Glass filament fibres in woven fabric are used for heavy coated laminated fabrics where flame retardance, heat resistance, resistance to chemicals and good weathering properties are required. Glass fibre has low resistance to flexing fatigue, due to its brittle nature, and can only be applied where flexing, crumpling and folding are kept to a minimum. Semipermanent covers for sports stadia, tensioned structures, thermal insulation and endless conveyor belts are typical examples.

Aromatic polyamide (Aramid) fabrics, both woven and knitted, are used where flame retardance and non-melting behaviour are required. Blends of poly(*meta*-phenylene isophthalamide) and poly(*para*-phenylene terephthalamide) (Nomex® and Kevlar®) provide fabrics of great strength and abrasion resistance. Other Aramid fibres used include Conex®, Apyeuil®, Kermel® and Twaron®.

7.3 Techniques for coating fabrics

7.3.1 *Basic techniques*

There is a wide range of techniques in industrial use, but they all require similar basic approaches.

(a) The textile fabric to be coated or laminated is supplied full width on a roll.
(b) The fabric is fed under careful tension control to a coating or laminating head or zone.
(c) After application, the coated fabric is passed through an oven to cure the composite and remove volatile solvents before cooling and rolling up.
(d) The film or coating can range in weight from about 10 g/m^2 up to several hundred g/m^2, according to usage.

7.3.2 *Application systems (Smith, 1971)*

(*a*) *Knife over roller.* This is one of the most common coating techniques in use. The diagram (Figure 7.2) shows a metal Doctor blade which is suspended above a steel or rubber-covered roller. The gap between the roller and blade can be set very accurately to about 0.01 mm to control the thickness of the coating. The polymer to be applied is in the form of a viscous solution or gel which rolls against the Doctor blade as the fabric passes underneath and leaves a layer of polymer on the fabric.

(*b*) *Floating knife.* In this system the Doctor blade is fixed above the fabric, as shown in Figure 7.3. The coating weight and thickness are controlled by the tension in the fabric and the depression of the knife. This process tends to be used for impregnating fabrics with small amounts of coating.

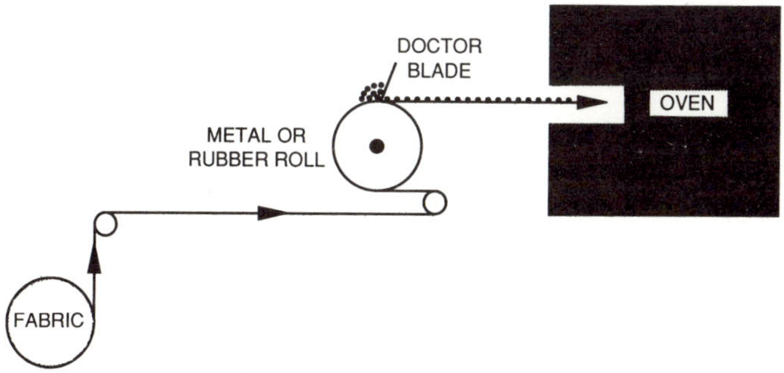

Figure 7.2 'Knife over roll' system.

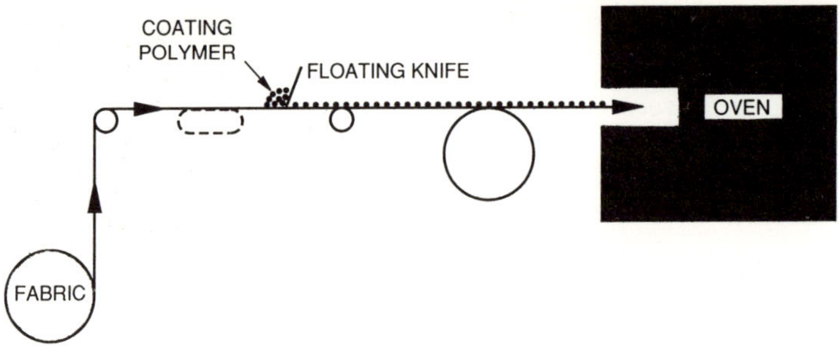

Figure 7.3 Spreading by 'floating knife' system.

(*c*) *Back coating.* In this system the coating polymer is contained in a trough where it is picked up by a revolving gravure roller and transferred to the underside of the fabric, as shown in Figure 7.4.

(*d*) *Reverse roll.* In this process, one slow-moving precision roll is used to meter the thickness of the coating composition on to another precision roll, by accurately adjusting the distance between the two. The second roll is allowed to contact the moving substrate which is supported from below by the main rubber roll. The coating applied is such that its linear weight remains constant however uneven the surface of the substrate may be. This is thus a precision process (see Figure 7.5).

(*e*) *Transfer coating.* This is a sequential process in which the coating composition is metered by Doctor blade on to a silicone release paper, in a process similar to process (a), above. The prepared non-tacky and smooth film is then coated with a layer of an adhesive polymer (which may be the same as the original coating formulation) known as a tie-coat. Immediately

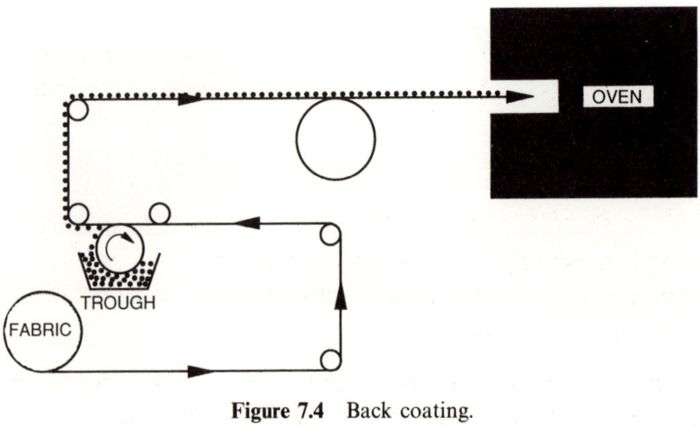

Figure 7.4 Back coating.

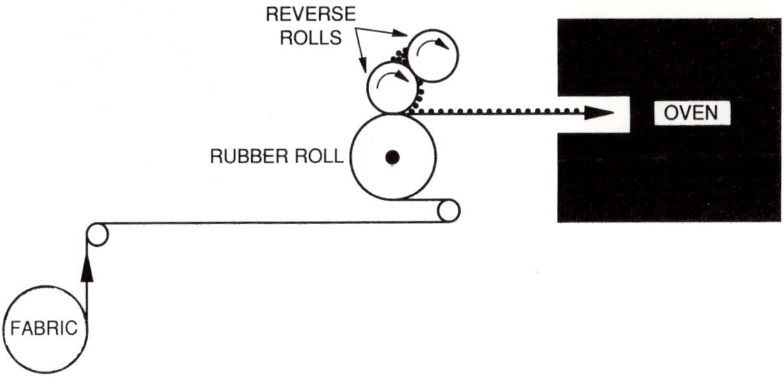

Figure 7.5 'Reverse roll' system.

after this a suitable textile substrate is combined with the tacky adhesive by passage through laminating rolls (see Figure 7.6a). After drying and curing in an oven, the release paper is peeled away, leaving a smooth compact coated surface (see Figure 7.6b).

(*f*) *Lamination.* This is a process similar to transfer coating, in that a pre-prepared film (without release paper) is coated with adhesive prior to combination with one or more textile fabrics under pressure between laminating rolls (see Figure 7.6b).

(*g*) *Flame lamination of foams.* This process is similar to adhesive lamination (f), except that a pre-prepared thin, thermoplastic foam sheet is exposed to a wide slot flame burner situated before the laminating rolls. A thin layer of foam is melted and adheres to the fabric during laminating and cooling. No drying or curing oven is required in this process.

(*h*) *Screen printing.* This is an ideal process for applying even coatings to fabrics with uneven surfaces. Rotary screens dispense the coating composition on to the fabric. The composition may be a stable foam or paste or even an unstable foam. A Doctor blade then controls the coating thickness precisely (Matthews, 1993).

7.3.3 *Control parameters in coating*

The coating process requires a delicate balance between adhesion efficiency and parameters such as tear-strength performance and handle of the finished fabric. It is important that:

(a) Specially formulated tie-coats may need to be applied on certain substrates to prepare the fabric surface for subsequent coating layers.

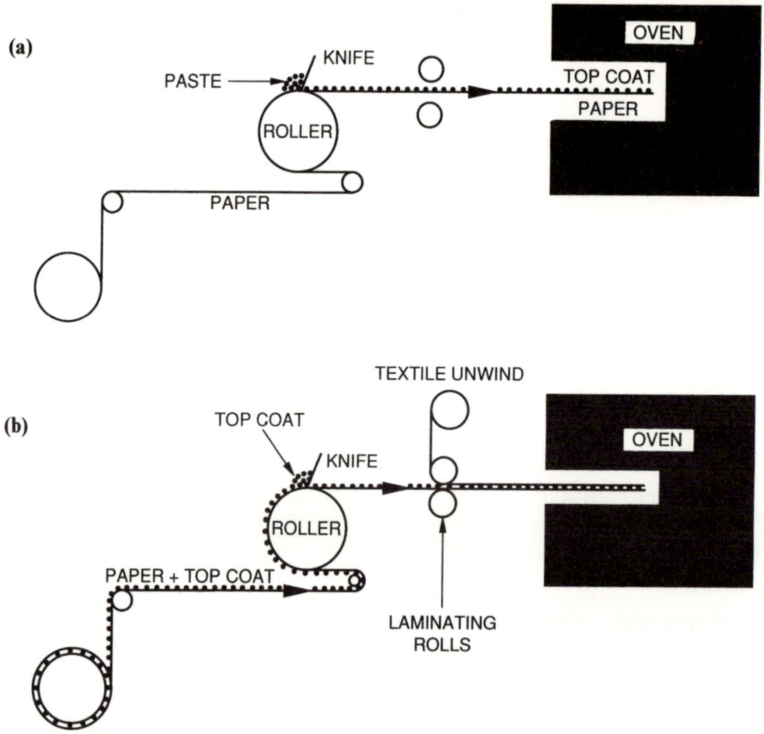

Figure 7.6 Transfer coating: (a) stage 1: film preparation; (b) stage 2: lamination of the textile to the film.

(b) If the coating is applied in several passes or stages, the barrier properties are improved, as imperfections such as pin-holes are covered.
(c) The polymer coating should not penetrate through the fabric warp and weft yarns as this locks the structure, leading to loss of tear strength, poor handle and drape, and increased noisiness.
(d) The coating should be of such a consistency that it sits on the surface of the fabric substrate, but is adequately bonded to it to prevent delamination or coating loss in use.
(e) Many waterproof coated and laminated fabrics are treated with a water-repellent finish (silicone, fluorocarbon or wax) which is applied to the side of the fabric exposed to the rain. These finishes can also improve abrasion resistance and tear strength. This finish is applied after coating/laminating and curing, otherwise it interferes with the bonding process.
(f) Some heavy rubber-coated fabrics are cured by pressure contact with a heated metal calender, to produce a compact surface.

7.4 Chemistry of coating polymers

7.4.1 *Polyurethanes*

The polyurethanes are a wide class of polymers ranging from rubbers and elastomeric fibres to surface coatings, adhesives and flexible or rigid foams. They are not direct polymers of urethanes, but are derived from the reaction of polyesters or polyethers with di- or poly-isocyanates, to produce complex structures containing urethane linkages (Roff and Scott, 1971b).

The fundamental unit is based upon the urethane group

—N—C—O—
   | ‖
  H   O

Polymer structure. Polyurethanes for coating are complex polymers. The major component of these is a segmented prepolymer, composed of a linear polyester or polyether that has been extended severalfold in chain length by coupling through urethane linkages. The prepolymer molecule (—PP—) can be further extended and crosslinked with a multifunctional isocyanate (—IC—) to produce the following structure:

—(PP1)—IC—(PP2)—IC—
|
(PP3)
|
IC
|

7.4.2 *Preparation of polyurethanes*

The isocyanate (R—NCO) compound is prepared by the condensation of primary amines with phosgene.

$$\underset{\text{Primary amine}}{R{-}NH_2} + \underset{\text{Phosgene}}{COCl_2} \rightarrow \underset{\text{Urethane}}{R{-}NCO} + 2HCl$$

The isocyanate group readily reacts with amino (—NH_2) or hydroxyl (—OH) groups.

$$R{-}NCO + {-}OH \rightarrow R{-}NHCO \cdot O{-} \qquad \text{(urethane linkage)}$$

$$R{-}NCO + {-}NH_2 \rightarrow R{-}NHCO \cdot NH{-} \qquad \text{(urea linkage)}$$

When excess isocyanate is present, the above products undergo further reaction, yielding branched structures that lead to crosslinkages of two main types.

$$\underset{\text{Urethane}}{\text{R—NHCO·O—}} + \underset{\text{Isocyanate}}{\text{—NCO}} \rightarrow \underset{\text{Allophanate linkage}}{\text{R—N(CONH—)CO·O—}}$$

$$\underset{\text{Urea}}{\text{R—NHCO·NH}} + \underset{\text{Isocyanate}}{\text{—NCO}} \rightarrow \underset{\text{Biuret linkage}}{\text{R—N(CONH—)CO·NH—}}$$

The urethane linkage (—NHCO·O—) is unstable in the presence of water (—OH groups) or carboxylic acid and breaks down to form amines or amides, with evolution of carbon dioxide.

7.4.3 *Production of complex polymers*

The first production step is to prepare segmented prepolymers by coupling a hydroxyl-terminated polyester or polyether with a polyfunctional isocyanate. This extends the molecular size through the urethane linkages.

A typical polyester could be poly(diethyleneglycol adipate) having a molecular weight between 2000 and 3000. A typical polyether is obtained by polymerisation of propylene oxide, usually in the presence of a small proportion of glycerol or sorbitol to provide branched structures. Two of the most commonly used isocyanates are 2,4- and 2,6-tolylene di-isocyanate (TDI). These are chosen because of their reactivity, cheapness and relative low toxicity.

The resulting prepolymer has —OH terminal groups when a deficiency of isocyanate is used, but —NCO terminal groups if an excess of isocyanate is employed (Scheme 7.1).

$$\underset{\text{Deficiency}}{(n-1)\text{R—(NCO)}_2} + n\text{HO—(PP)—OH} \rightarrow$$

$$\text{HO—(PP)—O—[CONHRNHCO·O—(PP)—O]}_{(n-2)}\text{CONHRNHCO·O—(PP)—OH}$$

Hydroxyl end group

or

$$\underset{\text{Excess}}{(n+1)\text{R—(NCO)}_2} + n\text{HO—(PP)—OH}$$

$$\text{OCNRNHCO·O—(PP)—O—[CONHRNHCO·O—(PP)—O]}_{(n-1)}\text{CONHRNCO}$$

Isocyanate end group

Scheme 7.1

7.4.4. *Isocyanates*

Table 7.1 lists some common di-isocyanates used.

Table 7.1 Common di-isocyanates

Abbreviation	Chemical name
DADI	Dianisidine di-isocyanate
HDI	1,6-Hexamethylene di-isocyanate
MDI	4,4-Diphenyl methane di-isocyanate
NDI	1,5-Naphthalene di-isocyanate
PDI	Phenylene di-isocyanate
TODI	Tolidine di-isocyanate
XDI	Xenylene di-isocyanate
TDI	2,4- and 2,6-Tolylene di-isocyanate

7.4.5 *Coating formulations*

(*i*) *Two-component systems.* Insoluble films can be prepared by mixing a solution of a hydroxyl-terminated polyester with a non-volatile di-isocyanate just before application to the fabric. The hardness of the film depends on the complexity of their composition and the degree of branching. Rubbery prepolymers, solvent-based and curable with di-isocyanates at room temperatures or by heat, adhere well to metals and textiles. Alternatively, isocyanate-ended polymers can be cured by intermediates of epoxy resins containing hydroxyl groups or fatty acid polyamides.

(*ii*) *One-component systems.* A branched isocyanate-ended polymer, applied from solution, will cure to a hard coating by reaction with atmospheric moisture. The adhesion of rubbers and polyurethanes to textile surfaces can be enhanced by simple treatment with 4,4,4-triphenylmethane di-isocyanate, which probably reacts with surface layers of hydroxide or moisture.

Another type of coating medium, known as a urethane oil or urethane alkoyd, is made by treating a drying oil such as linseed oil (a polyunsaturated glyceride) with glycerol, to yield a mixture of mono- and di-glycerides. An excess of these is then reacted with a di-isocyanate. This gives a product which is highly unsaturated, and which can dry rapidly by atmospheric oxidation.

(*iii*) *Thermoplastic polyurethane rubbers.* There are two types of product which are processible on plastics processing equipment. These materials undergo allophanate-type crosslinkages during processing. One is based upon adipic acid, 1,4-butanediol and MDI, with the trade name Estane. This is thermoplastic but is virtually crosslinked, and has good extensibility and elasticity. The other is a heat-cured rubber prepared from hydroxyl-terminated polyesters, MDI, and a diol which has the trade name Texin. Thermoplastic polyurethanes can be applied from the melt, and avoid the use of solvents.

(*iv*) *Polyurethane foams.* Flexible foams for coating fabrics can be made by reacting a polyfunctional isocyanate with a prepolymer and water. This yields a rubber and carbon dioxide gas sufficient to effect the expansion of a cellular structure.

Flexible foams are usually made from a branched polyether of molecular weight 2000 to 5000 crosslinked with TDI. Polymerisation is catalysed by an organometallic compound (e.g. stannous 2-ethylhexoate) and a tertiary amine catalyst (e.g. triethylene diamine). These act at room temperature and also promote the evolution of the blowing agent, carbon dioxide. A surface-active agent can be included to promote a uniformly textured foam. Flexible foams contain open, interconnected cells, whereas rigid foams have predominantly closed cells.

(*v*) *Water-based polyurethanes* (*Goldsmith, 1988*). Conventional polyurethanes tend to be applied from organic solvents such as di-methylformamide (DMF), methyl ethyl ketone (MEK) and mixtures containing toluene, xylene and chloro solvents. The solutions can contain up to 40% by weight of solvents. Large quantities of these expensive solvents are discharged into the atmosphere during drying. The economic aspects of this waste have caused concern in the past, but it is the environmental impact of the pollution caused by these organic solvents which is paramount, and is forcing governments to legislate against atmospheric pollution. This forces the manufacturers either to reclaim the solvent, or to move to polymer systems which limit or avoid their use. Solvent reclamation plants are extremely expensive systems which need to be built into the coating plant at the drying oven stage. They are, however, justified when solvent prices are high and local legislation is strict.

Aqueous polymer dispersion systems were introduced in the 1970s and 1980s. They contain up to 50% solids suspended in water, and although they avoid the problem of solvents, they require larger amounts of heat energy to drive off the water to dry and fix the polyurethane film which results. Polyurethane dispersions tend to have inadequate resistance to water. It is probable that surface-active interfaces remain in the solidified and dried polyurethane. These may provide routes for the diffusion of water into the film in moist conditions, leading to structural weaknesses which show when the polymer is strained. Bayer AG have attempted to overcome this by creating chemical crosslinks between the agglomerated microscopic polyurethane particles. An isocyanate has been developed which is stable in water for up to 2 h.

A typical aqueous polyurethane polymer would be prepared by the conventional first stages by reaction of a polyester or polyether diol with an isocyanate. This would be followed by the addition of an ionic functional group into the backbone, before neutralisation of the ionic groups, and finally dispersion in water. The neutralised ionic groups are hydrophilic, thereby eliminating the use of external surfactants to facilitate dispersion and stabilisation.

The advantages of water-based coating polymers are listed in Table 7.2.

Table 7.2 Advantages of water-based coating polymers

Low environmental impact
Low odour during application
Low toxicity of discharged vapours
Surfactant-free — no adhesion problems
Easy to use — no reactants or catalysts needed
Can be thermoplastic or thermosetting
Heat-resistant coating
Low-temperature flexibility
Water-vapour permeable

7.4.6 *Polyurethane properties*

The advantages and disadvantages of different types of polyurethanes are listed in Table 7.3.

7.5 Natural and synthetic rubbers

Rubbers are widely used in coated fabrics for a variety of purposes. For the purposes of this chapter, the types are:

Natural rubber	*cis*-Polyisoprene
Neoprene	Polychloroprene
Butyl	Polyisobutylene
SBR	Styrene–butadiene rubber
Nitrile	Acrylonitrile–butadiene rubber

Table 7.3 Advantages and disadvantages of different types of polyurethanes

Polyurethane type	Advantages	Disadvantages
Polyether	Good hydrolytic stability Good microbiological stability Excellent softness and flexibility Cheaper than polyesters Good solvent resistance Range of molecular weights available	Reduced adhesion to low-energy surfaces UV and weathering resistance not good Not as tough as polyester Lower physical properties
Polyester	Excellent adhesion to low-energy surfaces Excellent toughness and abrasion resistance Excellent flexibility and softness Range of molecular weights available	Lower solvent resistance Lower hydrolytic stability Lower microbiological stability Higher raw material costs

7.5.1 *Natural rubber*

The simplest fundamental unit is 1,4-isoprene:

$$-CH_2-\overset{\displaystyle CH_3}{\overset{|}{C}}=CH-CH_2-$$

Formula = C_5H_8; molecular weight = 68; degree of polymerisation = 1500–5000; specific gravity = 0.93.

There are two distinct isomeric configurations of 1,4-polyisoprene (Roff and Scott, 1971a):

cis—where the chains substituent to the double bond lie on the same side, as in structure (**1**). This is natural elastomeric rubber used in coating formulations.

trans—where the chains substituent to the double bond lie on opposite sides as in structure (**2**). This form is the harder gutta percha, chicle and balata.

As rotation about a double bond is not possible, the two forms are not readily interconvertible. The chains occupy three-dimensional space, which cannot be represented fully in structures (**1**) and (**2**). The *trans*-form permits highly regular zig-zag chains which fit together well, whereas the *cis*-form does not (Morrison and Boyd, 1987). Both exhibit the long chain structures which typify rubbers.

(1) *cis* - polyisoprene

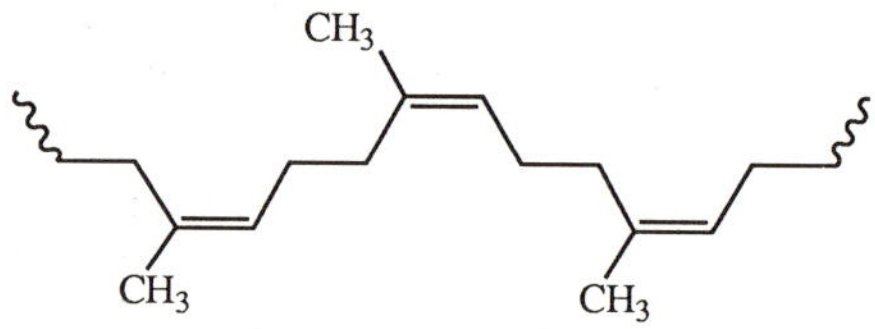

(2) *trans* - polyisoprene

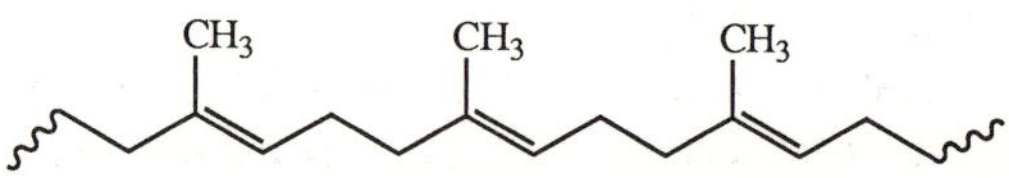

7.5.1.1 *Preparation.* Natural rubber is obtained by cutting (tapping) the bark of *Hevea braziliensis*, a wild tree occurring originally in South America. The rubber is in a milky form which contains about 35% of latex. The latex rubber can be separated by coagulation with an acid such as acetic acid. When washed and dried in air it becomes ‘crepe’, or if dried in wood smoke it becomes ‘smoked sheet’.

Rubber is soluble in a range of solvents, including hydrocarbons, chlorinated hydrocarbons, certain ketones, esters and carbon disulphide. Raw rubber is mechanically weak, deficient in elastic recovery, stiff when cold and soft when hot. It is necessary to blend it with other constituents and to vulcanise it to improve strength, elasticity and resistance to liquids, and to reduce the temperature sensitivity.

7.5.1.2 *Vulcanisation of rubbers.* Milling and masticating rubbers with sulphur, followed by heating, causes the formation of sulphur bridges between the chain molecules. This reaction involves reactive allylic positions, and thus depends on the double bond in the polymer:

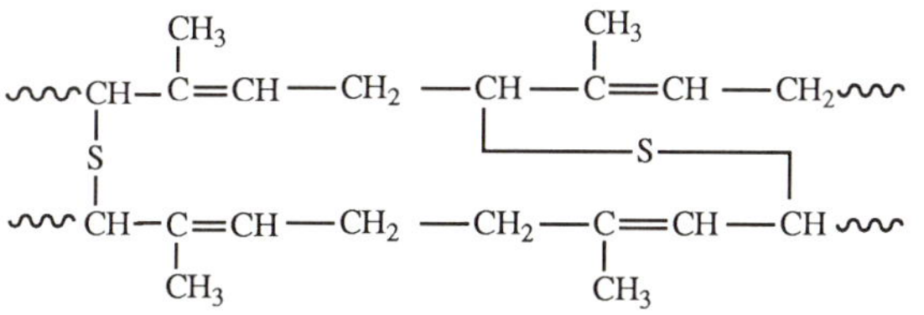

vulcanised rubber

7.5.1.3 *Mixing and blending constituents.* Practical rubbers are a complex mixture of organic and inorganic substances which must be mixed carefully to obtain uniformly high strength properties.

Rubber
Curing system { Vulcanising agent, Activator, Accelerator }
Diluting fillers
Reinforcing fillers
Plasticisers
Anti-oxidants
Colour/pigments

(a) The *vulcanising agent* is usually sulphur.
(b) The *accelerating agents* are usually metal oxides (MgO, PbO) or organic substances such as thiazoles, mercaptobenzthiazoles (MBT) or sulphenamides. The latter have a delayed action, in that they have no activity below a certain temperature.
(c) The *activators* are required to make the organic accelerators function effectively. They can be substances like zinc oxide and stearic acid. Stearic acid is also a lubricant, softener and plasticiser.
(d) *Diluting fillers* such as clays, talc and barytes act as diluents.
(e) *Reinforcing fillers* such as finely divided carbon black increase the tensile and tear strength of the mix.

(f) *Anti-oxidants* such as ketone–amine condensation products or phenols are used to prevent natural ageing and atmospheric (ozone) cracking.
(g) *Colours* in the form of pigments are added to colour the mix.

7.5.2 *Polychloroprene rubbers* (*neoprene*)

The simplest fundamental unit is:

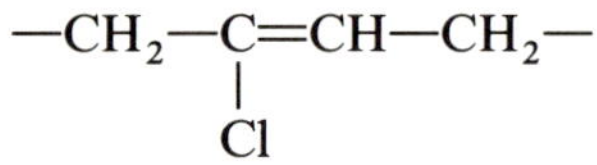

Formula is C_4H_5Cl; molecular weight = 88.5; degree of polymerisation = 1000–3500; specific gravity = 1.20–1.25.

Polychloroprene consists of about 85% *trans*-1,4 units and about 10% *cis*-1,4 units. The rest is composed of two other isomers.

cis-Polychloroprene

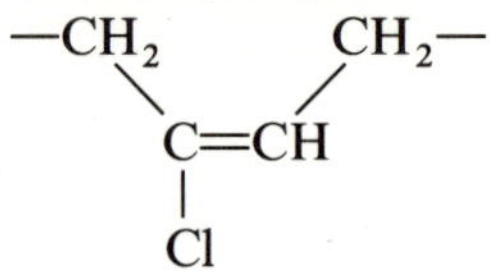

trans-Polychloroprene

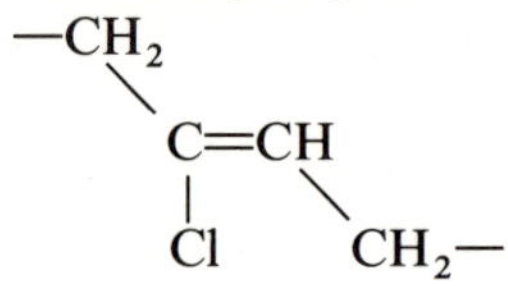

7.5.2.1 *Preparation.* Polychloroprene is normally prepared by dimerisation of acetylene, followed by addition of hydrogen chloride:

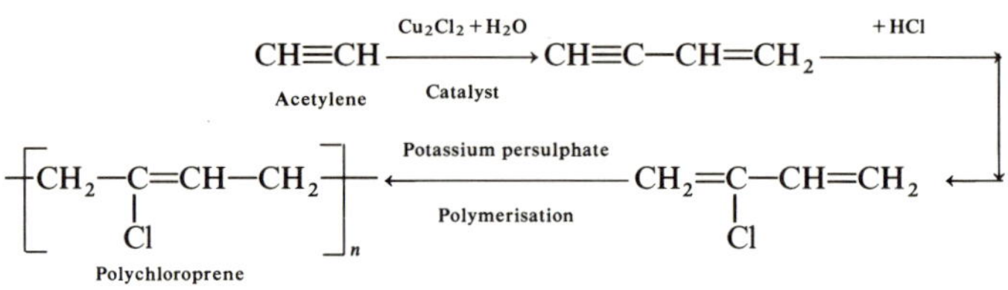

A more recent process involves the chlorination of C4 petroleum fractions to obtain 3,4-dichlorobut-1-ene. Dehydrochlorination of this gives the required monomer.

The forms of polychloroprene vulcanised using sulphur are known as 'neoprene' rubber. Vulcanisates resemble those of natural rubber, but have superior resistance to heat and flame, are self-extinguishing, and resistant to oils and ozone. These properties make neoprene a versatile and valuable rubber in coated fabric form.

Copolymers have been prepared commercially. One such is neoprene ILA, which is a copolymer of chloroprene and acrylonitrile, which has exceptional oil resistance. A carboxylated type of latex is manufactured in Germany (Baypren).

7.5.3 *Polyisobutylene rubbers (butyl)*

The simplest fundamental unit is:

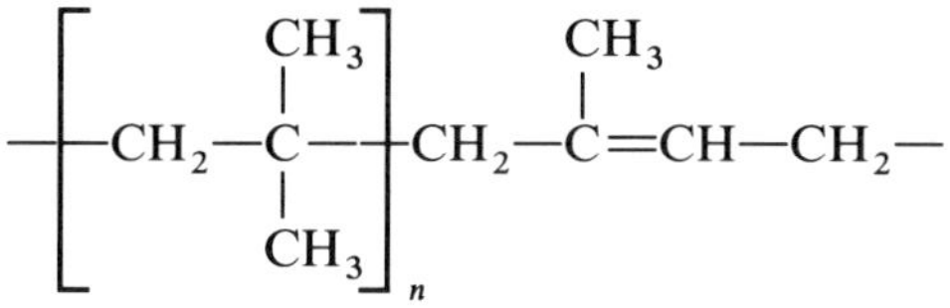

Isobutylene formula = C_4H_8; molecular weight = 56; 1,4-Isoprene formula = C_5H_8; molecular weight = 68; the value of n varies from 20 to 200; specific gravity = 0.91–0.98.

The isoprene units are randomly distributed and provide unsaturated bonds which allow the copolymer to be vulcanised. The small number of double bonds remaining in the vulcanisate means that it is resistive to oxidation and chemical attack.

7.5.3.1 *Preparation.* Can be prepared from C_4 petroleum fractions such as butane or butylene. If the starting point is butane, this is dehydrogenated to form butylene. A small amount of isoprene is mixed with isobutylene in an inert solvent (methylene chloride or liquified ethylene). This is made to undergo a Friedel–Crafts reaction using catalysts such as $AlCl_3$ or BF_3, together with a trace of a cocatalyst such as water. Polymerisation occurs at a very high rate, and cooling of the monomers is required to control a continuous process.

$$\underset{\text{Butane}}{C_4H_{10}} \xrightarrow[\text{Heat}]{-H_2} \underset{(n\text{- and iso-butylene})}{C_4H_8} \xrightarrow[\text{In } H_2SO_4]{\text{Separate}} \underset{\text{isobutylene}}{C_4H_8}$$

$$\underset{\text{Polyisobutylene}}{[C_4H_8]_n} + \underset{\text{Isoprene}}{C_5H_8} \xrightarrow[AlCl_3 + H_2O]{\text{Friedel–Crafts}} \underset{\text{Butyl rubber}}{-[C_4H_8]_n-C_5H_8-}$$

7.5.3.2 *Properties.* Butyl rubber is unaffected by oxygen, alkalis, hydrogen peroxide, alcohols, phenols, animal and vegetable oils, and some ketones, ethers, esters and fatty acids. Its compact structure gives it excellent resistance to gas and liquid permeation. Hence its use in tyres, balloons, tank and pool linings. It has specialist military uses in chemical warfare agent protective clothing, covers and shelters.

Butyl rubbers can be rendered heat-resistant and flame-retardant by combining small proportions of chlorine or bromine in the milling process.

7.5.4 *Styrene–butadiene rubbers (SBR)*

The simplest fundamental units are shown in structures **3** to **6**.

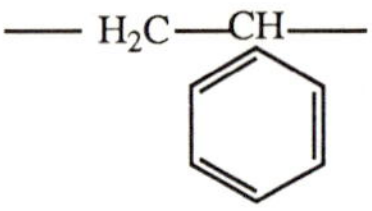

styrene

(3)

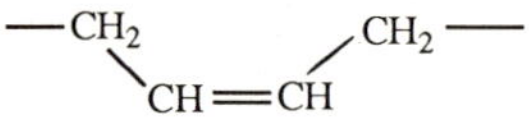

cis-1,4 ,butadiene

(4)

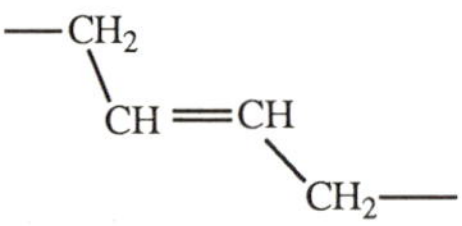

trans-1,4,butadiene

(5)

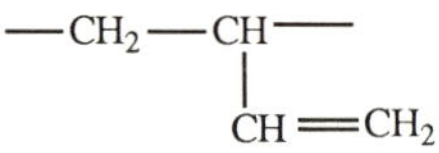

1,2 or vinyl butadiene

(6)

Styrene formula = C_8H_8; molecular weight = 104; butadiene formula = C_4H_6; molecular weight = 54; specific gravity = 0.91–1.00.

SBR is a mixture of styrene (**3**) with various molar proportions of the isomers **4**, **5** or **6**. The proportions depend on the method and temperature of polymerisation. Table 7.4 shows the proportions resulting from different processes.

Units **3** to **6** are randomly arranged in emulsion copolymers, but in solution polymers may occur in uniform blocks of varying length. Such block copolymers of Sty–Buta–Sty type have reversible thermoplastic–elastomeric properties. They are resilient and rubber-like at room temperature, but are thermoplastic at higher temperatures.

7.5.4.1 *Preparation*

Emulsion polymerisation route. Styrene and 1,3-butadiene are emulsified in water containing soap or other surface-active agent. They are polymerised in two different ways:

Table 7.4 Proportions of butadiene isomers in SBR from different processes

Type of polymerisation process	Percentage of the butadiene units		
	(4)	**(5)**	**(6)**
'Hot' emulsion polymerisation	18.3	65.3	16.3
'Cold' emulsion polymerisation	12	72	16
Solution polymerisation	40	54	6

(a) 'Hot' rubber is produced at about 50°C using an initiator such as potassium persulphate or an organic peroxide.
(b) 'Cold' rubber is produced at about 5°C using a redox initiating system such as ferrous sulphate plus a peroxy compound, sometimes with sodium formaldehyde sulphoxylate. Polymerisation is terminated at about 70% conversion by adding a 'short stopper' such as quinol.

Solution polymerisation. The two monomers, diluted with a hydrocarbon solvent, are polymerised using a stereospecific catalyst such as a lithium alkyl.

7.5.4.2 *Properties.* SBRs are similar to natural rubber in that they are susceptible to atmospheric oxidation and ozone cracking when stretched in air. They are swollen and weakened by hydrocarbons and halogenated hydrocarbons. Their abrasion resistance and ageing resistance are superior to those of natural rubber.

7.5.5 *Nitrile rubbers*

The simplest fundamental units are structures (**4**) to (**6**) (see section 7.5.4) and structure (**7**).

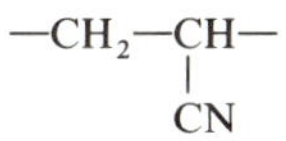

Acrylonitrile

(**7**)

Acrylonitrile formula = C_3H_3N; molecular weight = 53; butadiene formula = C_4H_6; molecular weight = 54; specific gravity = 0.95–1.02.

Nitrile rubbers are classified according to the content of acrylonitrile: low acrylonitrile = approximately 20%; high acrylonitrile = approximately 40%; very high acrylonitrile = approximately 45%. The butadiene units are predominantly in the *trans*-1,4 form (**5**).

7.5.5.1 *Preparation.* Acrylonitrile and freshly distilled 1,3-butadiene are copolymerised in emulsions by methods analogous to those used to prepare styrene–butadiene rubbers detailed earlier in section 7.5.4.1. Both 'hot' and 'cold' rubber processes can be used. The initiator in the hot process can be potassium persulphate. A redox type of hydroperoxide–dextrose sequestering agent is used for the 'cold' process.

7.5.5.2 *Properties.* Nitrile rubbers have low tensile strength unless filled with reinforcing carbon blacks. Resilience is low, and low temperature flex-cracking is worse than that of natural rubber, especially with rubbers containing large proportions of acrylonitrile. They do, however, have better resistance to heat ageing and light than natural or SBR rubbers.

Nitrile rubbers have good resistance to oils, greases and other hydrocarbon liquids, and are thus used for products where this property is of primary importance.

7.6 Coating polymers and elastomers

7.6.1 *Poly(vinyl chloride) (PVC)*

The fundamental unit is:

$$-CH_2-\underset{\displaystyle Cl}{\underset{|}{CH}}-$$

Formula is C_2H_3Cl; molecular weight = 62.5; degree of polymerisation = 800–2000; specific gravity (plasticised) = 1.1–1.7.

7.6.1.1 *Preparation.* PVC is an additive polymer made from the monomer vinyl chloride (CH_2-CHCl). It can be formed by the following methods.

(a) The addition of hydrogen chloride to acetylene:

$$\underset{\text{Acetylene}}{CH{\equiv}CH} + HCl \xrightarrow{HgCl_2 + BaCl_2} \underset{\text{Vinyl chloride}}{CH_2{=}CHCl}$$

(b) Chlorination of ethylene, followed by dehydrochlorination by treatment with alkali or by cracking at 300–600°C (Woolfe process):

$$\underset{\text{Ethylene}}{CH_2{=}CH_2} + Cl_2 \xrightarrow{\text{Catalyst}} CH_2Cl-CH_2Cl$$

$$CH_2Cl-CH_2Cl \xrightarrow[-HCl]{300\text{–}600°C} \underset{\text{Vinyl chloride}}{CH_2{=}CHCl}$$

(c) Another process is oxyhydrochlorination of ethylene:

$$CH_2{=}CH_2 \xrightarrow[500°C]{Air + CuCl_2 + KCl} CH_2{=}CHCl$$

The vinyl chloride monomer can be polymerised by high-energy radiation or free-radical initiators. The most common polymerisation techniques are emulsion and suspension processes (Roff and Scott, 1971b).

Emulsion polymerisation is carried out in aqueous emulsions using surface-active agents. The initiator is contained in the aqueous phase.

Table 7.5 Properties conferred by different plasticisers for PVC

Plasticiser types	Properties conferred
Di-octyl phthalate	General purpose plasticisers
Trixylyl phosphate	Offers some flame-retardance
Di-octyl sebacate, polypropylene sebacate	Low-temperature plasticisers

Suspension polymerisation utilises water-soluble suspension agents such as poly(vinyl alcohol) or methyl cellulose. The initiators used can be benzoyl or lauroyl peroxides.

7.6.1.2 *Plasticisers for PVC.* Coating formulations are plasticised by using high boiling point esters of C_8 to C_{10} alcohols (phthalates, phosphates, sebacates), as shown in Table 7.5.

7.6.1.3 *Properties.* PVC is dissolved by tetrahydrofuran (THF), dimethyl formamide (DMF), ketones, carbon tetrachloride, and other chloro-solvents. It will not resist petrols and light oils. All PVC polymers are degraded by heat and light, although PVC is flame-retardant. The degradation is thought to proceed as a chain reaction, initiated and propagated by activated chlorine. The polymer decomposes by an unzipping mechanism. Stabilisers such as lead carbonate, lead silicate, salicylates or stearates interfere with the chain reaction.

PVC-coated fabrics stiffen and can crack if flexed at low temperatures (colder than $-10°C$), although choice of plasticisers can extend the utility range downwards to lower temperatures.

7.6.1.4 *Application in coated fabrics.* PVC can be applied from mixed solvents, or more recently as plastisols, which are creamy pastes. The plastisols consist of very finely divided PVC dispersed in a plasticiser. This method was mainly used for heavy coatings such as artificial leathercloth and heavy-duty industrial clothing, but it is now used for a wider range of products. It has the advantage that no solvents are used, and thus avoids the pollution or recycling problems that solvent can cause.

7.6.2 *Polyacrylate elastomers*

The simplest fundamental unit is:

$$-CH_2-\underset{\displaystyle COOR}{\underset{|}{CH}}-$$

Acrylate polymer

where R can be an alkyl (C_1 to C_{12}) or an aryl group, or, in the case of water-soluble polymers, a monovalent cation. The formula is $C_3H_3O_2$—(R); molecular weight = 71 + (R); degree of polymerisation = 5000–10 000; specific gravity = 1.05–1.11.

7.6.2.1 *Preparation.* There are two main methods of preparation.

(a) By the oxidation in air of ethylene, followed by conversion to ethylene cyanohydrin. This intermediate is then dehydrated and hydrolysed in the presence of alcohol, as shown in Scheme 7.2.

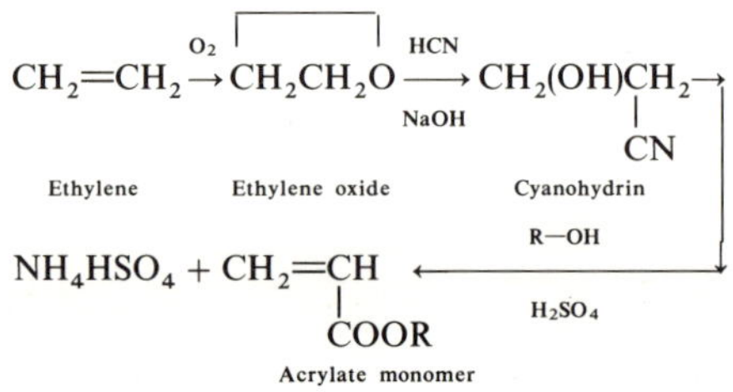

Scheme 7.2

(b) By the Reppe synthesis from ethyne, carbon monoxide, and the appropriate alcohol:

$$\underset{\text{Ethyne}}{CH{\equiv}CH} + \underset{\text{Carbon monoxide}}{CO} + \underset{\text{Alcohol}}{ROH} \xrightarrow[\text{Heat, Pressure}]{\text{Catalyst}} CH_2{-}\underset{\displaystyle COOR}{\underset{|}{CH}}{-}$$

The monomer is polymerised as an emulsion to give latexes for the impregnation and coating of leather, paper and textiles. Molecular weight and reaction rate are controlled by initiators such as peroxides, or a redox system such as persulphate–bisulphite. Polyacrylates can be cross-linked by diamines and peroxides to produce acrylic rubbers.

7.6.2.2 *Properties.* Acrylates are soluble in most chlorinated hydrocarbons, esters, ketones, and aromatic hydrocarbons. They can be plasticised by esters similar to those used for PVC. As the length of the alkyl chain (R) increases, the acrylate polymers increase in softness.

Acrylate polymers used for coatings possess excellent fastness to light, weathering, and microbiological attack. The rubbers have good resistance to flex-cracking, oxidation, ozone, ultraviolet radiation and lubricating oils. They retain light colours, including white, in sunlight. They tend to be used for items such as sun blinds, white hat covers, and snow camouflage for these reasons.

7.6.3 *Silicone elastomers*

The simplest fundamental units of silicone are elastomers:

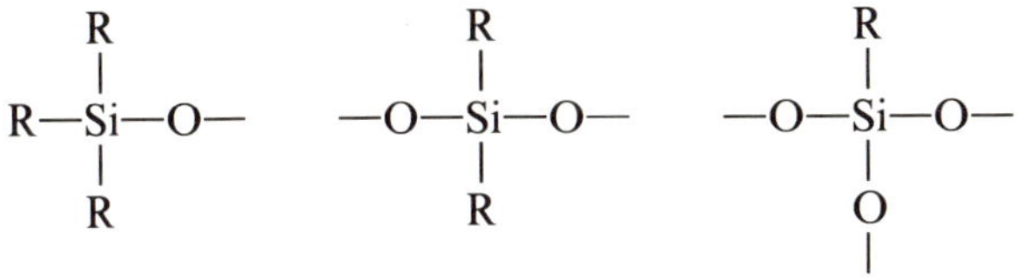

Monofunctinal Difunctional Trifunctional

These are polysiloxane units, where R is an organic substituent, usually an alkyl group, and often CH_3—.

Silicone rubbers are primarily linear chains of the difunctional units with monofunctional end-groups. General-purpose rubbers use methyl radicals, where low-temperature rubbers utilise about 5–15% molecular percentage of phenyl groups:

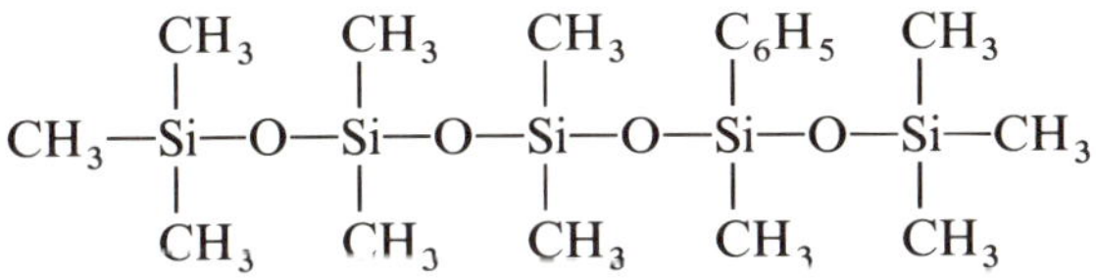

7.6.3.1 *Preparation.* Silicone monomers can be prepared by the 'direct' process whereas silicon is reacted with methyl chloride at high temperature in the presence of copper. This produces a large proportion of dimethyl dichlorosilane. Hydrolysis in dilute HCl first replaces Cl with OH, which is followed by condensation polymerisation at 175°C to give a mixture of mono- and di-functional units.

The monofunctional units in small quantities control the molecular weight of the polymer chains by termination. The trifunctional units need to be absent, since they produce chain branching and crosslinking (Roff and Scott, 1971b).

7.6.3.2 *Properties.* Silicones are soluble in hydrocarbons, ketones, esters and ethers. They are unaffected by water, some mineral oils, petrol, cold acids and alkalis.

Silicone rubbers and resins remain flexible at lower temperatures than organic rubbers (down to −70°C) and resist heat-ageing well. When used as thin coatings on lightweight fabrics they maintain high fabric tear strengths with soft handle. When vulcanised (with organic peroxides) they resist microbiological attack. They are odourless, tasteless, physiologically inert, and water-repellent.

They are used for outdoor purposes such as tents and covers, and in the food, engineering and medical-product industries. Their inertness makes them suitable for surgical implants.

7.6.4 *Poly(tetrafluoroethylene) (PTFE)*

The simplest unit is:

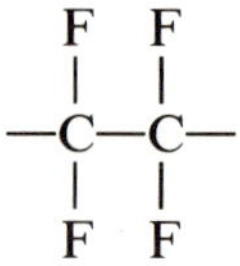

Formula is C_2F_4; molecuclar weight = 100; degree of polymerisation = 10 000; specific gravity = 2.1–2.3.

7.6.4.1 *Preparation.* An industrial synthesis involves the reaction between chloroform and anhydrous hydrogen fluoride to produce a mixture of chlorofluoromethanes. Chlorodifluoromethane is then cracked thermally to produce the monomer, as shown in Scheme 7.3. The monomer is then polymerised in the presence of a radical initiator (Roff and Scott, 1971b).

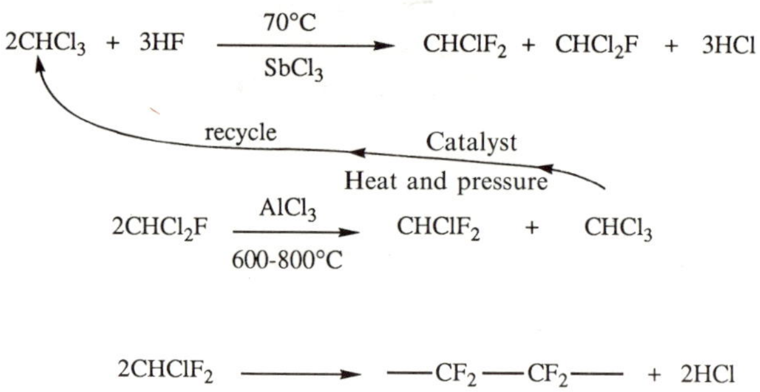

Scheme 7.3

7.6.4.2 *Properties.* PTFE is extremely stable and inert. It is not soluble in any known liquids, and is unaffected by concentrated acids and alkalis.

It will not burn in air, and melts only slowly in a flame. It is flexible down to −80°C and dimensionally stable up to +250°C. PTFE has an unusually low coefficient of friction and its abrasive resistance is good. Its remarkable non-adhesive properties mean that it sheds or repels liquids, especially water, readily. Processed thin, light films can be laminated to textiles (see section 7.7 on waterproof (breathable) fabrics).

7.6.5 *Polyethylene*

Polyethyelene has one of the simplest fundamental units in organic chemistry:

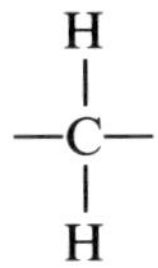

Formula = CH_2; molecular weight = 14; degree of polymerisation = 200 000; specific gravity = 0.915 to 0.94 (low density), 0.94 to 0.97 (high density).

7.6.5.1 *Preparation.* Both low- and high-density polyethylenes are prepared from ethylene ($CH_2{=}CH_2$) which is obtained by catalytic cracking of natural gases, such as propane, or light oil fractions. The monomer has to be free from impurities such as other hydrocarbons, water, oxygen, etc.

(*a*) *Low-density polyethylenes.* Ethylene is passed over heated copper to remove oxygen. It is then compressed to 2000 to 3000 atmospheres, and a catalyst such as a peroxide compound, and a chain modifier, such as an aldehyde or ketone, are added in controlled amounts. Polymerisation is exothermic, and has to be controlled at temperatures up to 200°C. Propagation of the polymer radical can be terminated by combination, chain transfer, or disproportionation. Both solution and emission polymerisation processes have been used.

(*b*) *High-density polyethylenes.*

(i) *The Ziegler process* involves feeding ethylene into an inert hydrocarbon solvent containing a Ziegler–Natta catalyst (i.e. triethyl aluminium and titanium tetrachloride). Polymerisation takes place at atmospheric pressure and at about 75°C. Chain terminators such as oxygen or hydrogen are added to control molecular weight.

(ii) *The Phillips process* involves feeding ethylene into cyclohexane solvent which contains a catalyst of partially-reduced chromium oxide on activated silica or alumina. Polymerisation occurs at 25–35 atmospheres and at up to 175°C.

(iii) *The Du Pont process* is another high pressure process (over 7000 atmospheres) which uses a catalyst of α,α'-azobisisobutyronitrile. Polymerisation occurs in solutions or emulsions, similar to the low-density polyethylene process above.

7.6.5.2 *Properties.* Both low- and high-density polyethylenes are generally insoluble in hydrocarbons and chlorinated hydrocarbons below 50–70°C. They are relatively unaffected by polar solvents, such as alcohols, phenols, esters and ketones. They will resist attack by alkalis, most concentrated acids, vegetable oils and ozone. Generally, chemical resistance is highest in linear high-density polyethylenes.

Polyethylenes are thermoplastic and melt at around 130–140°C. They burn, melt and drip when exposed to flame. They are susceptible to photo-oxidation by ultraviolet light, which limits their outdoor use, unless protective agents are used.

Polyethylenes are resistant to microbiological attack, water and gas penetration. They are utilised as thin abrasion-resistant films on textiles, predominantly on flash-spun polyolefin non-woven fabrics for use as cheap disposable materials for packaging, protective clothing and linings (e.g. Tyvek).

7.6.6 *Chlorinated and chlorosulphonated polyethylenes*

Chlorinated polyethylenes are prepared by passing gaseous chlorine through a hot suspension or emulsion of polyethylene in carbon tetrachloride, in the presence of a catalyst. Useful polymers contain 25–30% of chlorine.

Chlorosulphonated polyethylenes are prepared by a process similar to that above, except that gaseous sulphur dioxide and chlorine are introduced simultaneously into a hot emulsion or suspension of polyethylene in carbon tetrachloride.

The chlorosulphonyl groups ($—CH_2CH(SO_2Cl)—$) make up about 1–2% of the ethylene polymer, whereas the chlorinated groups ($—CH_2CH(Cl)—$) make up about 25–45% of the ethylene units. The SO_2Cl groups provide reactive sites for crosslinking, very much as in general purpose rubbers.

7.6.6.1 *Properties.* Compared with straight polyethylene, these polymers possess outstanding resistance to heat ageing, and they have a degree of flame retardance. They are highly resistant to weathering, ozone, abrasion and general chemical attack.

They are utilised where flame retardance, good weathering properties, and chemical resistance are required. Coated fabrics are used for tentage, chemical protective clothing, collapsible fuel tanks, roofing and flooring, amongst other uses.

7.7 Liquid proof/vapour permeable coated fabrics

Coated and laminated fabrics are widely used in functional clothing and covers to protect the wearer from rain, water and other liquids. When such garments or covers are worn by humans or animals, there is a physiological problem associated with the inability to allow the free passage of perspiration vapour through the clothing. The consequence of this for active wearers is progressive dampening of the underclothing. In the mildest cases this can cause discomfort, but in the worst survival situations it can result in injury, incapacitation or death by hypothermia or hyperthermia. These conditions are explained in Table 7.6.

Table 7.6 Effects of waterproof clothing in different conditions

Conditions	Activity	Consequence
Cold/wet climate	Medium activity	Discomfort
Cold/wet climate in sweat-wetted waterproof clothing	High activity followed by low activity	Hypothermia (cold stress)
Hot/moist climate (when protective clothing required)	High activity	Hyperthermia (heat stress)

The most demanding situations tend to be faced by those with professional outdoor occupations such as military personnel, police, fire services, and those involved in professional sport and expeditions.

One of the biggest growth areas in coated fabrics is that of waterproof/vapour permeable materials, commonly called waterproof/breathable fabrics.

Historically, the first such fabrics consisted of cotton fabrics impregnated with waxes and oils to provide a degree of waterproofness and vapour permeability.

The first microporous polymeric membranes were commercialised by Sartorius-Werkes in Germany in 1927. They were used in microfiltration of liquids and were based upon nitrocellulose polymers. This technology has been developed and modified over the years to produce a wide range of advanced polymeric films and coatings with high vapour permeability and high waterproofness.

The Second World War saw the urgent need to provide comfortable waterproof clothing for RAF Hurricane and Spitfire pilots escorting convoys to Northern Russia. Many pilots were lost through ditching in the sea and dying of hypothermia. A fabric was developed using the finest long staple (Sea Island) cotton, which was carefully spun and doubled before being woven into a dense parchment-feel Oxford weave fabric. The fabric was then finished with a permanent water-repellent finish based upon stearamido derivatives. Such 'Ventile' fabrics (Anon, 1993) have a high vapour permeability in air but, when immersed in water, swell to block the interstitial voids between cotton threads, thus providing a degree of waterproofness. Production started in 1943, and it was still widely used in the 1990s for military aviation, expedition clothing and tentage.

7.7.1 *Physiological aspects*

Comfort is difficult to define; in fact, it is often referred to as 'the lack of discomfort'. Comfort is highly subjective, but in clothing is directly related to the temperature and humidity conditions inside garments. The balance between a person's energy production and its exchange with the environment can be controlled by clothing. Woodcock's formula (Woodcock, 1962) shows this relationship:

$$H = \frac{T_s - T_a}{I} + \frac{p_s - p_a}{E}$$

where: H = total rate of heat transfer per unit area; T_s = skin temperature; T_a = ambient temperature; I = insulation of clothing plus overlying air layers; p_s = water vapour pressure at skin boundary; p_a = water vapour pressure of ambient air; E = resistance to moisture vapour transmission of clothing and overlying air layer.

In this formula H, T_s and p_s are physiological parameters controlled by the individual. T_a and p_a are controlled by the environment. I and E are controlled by the clothing, including air layers. The levels of perspiration produced by a human are as shown in Table 7.7.

Table 7.7 indicates that the human skin exudes a small amount of insensible perspiration even when inactive and in a cool environment. Under continuous strenuous conditions almost 1.5 l/h of sweat can be liberated. For it to cause cooling of the skin, and hence the core of the body, the sweat must leave the skin as a vapour. In impermeable clothing, evaporative cooling cannot take place and overheating results. At the cool end of the climatic scale, sweat-wetted clothing loses up to 70% of its dry thermal insulation. During periods of low activity, chilling and rapid cooling can occur, leading to hypothermia and finally death.

7.7.2 *Types of waterproof/vapour permeable fabrics*

There are four main types of 'breathable' fabric:

(a) High-density woven fabrics
(b) Microporous coatings and films
(c) Hydrophilic coatings and films
(d) Combined microporous/hydrophilic layers

For the purposes of this chapter, we shall concentrate on (b), (c) and (d).

Table 7.7 Sweating rates for humans

Activity	Sweat rate (ml/h)
Sleeping (cool, comfort zone)	15–30
Walking at 5 km/h (comfort zone)	200–500
Hard physical work (hot, humid conditions)	400–1000
Maximum sweat rate (only possible for a few minutes)	~1800

7.7.3 *Microporous coatings and films (poromerics)*

Microporous membranes are defined as having a narrow pore-size distribution, usually in the sub-micrometre range, although they can range from 0.1 to 10 µm. It is common for microporous films to have 1–2 billion pores per square centimetre in a film 10–50 µm thick. The most widely used polymers for microporous fabrics are as follows.

Polyurethanes, which offer toughness, flexibility and wide variation in formulation, allowing 'tailor-made' polymers.
PTFE, which offers an inert, extremely hydrophobic polymer with very open porous structures.

Other polymers used include:

Acrylics.
Polyamino acids.
Polyolefins

7.7.3.1 *Microporous polyurethanes.* There are several methods used to render polyurethane films and coatings microporous. The main commercial techniques are as follows.

(*a*) *Wet-cast coagulation.* A large proportion of polyurethene microfiltration membranes are produced by this technique (at the time of writing). A polyurethane polymer, usually thermoplastic, is dissolved in an organic solvent which is also water-soluble.

The resulting viscous solution is coated on to a fabric directly, or cast as a thin film on a release paper. Whilst still liquid, the film is exposed to water vapour or steam in a controlled manner. The water vapour dissolves in the film, causing phase separation of the polymer, producing a coagulated gel with microporous voids, channels and blow holes. The film is then washed and dried (see Figure 7.7).

The process requires careful control, and is relatively slow. It produces coatings or films with pores in the 0.1–10 µm range. Examples of commercial products are Cyclone®, Entrant® and Keelatex®.

(*b*) *Salt dissolution process.* In this process a polyester or polyether urethane polymer is cast as a film from an organic solvent such as dimethyl formamide (DMF). The polyurethane film contains a water-soluble finely-divided inorganic crystal substance such as sodium chloride. This is subsequently leached out of the film using water. This leaves microporous voids in the film. Examples of commercial products made by this process are Porvair Porelle® and Permair® (Threlfall and Clarke, 1989).

Microporous polyurethanes, having a high permeability to water vapour, tend to suffer from a corresponding decrease in resistance to water ingress. Waterproofness can also suffer during use when contaminants, including

Figure 7.7 Microporous polyurethane coated fabric.

sweat residues, migrate into the pores, overcoming surface tension forces. These effects can be offset by applying a surface skin, either directly or indirectly. A thin continuous solid skin must decrease the vapour permeability by a significant amount, but it is a compromise which results from seeking the optimum performance in rainwear. Some microporous products are finished with a thin hydrophilic polyurethane top coat, which in itself has a relatively high water-vapour permeability (see Figure 7.8).

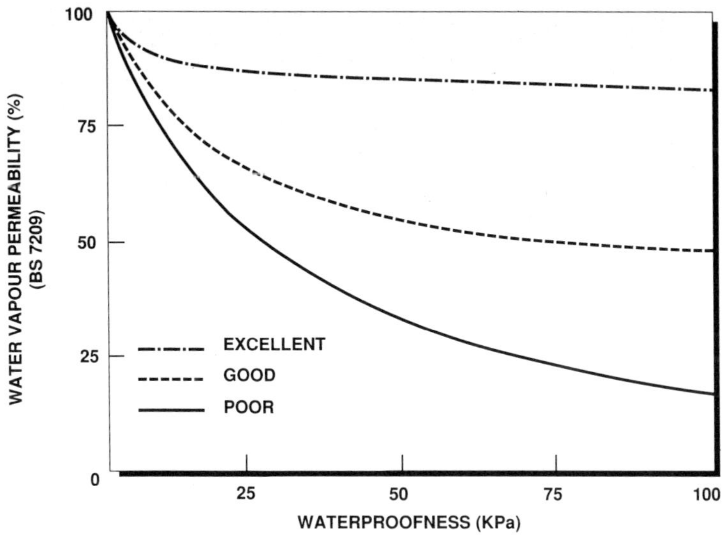

Figure 7.8 Typical curves of water-vapour permeability versus waterproofness.

7.7.3.2 *Microporous poly(tetrafluoroethylene) (PTFE).* Very thin films of PTFE (5–15 μm) can be made by extruding a paste of PTFE dispersion and oils through a slit die. The extrudate is leached to remove the oil and biaxially stretched. This causes microporous tears or fibrils in the film, which have effective pore sizes of 0.1–1 μm (Gohlke and Tanner, 1976) (Figure 7.9). The material has a very high void volume, but is mechanically weak, and has to be laminated to one or more textile fabrics to be used for rainwear. Certain PTFE films also incorporate thin continuous layers of a hydrophilic/oleophobic polymer on the side facing the sweat. This resists sweat contamination and improves the penetration resistance to low surface tension organic liquids. At the time of writing, this hybrid microporous/hydrophilic membrane is patented.

Such films are adhesively laminated using discrete dots of adhesive applied by a very fine gravure roll. The technique is highly critical, in that there must be sufficient adhesive to bond the hydrophobic 'non-stick' film to a textile fabric, but the adhesive dot coverage has to be kept low to minimise the area of blocked micropores. One of the most widely recognised PTFE laminates is Gore-Tex®.

7.7.3.3 *Ultraviolet-beam polymerised microporous films.* Traditional processes for producing microporous films and coatings tend to be relatively slow, and hence expensive. A process was developed in the late 1980s which was based upon long-standing non-porous coating technology, where monomers and

Figure 7.9 Microporous PTFE membrane.

oligomers are crosslinked under a radiation source and cured in milliseconds. In the Sunbeam® process the film is cured by an ultraviolet beam to produce controlled pores of about 0.2 μm in very thin films of acrylic-based polymers (Lovell, 1984; Gregor *et al.*, 1988). The process can run at speeds of more than 350 feet/min and in widths of up to 100 cm.

Such ultraviolet cured films are very cheap and can be laminated to non-woven substrates for disposable protective clothing. The materials possess high water-vapour permeabilities and good barrier properties. A commercial product is marketed as Gelman Repel®.

7.7.4 *Hydrophilic coatings and films*

In contrast to microporous films, the hydrophilic films are continuous, pore-free solid films. As such, they have the potential for a high resistance to ingress of liquids, provided that they are free from pin-holes and that the film has a certain minimum thickness. However, water-vapour permeability rates through hydrophilic films are inversely proportional to their thickness (Lovell, 1984; Lomax, 1990).

Diffusion of water vapour through a solid film is achieved by the incorporation of hydrophilic functional groups along the molecular chain.

Such groups can be —O—, —CO, —OH or —NH_2, all of which can form reversible hydrogen bonds with water molecules. These bonds are comparatively weak, and can readily be broken down by thermal motion. Water molecules diffusing through the film pass stepwise along the molecular chains, a process which is facilitated by introducing pendant side groups, which prevent close-packing of adjacent chains.

The passage of water molecules through solid films is governed by Fick's Laws of Diffusion (Crank and Park, 1968), such that the rate of diffusion is proportional to the concentration gradient of water vapour between the two surfaces of the film. Expressed mathematically this is:

$$\frac{\mathrm{d}s}{\mathrm{d}t} = -DA(C_1 - C_2)$$

where: $\mathrm{d}s/\mathrm{d}t$ is the diffusion rate across an area A of film; $(C_1 - C_2)$ is the concentration gradient of vapour between the two surfaces; and D is the absolute diffusion coefficient.

7.7.4.1 *Hydrophilic polyurethanes.* Of the various possible routes to hydrophilic coatings and films, segmented copolymers of polyester or polyether polyurethanes with poly(ethylene oxide) are the most widely used for breathable apparel. The hydrophilicity of these products can be varied, either by increasing the overall content of the hydrophilic component, or by altering the length of the hydrophilic segments (Lovell, 1984; Lomax, 1990).

An important feature of hydrophilic films and coatings is that there has to be a balance between swelling of the membrane, which encourages the passage of water vapour, and dissolution of the polymer in water. Constant swelling and shrinking of the polymer coating can also lead to delamination from the textile substrate. Modern formulations are designed to provide low swelling with high vapour transmission rates. An example of a commercial polymer product is Witcoflex Staycool®, which can be applied as a direct coating, or as a transfer film.

7.7.4.2 *Hydrophilic polyesters.* Another type of hydrophilic copolymer is based on a modified type of polyester into which polyether groups have been incorporated to impart hydrophilic properties with a limited degree of swelling (Drinkmann, 1992). Commerical products such as Sympatex® are available as pre-prepared films which are 10–25 μm thick (Figure 7.10), The thinner films are used for apparel, and can be laminated to textiles in a variety of ways such as two-layer or three-layer laminates, but can also be used as lightweight linings or inserts in waterproof clothing. As with hydrophilic polyurethanes, the water vapour transmission rates are inversely proportional to the film thickness (Figure 7.11).

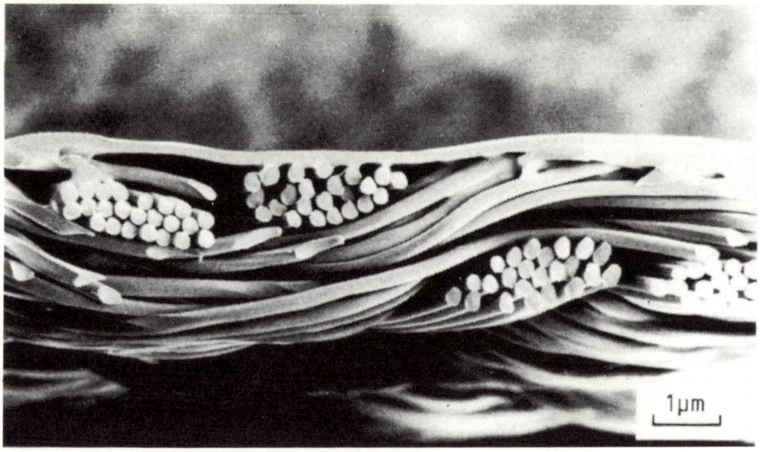

Figure 7.10 Solid hydrophilic polyester coated fabric.

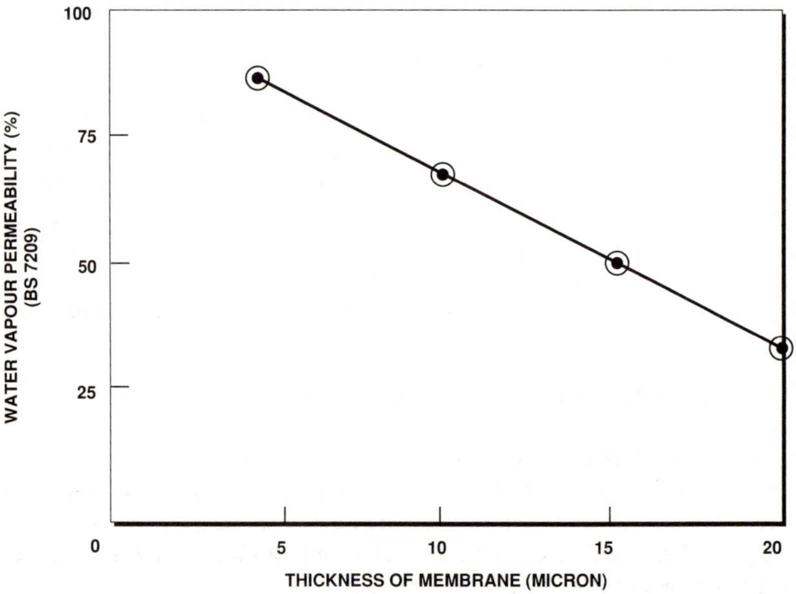

Figure 7.11 Typical graph of water-vapour permeability versus membrane thickness.

Table 7.8 Star rating of breathable barrier fabrics. One star = poor; five stars = excellent

Type of breathable barrier fabric	Water-vapour permeability	Waterproofness	Cost	Comments
PTFE laminates	*****	*****	Very high	Market leader; versatile usage; expensive
Microporous polyurethanes	** to *****	** to *****	Medium to high	Widely used; reasonable durability; coated or laminated forms
Hydrophilic polyurethanes and polyesters	** to ***	*** to *****	Low to medium	Polyurethanes very cheap; some delamination can occur in use
Conventional impermeable coatings	Not significant	** to *****	Low to medium	Waterproof but uncomfortable

7.7.5 *Relative performance of breathable barrier fabrics*

A wide range of coated and laminated products is available worldwide, and their performance improves with development (Scott, 1991). It is not possible to give detailed specific performance figures for all products in this chapter, but they can be compared in general representative groups using a star-rating (shown in Table 7.8).

7.8 Test methods for coated and laminated fabrics

The performance of coated and laminated fabrics is quantified by a large range of laboratory tests. Such test methods are widely used by purchasers, suppliers and users of such fabrics. Many public bodies, such as government departments and private sector companies and corporations, purchase materials to detailed specifications which call upon national standard tests, and quote required minimum performance figures.

Table 7.9 lists contain details of those tests which are most widely used.

7.9 End-use applications for rubber-coated fabrics

Natural rubber has poor resistance to ozone, oxidation, sunlight, elevated temperatures, hydrocarbons and micro-organisms. This limits its usefulness, as the synthetic rubbers possess better performance in specific areas.

Butyl rubbers have excellent gas- and liquid-holding properties and can operate over a wide range of temperatures.

Neoprene rubbers have good oil and solvent resistance, and possess a degree of flame retardance.

Uses are summarised in Table 7.10. Gillette (1992) also discusses end-uses.

Table 7.9 Test methods for coated and laminated fabrics

Test method title	Standard test reference	Comments
Coated fabrics for use in the manufacture of water penetration resistant clothing	BS 3546: Part 1	Specification for polyurethane and silicone elastomer fabrics
	BS 3546: Part 2	Specification for non-water vapour permeable fabrics
	BS 3546: Part 3	Specification for natural and synthetic rubber polymer fabrics
	BS 3546: Part 4	Specification for water vapour permeable fabrics
Mass per unit area	BS 3424: Part 3 ISO 2286	Includes measurements on total mass, mass of coating and mass of substrate
Breaking strength and elongation at break	BS 3424: Part 4 ISO 1421	Test methods for CRL, CRT and CRE machines
Methods for the determination of tear strength	BS 3424: Part 5	Three methods of test, and advice on analysis of autographic traces
Coating adhesion strength	BS 3424: Part 7 ISO 2411	Includes several test procedures
Low temperature performance	BS 3424: Part 8 ISO 4646	Cold crack tests — three methods listed
Resistance to damage by flexing	BS 3424: Part 9 ISO 7854	Four methods to assess dynamic flex fatigue properties
Resistance to accelerated ageing	BS 3424: Part 12 ISO 1419	Ageing in dry heat, tropical conditions and ultraviolet exposure
Resistance to blocking	BS 3424: Part 11 ISO 5978	Two methods to determine the resistance of polymeric coatings to self-adhesion
Fusion of PVC coatings and the state of cure of vulcanised rubber coatings	BS 3424: Part 22 ISO 6451	Checks for tackiness, cracking, or disintegration of the coating
Resistance to water penetration and surface wetting	BS 3424: Part 26	Hydrostatic head tests at different pressures and conditions. One method to measure surface wetting (repellency)
Method for the determination of surface wetting	BS 3702	Water repellency test
Resistance to wicking and lateral leakage	BS 3424: Part 18	Two methods for fabrics coated on one or both sides, or double textured or plied fabrics
Water entry pressure of coated fabrics	Federal Standard 191 Method 5512	Mullen Test; produces pressures up to 400 psi
Water vapour permeability index (WVPI)	BS 3424: Part 37 BS 7209	Water vapour permeability compared with that of an open polyester mesh fabric
Moisture vapour transmission rate	ASTM 96-66BW	Modified Inverted Cup Test (USA)
Water vapour resistance	CGSB CAN–4.2 (M77) Method 49	Canadian 'Turl' Dish Method; measures resistance in equivalent thickness of still air layer
Measurement of stationary thermal and water vapour resistance by means of a thermoregulatory model of human skin	DIN 54101 ISO/DIS 11092	Hohenstein skin model measures water vapour resistance

Table 7.10 Use of rubber coatings

Rubber coating	End-use applications
Natural rubber	Conveyor belts, inflatable balloons, surgical gloves, hoses, flooring, mats, groundsheets, immersion and diving suits, gaskets, aprons
Styrene–butadiene rubbers	Similar uses to natural rubber
Neoprene rubber	Waterproof clothing, oil- and fuel-resistant clothing, gloves, tents, shelters, inflatables, fuel tanks, conveyor belts, life jackets, gaskets
Butyl rubber	Gas-holding bags, chemical warfare protective clothing and shelters, pond and reservoir liners, water tanks, inflatable boats, covers, bellows, conveyor belts, air cushions, gaskets
Nitrile rubber	Oil-resistant clothing, gloves, aprons, fuel-cell liners, gaskets

Table 7.11 Uses of polymer coatings

Polymer coating	End-use applications
Polyurethanes	Lightweight waterproof clothing, covers, tents, bags, load carriage pouches, immersion clothing, inflatable boats and tanks, camouflage nets, acoustic insulation, vibration-absorbing materials
Poly(vinyl chloride)	Truck covers, tents, shelters, tarpaulins, waterproof clothing, leather-cloth upholstery, blinds, awnings, bags, shoes, belts, aprons
Silicones	Tents, covers, heating ducts, diaphragms, nonstick conveyor belts, gaskets, medical products
Polyethylene	Chemical protective clothing (disposable), disposable covers, shelters, gaskets, linings
Acrylates	Blinds, awnings, white hat covers, camouflage nets (snow)
Poly(tetrafluoro-ethylene)	Lightweight waterproof clothing, chemical protective clothing, tents, shelters, tensioned covers, conveyor belts, medical products, gaskets, seals
Chlorosulphonated polyethylene	Tents, covers, tarpaulins, shelters, inflatables, gaskets, ducting

Table 7.12 Properties of rubbers for coated fabrics

Properties	Natural rubber	Styrene–butadiene (SBR)	Butyl rubber	Nitrile rubber	Neoprene rubber
Adhesion to textiles	Excellent	Good	Good	Good	Excellent
Tear resistance (alone)	V. good	Fair	Good	Fair	Good
Abrasion resistance	Excellent	V. good	Good	Good	Excellent
Electrical insulation	V. good	V. good	V. good	Poor	Fair
Permeability to gases	Medium	Medium	V. low	Low	Low
Acid resistance					
Dilute	Medium	Medium	Excellent	Good	Excellent
Concentrated	Medium	Medium	Good	Good	Good
Solvent resistance					
Aliphatic hydrocarbons	Poor	Poor	Poor	Excellent	Medium
Aromatic hydrocarbons	Poor	Poor	Poor	Good	Fair
Ketones, ethers	Good	Good	Good	Poor	Poor
Resistance to swelling					
Mineral oils and gasoline	Poor	Poor	Poor	Excellent	Good
Animal and vegetable oils	Fair	Fair	V. good	V. good	Good
Water absorption	V. low	V. low	V. low	Low	Low
Resistance to					
Oxidation	Good	Fair	Excellent	Good	V. good
Ozone	Poor	Poor	Excellent	Fair	V. good
Sunlight ageing	Poor	Poor	V. good	Poor	V. good
Heat ageing	Fair	Medium	V. good	Good	Good
Low temperature	V. good	V. good	Good	Medium	Good
Flame	Poor	Poor	Poor	Poor	Good

7.10 End-use applications for polymer-coated fabrics

Many coating and laminating polymers have specific properties that make them suitable for specific purposes. Polyurethanes are used in lightweight waterproof fabrics which can remain flexible and durable over a wide range of operating temperatures and conditions.

PVC is a good general-purpose, low cost, heavy covering material with a degree of flame resistance.

Silicones produce very lightweight waterproof/liquid proof fabrics. They can operate at high and low temperatures, and can be used in the food and medical industries. Polyethylenes possess good chemical resistance.

7.11 Summary

Tables 7.12 and 7.13 summarise the properties of rubber and polymer coatings, respectively.

Table 7.13 Properties of polymers used for coated fabrics

Properties	Poly-urethanes	PVC	PTFE	Chloro-sulphonated polyethylene	Silicone elastomer	Poly-ethylene
Adhesion to textiles	V. good	V. good	Fair	Good	Excellent	Good
Tear resistance (alone)	Excellent	Good	Medium	Fair	Fairly low	V. good
Abrasion resistance	Excellent	Good	Good	Excellent	Fair	Good
Electrical insulation	Good	Good	Excellent	Good	Excellent	Excellent
Permeability to gases	Fairly low	Fairly low	V. low	Fairly low	Fairly low	V. low
Acid resistance						
Dilute	Fair	Excellent	Excellent	Excellent	Excellent	Excellent
Concentrated	Poor	V. good	Excellent	Good	Fair	V. good
Solvent resistance						
Aliphatic hydrocarbons	Fair	Good	Excellent	Fairly good	Poor	Fair
Aromatic hydrocarbons	Poor	Fair	Excellent	Fair	Poor	Fair
Ketones, ethers	Poor	Poor	Poor	Fair	Fair	Good
Resistance to swelling						
Mineral oils and gasoline	V. good	Poor	Excellent	Good	Fair	V. good
Animal and vegetable oils	V. good	Poor	Excellent	Good	V. good	V. good
Water absorption	Low	Low	V. low	Low	V. low	V. low
Resistance to						
Oxidation	Excellent	Fair	Outstanding	Excellent	Excellent	Poor
Ozone	Excellent	Fair	Outstanding	Outstanding	Excellent	Good
Sunlight ageing	V. Good	Poor	Outstanding	Outstanding	Excellent	Fair
Heat ageing	Fairly good	Fair	Outstanding	V. good	Outstanding	Medium
Low temperature	Excellent	Poor	Fairly good	Good	Outstanding	V. good
Flame	Fairly poor	V. good	Excellent	V. good	Fair	Poor

References

Anon (1993) Ventile, a natural choice. *Manufacturing Clothier*, January 1993, 18–20.

Crank, J. and Park, G.S. (1968) *Diffusion in Polymers*, Academic Press, New York.

Drinkmann, M. (1992) Structure and processing of Sympatex® laminates. *Journal of Coated Fabrics*, **21**, January, 199–211.

Gillette, S.M. (1992) End-use application of coated fabrics. *Journal of Coated Fabrics*, **22**, July, 75–79.

Gohlke, D.J. and Tanner, J.C. (1976). Gore-Tex waterproof breathable laminates. *Journal of Coated Fabrics*, **6**, July, 28–38.

Goldsmith, J. (1988) Water-based polyurethanes. *Journal of Coated Fabrics*, **18**, July, 12–25.

Gregor, E.C., Tanny, G.B., Shchori, E. and Kenigsberg, Y. (1988) Sunbeam® process for microporous membranes; A high performance barrier for protective clothing. *Journal of Coated Fabrics*, **18**, July, 26–37

Lomax, G.R. (1990) Hydrophilic polyurethane coatings. *Journal of Coated Fabrics*, **20**, October, 88–107.

Lovell, K.V. (1984) Studies of novel polymers and composites. Royal Military College of Science Report on CDE Agreement No. IE 1139/11, pp. 5–8.

Matthews, T. (1993) Polymer coatings for textiles. *Technical Textiles International*, June, 14–20.

Morrison, R.T. and Boyd, R.N. (1987) *Organic Chemistry*, 5th edn, Allyn & Bacon, Boston, p. 1255.

Roff, W.J. and Scott, J.R. (1971a) *Fibres, Films, Plastics and Rubbers*, Butterworths, London, pp. 313–435.

Roff, W.J. and Scott, J.R. (1971b) *Fibres, Films, Plastics and Rubbers*, Butterworths, London, pp. 446–457.

Scott, R.A. (1991) Waterproof/water vapour permeable textile materials. Unpublished MOD Seminar. Ministry of Defence Stores and Clothing Research and Development Establishment, Colchester, September.

Smith, J.C. (1971) Spooner–Edmeston pilot scale coating plant. *Shirley Institute Bulletin*, **44**(3), 74–78.

Threlfall, P. and Clarke, J. (1989) Research into water vapour permeable/waterproof microporous laminates for use in military clothing. Porvair Report on MOD Contract No. HW11B/1115, pp. 2–4.

Woodcock, A.H. (1962) Moisture vapour transfer in textile systems. *Textile Research Journal*, **32**, 628–633.

8 Scouring, enzymes and softeners

J. ELLIS

8.1 Introduction

Scouring is probably the single most important process in the wet processing of textile materials. Effective scouring is essential for the subsequent processing of any textile substrate, regardless of type. In early textile processing, soap, often in combination with soda (sodium carbonate) was the universal agent for cleaning textile substrates. In many cases its use persists to this day, particularly for the processing of natural fibres where its gentle action is of some benefit. However, the majority of cleaning now employs manufactured detergents, specially formulated for particular purposes and applications.

In general, the same principles apply in all scouring operations, that is the effective removal of oils, waxes, fats, sizes, etc., prior to further wet processing such as dyeing, bleaching, resin application, softening or shrink-resist processing. If such materials are not properly removed from the substrate, this will inevitably lead to problems such as patchy, uneven dyeing or bleaching, or failure of shrink-resistance, for example. Scouring may in some instances be the only wet process applied, when its use is to prepare the article to make it acceptable to the consumer.

It is not unusual for a material to be scoured more than once during its progression from raw fibre to the final made-up article, as many finishes must be applied for a particular process, and then removed again once their job is done. This is especially true of the natural fibres, and of wool in particular. For example, wool destined for woollen spinning will be scoured in its raw state to remove the natural oils and waxes present in the fleece as well as general soil. The scoured loose wool will be blended and carded, and a spinning lubricant applied to aid the spinning process and the cohesion of the yarn. Typically (in the UK), around 5–8% oil will be applied, although this can vary greatly depending on the quality of the wool being processed. Once the yarn is knitted, the oil must be readily removable for any subsequent processing of the article.

There are two means by which textile substrates may be scoured, namely in wet (i.e. aqueous) processes or in solvent processes ('dry cleaning'). Both should achieve the same end result, that is, a clean article prepared for subsequent processing or sale, but the choice of which process may depend on factors such as:

Cost — dry cleaning is generally more expensive than aqueous scouring;
Ease of removal — in many cases, dry cleaning is only used as a last resort on particularly difficult jobs;
Subsequent processing requirements — there are some processes which are carried out entirely in solvent;
Machinery availability — the machinery requirements for aqueous scouring and dry cleaning are specific and are not interchangeable.

There are additionally some aqueous systems which incorporate a solvent, sometimes even a dry-cleaning solvent such as perchloroethylene, but these are essentially aqueous scouring agents where the solvent has been added to boost the effect.

8.2 Aqueous scouring processes

As has already been mentioned, soap is the traditional material used in the aqueous scouring of textile substrates; before the introduction of synthetic detergents, its use was universal. Indeed, soap has been known for more than 2000 years. However, the major drawback with soap as a scouring agent is the tendency to form insoluble calcium salts or 'scum' in hard water. Synthetic detergents were first developed to overcome this particular problem.

Since soap is manufactured by the reaction of a fatty acid with an alkali-metal base, it could be regarded as 'synthetic'. However, it is customarily accepted that soap is not classified with modern synthetic detergents, which were first introduced in the late nineteenth century.

In a sense, the chemistry of both soap and synthetic detergent molecules is very similar, and certainly the basic principles underlying their action are the same. Synthetic detergents are, however, much more effective in their action, and their chemical structures can be substantially altered to achieve a variety of effects in a way that is not possible with simple soap molecules. Some understanding of basic surfactant science is necessary to appreciate why scouring agents are effective in the removal of oils and other materials from surfaces. In this discussion, no distinction is made between soap and synthetic detergents, since other than their chemistry, their modes of action are considered to be similar.

Surfactants — or surface-active agents — are materials that act at the interface between two (or more) immiscible substances. In the widest sense, the immiscible substances may be solids, liquids or gases, in any combination. In textile processing, of course, there is most often a bicomponent immiscible system, being, in the large majority of cases, water and an oily material. However, liquid–solid and liquid–gas interfaces are also important in textile processing, for example in the wetting-out or de-aeration of materials. The role of a surfactant can most simply be described as overcoming such natural

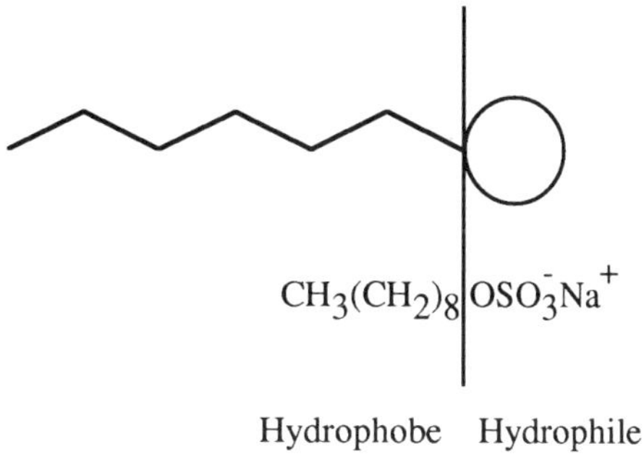

Figure 8.1 Schematic representation of the structure of a surfactant molecule.

immiscibility and allowing the two substances to come into close, intimate contact. In order to achieve this, all surfactants have a characteristic type of structure, although the exact structures employed depend upon the requirements of a particular application. Since we are considering aqueous processes, it can be assumed that the surfactant molecule will, in general terms, have one portion which is hydrophilic and one which is hydrophobic. For use as a scouring agent, the structure will be designed such that the water molecule has a high water solubility.

8.2.1 *The surfactant molecule*

The characteristic feature of all (aqueous) surface-active agents is the possession of a hydrophobic 'chain' and a hydrophilic 'head' (Figure 8.1). This is true of the simplest soap molecule right through to the most complex synthetic detergents. In the majority of cases, the hydrophobic portion is a long-chain hydrocarbon, while the hydrophilic portion is a highly polar or charged (ionic) group that will confer water solubility. When one considers the potential pairings that might be possible in designing such a molecule, the number of possibilities is almost limitless, and indeed the number of commercially available surfactants is very large. However, for practical purposes, there are limits on the structures which are useful; for example, the solubility of the molecule in the desired medium must be considered, and this may place a limit on the length of the hydrocarbon chain.

It is most convenient to consider the two distinct components of the surfactant molecule separately, namely the hydrophile (or lipophobe) and the hydrophobe (or lipophile). Surfactants are customarily classified on the basis of the nature of the hydrophobe, this classification relating to the ionic charge carried by the hydrophobe. There are only four classes of surfactants, using this classification scheme, that is anionic, cationic, nonionic and amphoteric. Notice that the term 'anionic' (and similarly 'cationic') refers to the charge carried by the hydrophobe on the surfactant molecule itself, not to the (non-surface-active) counterion. There are, of course, both anionic and cationic components in any charged, electrically balanced molecule.

The hydrophobe portion can vary widely, and though most are based on linear hydrocarbon chains, the actual number of structures possible is immense.

8.2.1.1 *Anionic surfactants.* Soap can be considered as an anionic surfactant, but there are many examples of synthetic anionic surfactants. Indeed, the first synthetic surfactant—Turkey Red Oil, a sulphonated castor oil—was an anionic material. The characteristic feature is the possession of a group carrying a negative charge, the most typical examples being carboxyl ($RCOO^-M^+$), sulphate ($RSO_3{}^-M^+$) and sulphonate ($ROSO_3{}^-M^+$).

8.2.1.2 *Cationic surfactants.* The hydrophile carries a positive charge, the most common examples being quaternary ammonium compounds ($R_4N^+Cl^-$). Cationic surfactants are rarely used in textile scouring. They have a pronounced germicidal/batericidal activity which is particularly important in personal-care and hard-surface cleaning products.

8.2.1.3 *Nonionic surfactants.* As their name implies, these carry no electrical charge, and water solubility is conferred by the presence of highly polar groups. By far the large majority of nonionic surfactants contain a polyoxyethylene chain —$(OCH_2CH_2)_n$— of varying length. An increase in the value of n tends to increase the water solubility, but solubility is also influenced by the nature of the hydrophobe group.

8.2.1.4 *Amphoteric surfactants.* These contain both negative and positive charges within the same molecule, although the particular ionic charge under given conditions will be influenced by solution pH. Thus, under alkaline conditions the molecules act as anionic species, while under acidic conditions they behave as cationic species. Between these extremes of pH lies the isoelectric point, where the ionic charges are balanced and the molecule carries no charge. The most simple examples of amphoterics are the alkylamino acids ($RN^+H_2CH_2COO^-$). Other common examples include the betaines, trisubstituted alkylamino acids:

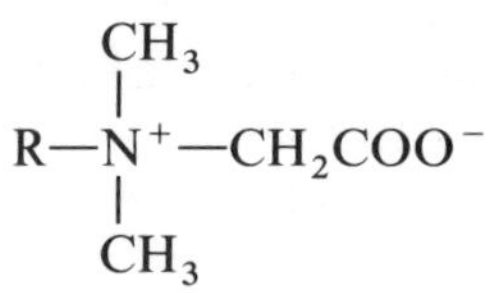

8.2.1.5 *Hydrophobes.* The range available is very large, but is substantially limited to the straight-chain hydrocarbons of chain length from eight to eighteen carbon atoms in the majority of cases. The chain may be saturated and unsaturated, and of natural (animal or vegetable fats and oils) or synthetic (crude oil) origin. The naturally occurring oils and fats contain carbon chains with an even number of carbon atoms only; chains with odd numbers of carbon atoms must be manufactured.

Other more or less common hydrophobes include:

- Branched-chain alkyl groups
- Alkylphenols of various configurations, which includes the important nonylphenol derivatives
- Polyoxypropylene derivatives
- Polydimethyl siloxanes
- Fluorinated derivatives, generally partially or fully fluorinated alkyls

8.2.2 *The nature of surface activity*

Surface activity is a highly complex field of study, and is of great significance in a surprising variety of aspects of our daily lives. However, we can confine ourselves to a very brief description of the most basic principles of surface activity, and its importance in understanding wetting and detergency. For those requiring a more detailed description of surfactants, there are a number of excellent reference books available, such as Myers (1992).

Surface activity is concerned mainly with the behaviour of materials at interfaces, that is, at the boundary between two or more immiscible phases. It should be clearly understood that the bulk properties of each individual material may be quite unaffected by such surface effects, but that the properties of a system as a whole (i.e. of the two or more immiscible materials) are governed entirely by the behaviour at such an interface. The principal feature of a surfactant is the ability to modify the behaviour of materials at such interfaces. The most obvious example of this is of course the use of a detergent to emulsify water and oil. The surfactant acts as an agent for overcoming the natural tendency for the boundary to be minimised; in aqueous systems this is governed by the relatively high surface tension of water. The net result of emulsification is a vast increase in the mutual surface area of the two phases as the oil is dispersed as small droplets of water (or *vice versa*). A surfactant molecule is able to achieve this because it contains within its structure an oil-soluble portion and a water-soluble portion. As such, it can act as a bridge at the interface of two phases, where the hydrophilic end of the surfactant is oriented towards the water while the hydrophobic end is oriented towards the oil phase. It is thus able to bring two immiscible phases into intimate contact.

In an aqueous system this orientation of the surfactant molecule between the aqueous and oil phases is controlled by the balance between the hydrophobic and hydrophilic components. Thus, the balance determines the characteristics of a surfactant, and hence its usefulness in a particular situation.

Although there are a number of schemes for relating a surfactant's properties to its physical structure, the most commonly used, and perhaps the most useful, is the hydrophile–lipophile balance (HLB) first proposed by Griffin (1949). Although they are of an empirical nature, in many ways HLB values can be used quantitatively for determining the characteristics of a surfactant.

Within this classification, surfactants are assigned an HLB value between zero and twenty or higher. Low values indicate surfactants with high lipophilicity (i.e. oil solubility), while a high HLB denotes a high water solubility. The HLB classification is especially useful for the nonionic ethoxylate surfactants, where the degree of ethoxylation readily identifies a surfactant as being of low, medium or high HLB. Thus, low degrees of ethoxylation result in low water solubility/high oil solubility and hence a low HLB. Increasing the degree of ethoxylation increases the water solubility (and subsequently decreases oil solubility), and hence the HLB increases. Table 8.1 lists typical applications related to HLB range for nonionic surfactants, while Table 8.2 gives the approximate behaviour in water of surfactants of varying HLB. From this behaviour, an estimation of HLB can be gauged for an unknown surfactant.

It is possible to calculate the HLB of a pure surfactant of unknown HLB. For example, the HLB of an ethoxylated nonionic surfactant is calculated from the formula:

$$\text{HLB} = \frac{\%\ \text{by mass of the hydrophilic group}}{5}$$

which gives a value in the range 0–20.

Table 8.1 Approximate HLB values *versus* surfactant characteristics

HLB range	Application
3–6	W/o emulsifier
7–9	Wetting agent
8–18	O/w emulsifier
13–15	Detergent
15–18	Solubiliser/disperser

Table 8.2 Approximate HLB from water solubility

Behaviour in water	Approx. HLB
No dispersibility	1–4
Poor dispersion	3–6
Unstable milky dispersion	6–8
Stable milky dispersion	8–10
Translucent/semi-clear dispersion	10–13
Clear solution	13+

For fatty esters derived from polyhydric alcohols, e.g. glycerol monostearate, the following relationship applies:

$$\text{HLB} = 20[1 - (S/A)]$$

where S is the saponification number of the ester and A is the acid number of the fatty acid.

Table 8.3 gives the HLB values of typical nonionic surfactants.

The most valuable use of HLB values is in determining the effectiveness of a surfactant to stabilise an emulsion of either oil-in-water (o/w) or water-in-oil (w/o). In order to make a stable emulsion it is necessary to match carefully the HLB of the surfactant to that of the oil, bearing in mind the requirements of the application. However, when choosing surfactants for a particular situation, it is not always necessary to match the HLB of the oil with that of a single surfactant, and indeed in many instances it is preferable and more effective to choose a blend of two or more surfactants. The HLB of a mixed surfactant system is simply the algebraic mean of the individual HLBs. For example:

$$\text{HLB}_{\text{mix}} = F_a \times (\text{HLB})_a + (1 - F_a) \times \text{HLB}_b$$

where F_a is the weight fraction of component a in the blend and HLB_a, HLB_b are the HLB values of components a and b respectively.

It is of course possible to use such a relationship to calculate a precise blend given the individual HLBs of the components and the desired HLB of the blend.

When designing an emulsification system for an oil of unknown HLB, the most convenient method is to take two surfactants a and b of widely differing

Table 8.3 HLB values for typical nonionic surfactants

Surfactant	HLB
Cetyl alcohol, 10-ethoxylate	12.9
Cetyl alcohol, 20-ethoxylate	15.7
Tridecyl alcohol, 6-ethoxylate	11.4
Tridecyl alcohol, 12-ethoxylate	14.5
Tridecyl alcohol, 15-ethoxylate	15.4
Nonyl phenol, 1-ethoxylate	4.5
Nonyl phenol, 6-ethoxylate	10.9
Nonyl phenol, 9-ethoxylate	12.8
Nonyl phenol, 12-ethoxylate	13.9
Sorbitan trioleate	2.3
Sorbitan mono-oleate	4.6
Glyceryl monostearate	4.4
PEG 200 monostearate	7.9
PEG 300 monostearate	10.2
PEG 400 monostearate	11.5
PEG 400 mono-oleate	11.3
PEG 400 monolaurate	13.1

HLB values and combine them in such proportions as to give a range of HLB values, using the calculation above. The stabilities of the emulsions produced from these various surfactant blends are compared visually, and the most stable indicates the effective HLB for that particular oil. Further tests can then be carried out to optimise the HLB around this value, perhaps using blends of differing surfactants to achieve the same HLB value.

8.2.3 *Critical micelle concentration (CMC)*

In aqueous solution, all surfactants tend to agglomerate in small particles. At low concentrations, surfactants will tend to orient themselves at interfaces, resulting in effects such as the lowering of surface tension. As the surfactant concentration is increased, these interfaces become saturated, and the surfactant molecules tend to form discrete clusters, called micelles, in the bulk of the solution. This concentration is called the critical micelle concentration or CMC (see Figure 8.2). Above this concentration there is no further reduction in surface tension and other properties of the surfactant solution, such as wetting power or foaming, are unchanged. However, as we shall see below, the formation of such micelles is crucial to the usefulness of surfactants as detergents in scouring operations.

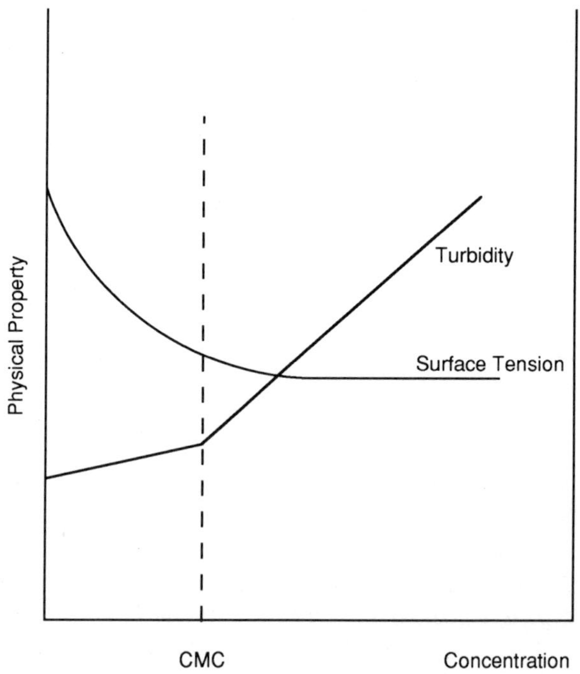

Figure 8.2 Critical micelle concentration.

The CMC of most surfactants is low, of the order of 0.2–0.5 g/l, the exact value depending upon the chemical structure of the surfactant. As a general rule, it can be said that the CMC increases with hydrophobe chain length and that nonionic surfactants have a lower CMC than ionic surfactants. The CMC is also very sensitive to temperature, with many ionic surfactants showing a minimum around 60–70°C. This is believed to be related to the ordering of water (hydration) around the ionic head group. Nonionic and amphoteric materials do not show such predictable CMC minima, but many polyoxyethylene nonionic surfactants do exhibit a cloud point.

8.2.3.1 *Cloud point.* Nonionic polyoxyethylene surfactants show an inverse temperature–solubility relationship, that is as the temperature of a solution is increased, the surfactant may become less soluble in water. The cloud point is the temperature at which a surfactant (measured as a 1% solution) becomes cloudy as a result of a decrease in water solubility. At this temperature it may no longer perform as a surfactant in some or all respects. For a given hydrophobe, cloud point increases with degree of ethoxylation.

8.2.4 *Scouring of textile substrates*

As has been said, the purpose of scouring is to remove all extraneous matter from the textile substrate, either as a preparation to further processing or as a final treatment prior to sale of the goods. The scouring agent must perform a number of functions, including wetting out of the goods, emulsification and suspension or solubilisation of the removed material. As has been indicated, the precise action of a scouring agent will depend to a large extent on its HLB, and how it is matched to that of the material to be removed. There are also other considerations, such as the type of soil and the nature of the textile fibre. In addition, the use of a 'builder' or other auxiliaries such as sequestrants may be necessary, depending upon the conditions. Still further consideration must be given to the processing machine type, especially where there is a need for low or zero-foaming products, such as in jet machines with a high-shear action.

It was indicated in Table 8.1 that the characteristic performance of a surfactant is defined by its HLB value, to an extent. For example, surfactants which show good wetting properties are likely to be poor detergents and *vice versa.* In order to create a balanced product which is able to wet the goods effectively and remove and suspend soil, a blend of surfactants is the ideal. Commercially available scouring agents range from simple single-component systems to complex blends able to fulfil many functions more effectively than the simple systems. Where there are particular constraints on the scouring systems, these can usually be accommodated by careful formulation, perhaps by use of suitable surfactants, or by the use of additives in the blend. The free rinsing of surfactants after the scouring operation is important in many situations, where the presence of surfactant residues may interfere with subsequent processing. The printing of fabrics is a good example where surfactant residues lead to a loss of definition at the colour edges of designs, due to bleeding from one area to another. Low foaming is of vital importance in jet machines, where excessive foaming can impede or even stop the free movement of the fabric around the machine.

The use of co-agents with the scouring agent may be necessary. Certainly the use of a 'builder' improves detergency significantly, and the most widely used builders are alkalis. In industrial processing this is most likely to be sodium bicarbonate, carbonate or hydroxide. Although alkaline phosphates are also used, their chief application is in the domestic product laundry market. Alkalis improve scouring operations by saponifying oils, waxes and fats, in essence making 'soaps' from these materials, increasing their water solubilities and hence their removability from the textile substrate.

Sequestering agents may be employed in areas where hard water is a problem. Where alkali is added to form an oil soap, the presence of calcium or magnesium may lead to scum formation. Sequestrants act by forming soluble complexes with metal ions, preventing their reaction with other species. They are not formulated into the surfactant system, but are added as separate auxiliaries to the scouring bath. Some of the materials used as sequestrants may have other properties useful in a scouring operation; for example, polycarboxylic acids have high dispersing powers, and help to prevent the redeposition of solid materials onto the substrate.

Solvents are sometimes added to enhance the scourability of particularly difficult, greasy substrates. Typical solvents are perchloroethylene, xylene, white spirit, terpenes and pine oils. Most formulators tend now towards the less hazardous solvents, and the use of chlorinated and aromatic solvents is much reduced.

Oily materials are removed from textile surfaces by two mechanisms, these being emulsification, which is an entirely surface (interfacial) phenomenon, and solubilisation by micelles. In addition, we should first briefly consider the removal of solid soils. In order to remove a solid soil from the surface, it is necessary first to wet the surface and by so doing reduce the affinity (interfacial forces) of the soil for the surface. Often little more than wetting and mechanical agitation are required to remove solid soils. The surfactant must then be able to disperse the solid particles effectively and prevent their redeposition. In many cases, solid particles are attached to the substrate not by mechanical or electrostatic forces, but by an oily deposit, and this first must be emulsified, freeing the solid soil particles from the surface.

Oily materials may be removed by emulsification with a surfactant of a suitable HLB value. While it is true that emulsification is the probable first step in most such situations, in fact the surfactants used as detergents are not very effective emulsifiers, and the emulsions so formed are likely to be unstable and may coalesce and redeposit on the substrate. The solubilisation of oily materials is perhaps of more significance in textile scouring. It is found that oil removal is most effective only at surfactant concentrations significantly above the CMC, particularly for nonionics. At concentrations above the CMC the surfactant molecules form structures (micelles) of varying complexity and configuration, depending on the structure of the surfactant, its concentration and the bath temperature. Fairly simple spherical structures might be formed at relatively low concentrations (above the CMC), while at concentrations significantly above the CMC, large, complex structures are formed with the capacity to solubilise a large amount of oil. In typical scouring operations, however, the surfactant concentration is not generally significantly greater than the CMC, and hence large amounts of oil cannot be solubilised; emulsification may play a significant role in such situations. Another consideration is the temperature of the scouring bath, particularly when using nonionics. The efficacy of nonionics increases with temperature and is at its maximum when the cloud point is approached.

8.2.5 *Choosing the surfactant for a fibre type*

The complexity of processes carried out in various substrate preparations makes different demands on the surfactant. Also the relative cleanliness of the substrate has an impact. Raw wool, for example, may contain 20–60% of extraneous matter, while the synthetic yarns, by the nature of their manufacture, contain only small and relatively consistent amounts of oil or waxes. Cotton scouring agents must be able to withstand strongly alkaline conditions during processing. Obviously no single scouring agent is able to cope with all these situations. To follow is a brief summary of the requirements of different fibre types and the typical scouring agents which might be employed.

8.2.5.1 *Wool.* Generally nonionics are used, largely replacing the traditional soap-based methods. Nonionics have the advantage that they are highly effective in neutral solutions, resulting in less damage to the wool. They are stable to hard water, and of relatively low cost. Scouring consists of a series of bowls containing surfactant and alkali; the concentration and temperature decreasing from the first bowl, 0.8% surfactant and 0.3% sodium carbonate at 50°C, to just water at 40°C.

8.2.5.2 *Cotton.* Like wool, cotton contains a significant proportion of extraneous matter, up to 12%, composed of waxes, protein, pectins, hemicellulose and ash. Cotton is, however, remarkably stable to strongly alkaline conditions, such treatments being an essential part of the preparation processes. Scouring treatments vary according to the equipment used but follow the general regime of 2–4% sodium hydroxide, 0.2% surfactant, 100–140°C and 2–360 min. Surfactants must be stable to such conditions, and fatty alkyl sulphates or sulphonates are generally employed, often formulated to give both wetting and detergent characteristics. Sulphosuccinates, although excellent wetters, are avoided, since they readily hydrolyse under strongly alkaline conditions. The hydrophilic nature of a cellulosic surface also means that polar or hydrogen-bonding interactions with nonionic surfactants may result in an orientation of the surfactant on the fibre surface, thereby decreasing the soil-removing capability; in order to be effective a surfactant must orient itself with its hydrophobic tail in the oil phase and its hydrophilic head in the aqueous phase.

8.2.5.3 *Synthetic fibres.* Nylon and polyester are typically scoured with an alkyl ether sulphate-based products, although nonionics may be equally effective . Where a fibre is to be subsequently dyed with disperse dyestuffs, it is essential to scour with anionic surfactants, since nonionics, if carried over into the dyebath, may affect the stability of the dye dispersion. Acrylic fibres must be scoured with nonionic surfactants, and never with anionics, since any carry-over of anionic surfactants will affect the uptake of basic dyestuffs used on this fibre.

8.3 Enzymes

8.3.1 *Introduction*

There is a common misconception about the nature of enzymes, and this is perhaps reinforced by some of the terminology used when describing their action. Enzymes, despite their derivation from living materials, are not living organisms themselves. They are protein structures, of varying complexity, based on chains of amino acids linked by peptide linkages. There are a number of factors that may distinguish one enzyme from another, and these include the particular amino acids present, the order in which they are linked together, and the presence or absence of metal ions. Most important, however, is the conformation or shape of the protein structure. All enzymes are water-soluble; the amino acid chains are coiled in such a way that the hydrophobic portions are oriented inwards, allowing maximum hydrogen bonding intramolecularly and with water or other molecules. Indeed, water is an essential medium for the proper functioning of all enzymes. Enzymes are enormously effective catalysts which permit reactions, that under normal conditions would simply not occur, to proceed and at a greatly accelerated rate. They are true catalysts that are not consumed in the reaction, but provide a low-energy path for that reaction to proceed, and can then catalyse further reactions. Enzymes are highly specific and will catalyse only one reaction, without having any effect on the other components of a system.

Enzymes are present in all living organisms, where they serve to promote the reactions which are essential for life. They catalyse only very simple reactions, which proceed in a stepwise fashion towards the final objective; no single enzyme is able to catalyse highly complex, multistage reactions. In this sense they can be likened to computers, which carry out complex mathematical operations by breaking them down to their most simple operations, and then carry these out at lightning speed. Thus, an enzyme will catalyse only a particular reaction, and the reaction path from A to B may involve a number of enzymes, each catalysing a particular step in the sequence.

The working environment of an enzyme is very important to its functioning. Most have very specific requirements for temperature and pH, for example; all require water. This is probably where analogies to living organisms creep in. Enzymes, for example, are said to be 'killed off' by extremes of heat or acidity/alkalinity, in the way that bacteria might be killed. However, this is simply not the case, and the enzyme is often only deactivated when encountering an environment outside of its normal working range. When conditions are brought back within this range, enzyme activity may well recommence. There are means of permanently disabling enzymes, but these involve destroying the structure itself, for example by removing the central metal ion from certain types of enzyme molecules.

Since enzymes are present in all living organisms, there is, in theory, a plentiful supply. However, the economics of extraction may be prohibitive, and nowadays enzymes are specially produced from bacterial or fungal species in large fermentation vessels, from which they are subsequently extracted. These micro-organisms produce their own enzymes, for example cellulases to enable them to live on cellulosic materials such as wood. In the past many enzymes were extracted from animal pancreases, a rich store of enzymes, but now too expensive. Another useful source is plant material, for example malted barley.

Enzymes can be classified according to their action, the important classes of which are listed below:

- Lipases, which hydrolyse fats and oils into alcohol and organic acids
- Proteases, which split protein molecules, and in the extreme may break the protein down into the component amino acids
- Amylases, which convert starch to shorter chain sugars
- Cellulases, which break down cellulosic materials
- Pectases, which hydrolyse pectins.

It should be remembered that enzymes are highly specific in their action and that, for example, there is not one amylase enzyme, but a variety each of which has a different part to play in the conversion of amylose to simpler compounds. For example, malt extracts contain at least two enzymes, α-amylase and β-amylase, which each have different functions.

An enzyme is very rarely, if ever, a pure material, and the classification given to an enzyme is often of necessity simply no more than the activity ascribed for a particular application. Bacteria and fungi must produce a whole range of enzymes, for example cellulases to break down cellulose to glucose, each specific cellulase differing in its mode of action. An extracted 'cellulase enzyme' may contain from a few percent to eighty or more percent of various cellulases, and will undoubtedly have many other side-activities, for example, amylase activity. However, it is common practice to describe an enzyme preparation by the activity for which it has been assayed, even if it might equally be described as an amylase in different situations and for other applications.

8.3.2 *Enzymes in textile processing*

Although enzymes have been employed in textile processing for many years, there is currently renewed interest in the possible benefits offered by enzyme processing, and not just for the traditional desizing of cotton. Enzymes are becoming a relatively cost-effective, pollution-free means of treating cellulosic fabrics for dehairing, pill-preventing and handle modification (collectively, the so-called 'Biopolishing' technology, a trade mark of Novo Nordisk), as well as the well-known stonewashing applications. For wool fabrics, somewhat different enzymes can be used to dehair processed fabrics, and even to render wool shrinkproof. For the consumer, there are the 'biological' laundry powders and liquids which rely, in part, on the action of enzymes for their effectiveness.

The applications of enzymes in textile processing can now be considered in more detail.

8.3.3 *Desizing of cotton*

Desizing is an essential preliminary stage in the processing of cotton fabrics. The size itself may be natural (starch based) or synthetic (commonly poly(vinyl alcohol), PVA) in origin, and may be present with other materials such as paraffin waxes on the substrate. Waxes and synthetic water-soluble sizes are relatively easily removed from the fabric by scouring in warm water, and indeed would not respond to enzyme treatment alone.

Starch-based sizes are most frequently removed by enzymes, which have the advantage of not causing any physical chemical damage to the cotton. Although starch and cellulose are both composed of long chains of glucose units linked together, they are sufficiently different in the way that these units are linked that amylases have no effect on cellulose and cellulases no effect on starch.

Starch is a mixture of two similar carbohydrates of differing molecular size and shape, amylose and amylopectin. Amylose comprises about 20% of starch and is water-soluble; amylopectin, a branched, high molecular weight molecule which is largely responsible for the thickness of starch pastes, is insoluble. By mechanical action or treatment with hot water, starch can be made to swell in water to form gel. Thus, in order to remove starch effectively from a fabric, it must first be rendered completely water-soluble. This may be accomplished by the action of bacterial amylases, which hydrolyse the starch molecules stepwise to dextrin (a polysaccharide), maltose (a disaccharide) and finally to the monosaccharide glucose, the simplest unit in the starch molecule. The progress of size removal can most conveniently be followed by iodine coloration; the unhydrolysed starch gives an intense purple–blue colour, while glucose gives no colour.

8.3.4 *Biopolishing and stonewashing of cotton*

Originally devised as a means of introducing a clean, smooth surface appearance and soft handle to woven cotton fabrics, biopolishing is now a widely promoted technology in all areas of cellulosic fabrics. The process is essentially a degradative action, resulting in a loss of both weight and strength of the fabric, although these losses are only moderate, and are kept under control by proper operation of the processes. Stonewashing is a specific, more severe form of cellulase treatment that is restricted to vat-dyed fabrics, particularly denim. The general principles underlying each treatment are essentially similar, however, and they are considered together.

Cotton is composed almost entirely of pure cellulose, an insoluble linear chain made up of linked glucose units, of at least 1500 units length. It differs from starch in the way that these units are linked together. The cellulose chains also show hydrogen bonding between the chains to form fibres. Cotton forms both highly ordered, crystalline regions and amorphous regions.

Most commercial celluloses are derived from the fungal *Trichoderma* and *Penicillium* species. They may be produced both in powdered forms (often adsorbed on a support medium) and as concentrated liquids, typically 25% active in brine, for stability. It has already been said that there are many highly catalytic actions within a 'single' enzyme preparation. For cellulases, there are four main actions, each of which may be represented by a number of individual cellulases:

- Endocellulases cleave the long cellulose chains randomly to produce two shorter chains
- Exocellulases attack one end of the cellulose chain, cleaving one glucose unit at a time
- Cellobiohydrolases similarly act on one end of the cellulose chain, producing a cellobiose molecule
- Cellobioases degrade cellobiose molecules to the simplest glucose unit.

These actions are illustrated in Figure 8.3.

Because of the highly specific nature of enzyme activity, the effectiveness of enzyme preparations may vary widely, and it is essential to carry out proper dosage trials with any product. It should always be remembered that 'cellulase' is a collective name for a whole host of activities within a single

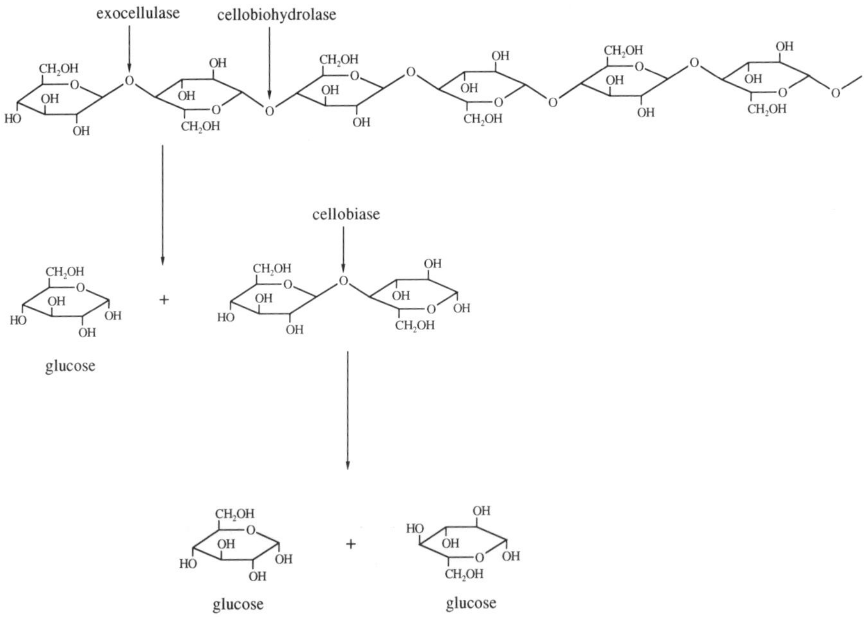

Figure 8.3 The actions of cellulase enzymes on a cellulose molecule.

preparation. Claims of total cellulase activity should always be closely scrutinised since the activity may relate to degradation of carboxymethylcellulose, a water-soluble derivative, quite dissimilar to cotton. To be meaningful for cotton processing, the enzyme should be assayed for activity on a substrate which more closely resembles cotton, such as filter paper or avicel, or preferably cotton itself.

The mechanisms of cellulase degradation of cotton are still not fully elucidated, but it appears that the first step involves attack by an endocellulase, which attaches itself to the cellulose surface, forming a complex. After cleaving a glucose unit the enzyme molecule may detach itself and move to another site, or may remain to degrade the same site further. Whether it moves to a fresh site or not may be governed by the nature of the cellulase, and even the fibre structure. The cellulase molecule itself has been found to be comprised of three identifiable regions: a *binding domain* which facilitates attachment to the cellulose molecule, a *core* which is the catalytic centre and a *linker*, which bridges these two regions.

8.3.4.1 *Operating conditions.* As with all enzymes, cellulases exhibit optimum properties within fairly narrow limits of pH and temperature. With regards to pH, there are two classes, the ‘acid cellulases’ which operate at pH around 4.5–5.5, and the ‘neutral cellulases’ which operate around pH 5.5–8.0. The acid cellulases also tend to have a slightly lower temperature range (45–55°C)

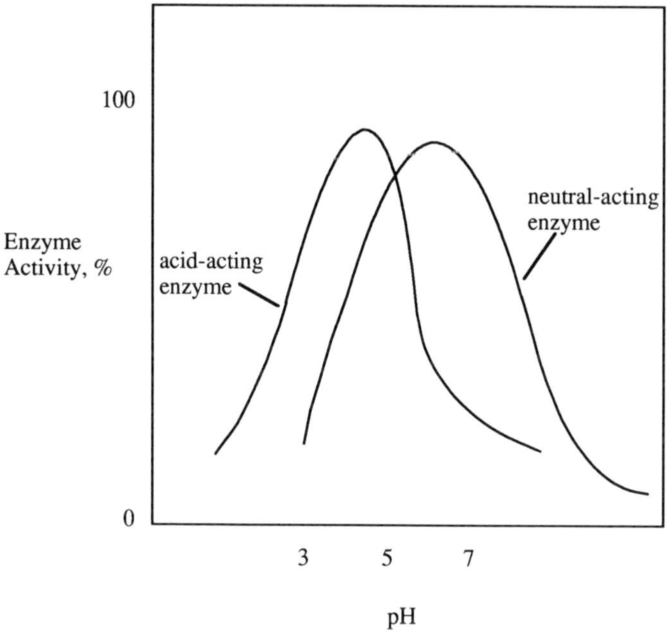

Figure 8.4 The influence of pH on the activity of cellulase enzymes.

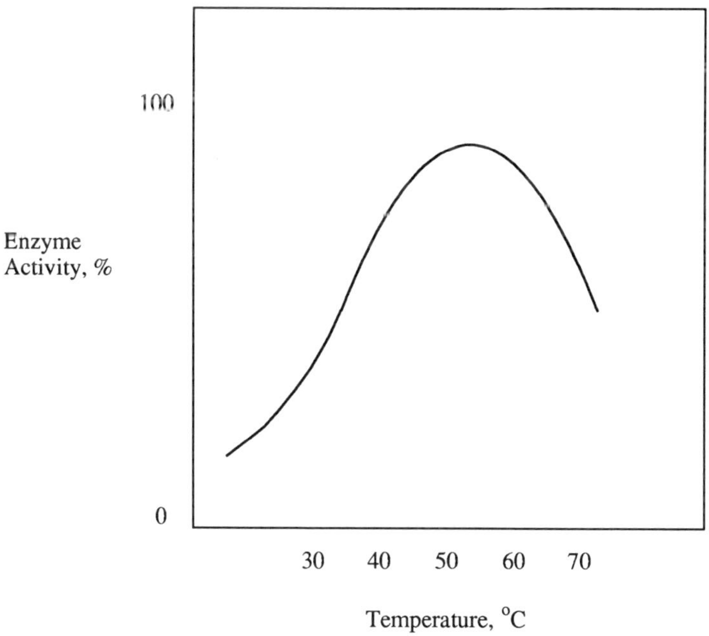

Figure 8.5 The influence of temperature on the activity of a cellulase enzyme.

than the neutral acting cellulases (50–60°C) (see Figures 8.4 and 8.5). The acid enzymes are available from a number of producers using various fungal strains, and will thus vary in effectiveness; neutral enzymes are currently produced exclusively by one manufacturer.

In stonewashing, the type of enzyme chosen has an important bearing on the process. Stonewashing effects as achieved by pumice stones can be duplicated by the action of cellulases which artificially ‘age’ the fabric by chemically degrading the cotton and thereby releasing indigo dye particles, giving an abraded, washed-out look. Cellulases may be used either in conjunction with a reduced load of stone or alone in a stoneless process. Acid enzymes are faster acting, and more aggressive, resulting in more chemical abrasion and resultant loss of fabric strength. Since the process is carried out under acid conditions, dyestuff redeposition is likely, giving backstaining on the reverse side of the fabric and less defined ‘pinpoints’ on the face side. By contrast, neutral enzymes are less aggressive, require longer cycle times, but give no backstaining.

8.3.5 *Wool processing*

One of the primary end-uses for enzymes in wool processing is to render the fibre shrink-resistant. Wool fabrics, both knitted and woven, have a natural tendency to shrink irreversibly unless treated; all current large-scale commercial anti-shrink processes for wool rely on a chemical modification of the wool fibre, in the majority of cases by the use of chlorine or permonosulphuric acid. This aspect is discussed in greater detail in chapter 6. Over many years a great deal of research has been directed toward the use of enzymes as anti-shrink treatments, these being seen as environmentally much cleaner than chemical processing, but to date the success in conferring worthwhile levels of shrink-resistance has been low.

Enzymes which act on wool (and proteins in general) are known as proteases. Like the cellulases, proteases are members of the class of enzymes known as hydrolases, that is, they catalyse hydrolysis reactions. As with cotton processing using cellulases, there are highly specific requirements for temperature and pH when processing wool with enzymes. Proteases attack polypeptide (protein) chains at the peptide (amide, CONH) linkage, giving rise to an organic acid and an amine. Again, as with cellulases, there are endoproteases which attack any peptide linkage in the chain and exoproteases which act only on the terminal linkage in each chain. However, in comparison with cellulose, proteins are relatively resistant to enzyme attack, and complete degradation to a protein fibre such as wool must usually be preceded by some form of denaturing.

Of course, there is a significant difference in the structure of wool compared with that of cotton. As mentioned above, cotton is composed solely of glucose molecules joined in a specific manner; wool, on the other hand, is composed of many amino acids in varying proportions depending upon both histology and the source of the wool fibre. Whereas cellulases always act on the β-1,4-glucosidic linkage between glucose molecules in the chain, proteases are much more selective in the exact linkage attacked. This is related to the spatial requirements of the enzyme protein — it was previously said that the conformation or shape of an enzyme structure is important to its functioning, and indeed enzymes can be thought of as acting like a key in a lock.

The specificity of proteases is seen in the hydrolysis of only those peptide links which satisfy the exact conformation of the enzyme. Indeed, by choosing proteases of closely controlled specificity it is possible to cleave a protein molecule effectively, and this has been used as a standard method of protein characterisation for many years. Thus, proteases for wool can be classified according to their action on particular peptide linkages, for example:

- Trypsin acts on lysine and arginine sites, at pH 7–9
- Chymotrypsin acts chiefly on tyrosine, phenylalanine, tryptophan and leucine, at pH 7–9
- Pepsin acts on linkages with an aromatic side chain at a much lower pH in the range 1.5–2
- A number of other less specific enzymes have activities in the pH range 5–8.5.

Enzyme attack on wool fibres results in loss of strength and weight; in extreme cases, up to 15% weight loss is seen. It is important therefore to select enzyme activity carefully, and to optimise reaction conditions (temperature, pH, enzyme concentration) such that the desired effect (for example, shrink-resistance or dehairing) is achieved without undue damage. However, one of the main considerations in the use of enzymes for the shrink-resist processing of the wool is the levelness of the treatment. Electron microscopic studies of wool fibres treated with enzymes show that some fibres are poorly treated while others are massively overtreated, with visible damage to the fibre cortex. In general, attack is related directly to enzyme concentration.

8.4 Softeners

8.4.1 *Introduction*

The processing of textiles to achieve a particular 'handle' or softness is one of the most important aspects of finishing technology. Very few fabrics with which people have contact are left in their natural state, and this includes not only apparel, but also, for example, furnishing fabrics. The exact nature of the finish to be applied may depend on any one of a number of factors, such as function, fashion or individual preference, and of course what is acceptable to one customer may be unacceptable to another.

The measurement of softness or handle is still largely a subjective decision on the part of the assessor, although there are some systems that are able to give some objective measure of the performance of a softener (these are described in section 8.4.3). The terms used when evaluating a softener tend therefore to be rather descriptive, attempting to convey an impression rather than hard fact; for example, 'dry' or 'greasy'; 'crisp', 'natural', 'warm'; and so on.

The basic function of a softener, then, is to impart a particular handle to a textile surface. This may be required to overcome the naturally harsh feel of the fabric (for example, linen), or to disguise harshness imparted by chemical processing of the fabric (for example, application of crease-resist resins to cottons). Alternatively, a softener may be applied simply to alter the handle for reasons of fashion. The range of materials that can be used as softeners is vast, but a relatively small number of basic chemical types are commonly in use. These are described in more detail in section 8.4.2.

It should be remembered that the performance of a softener will be altered by the nature of the substrate to which it is applied. This does not simply mean that, for example, the same softener will have a different feel on cotton and nylon; even on the same fibre, the handle may be different. The fabric construction is obviously important, and factors such as yarn fineness, yarn density, etc. in the fabric will play an important part in the overall impression of the feel of the fabric. In addition, the mechanical finishing of the fabric may alter the handle to a greater extent than will the application of a softener.

The durability of the softener should be taken into account. It is common to use fabric conditioners in the rinse cycle of domestic washing machines, therefore a durable effect from the mill-applied softener is perhaps unnecessary. What is then important is the point-of-sale handle; the softener is applied in order to make the garment or fabric feel more appealing, and so to encourage customers to make a purchase. Perhaps the most extreme example of this is the use of silicones in towels. Silicone softeners give a very soft, luxuriant handle to even the cheapest towel, but by their nature are hydrophobic, rendering the towel almost useless until the softener has been removed by washing.

The handle of a fabric may not be the only consideration in determining the 'comfort', particularly for fabrics which are in direct contact with the skin. In this category we can include underwear, sportswear and bedlinen. The most important properties a softener can impart to these fabrics is hydrophilicity and an ability to wick moisture away. Although not strictly concerned with the softness of the fabric, the softener plays an important role in maintaining the correct, comfortable environment close to the skin. A hydrophobic fabric of extreme softness is of no value in these situations, since moisture is allowed to build up as a film between the skin and the fabric, making the wearer sweaty and uncomfortable.

Softeners often have a multifunctional nature, depending upon chemical type. All quaternary ammonium containing softeners will show a degree of antistatic behaviour, for example; long-chain fatty acid-based materials may show yarn lubricating properties. Softeners may yield hydrophilic or hydrophobic finishes, either of which may be beneficial in different circumstances. Others may show some anti-soiling behaviour.

8.4.2 *Softener types*

Softeners can be grouped under a number of general headings. Each of these chemical types will be considered separately.

8.4.2.1 *Oils and waxes.* Although oils have largely been replaced as softeners for fabrics, they were once widely used, in the form of water-dispersible emulsions. Waxes are, however, still widely used, most commonly paraffin- or polyethylene-based. The waxes are manufactured and sold as emulsions in water, containing a proportion of surfactant. Paraffin wax emulsions tend to be sold as anionic formulations, and, as such, must be applied by padding techniques, since they will not exhaust. These materials will also act as effective lubricants during processing.

Polyethylene waxes may be produced as cationic, nonionic and anionic emulsions, although only the cationic and nonionic forms are encountered in textile processing. They also give good lubricating properties, and are often incorporated into formulations specifically to improve sewability of fabrics. They impart a rather greasy, full handle to fabrics, and are widely used on cotton and cotton-blend fabrics, especially twills and poplins. The durability of these waxes is not high, but can be improved by using a higher molecular weight wax.

8.4.2.2 *Cationic softeners.* This probably comprises the largest and most important group of softeners currently in use. Whilst many softeners have a cationic nature (for example, certain silicones and cationic polyethylene emulsions, mentioned in section 8.4.1); included in this group are the true cationic materials which contain a positive nitrogen atom in the structure. Perhaps the simplest, and certainly the cheapest and most widely available of this type, is dimethyldistearylammonium methosulphate (Figure 8.6).

Some of these types are based on the ethosulphate ($SO_4^-CH_2CH_3$) or chloride (Cl^-). The ethosulphates are less common, while the chlorides are not favoured in industrial use because of the more corrosive nature of the chloride ion. These materials are hard waxes with relatively low melting points and can be dissolved in water. Because they contain a quaternary

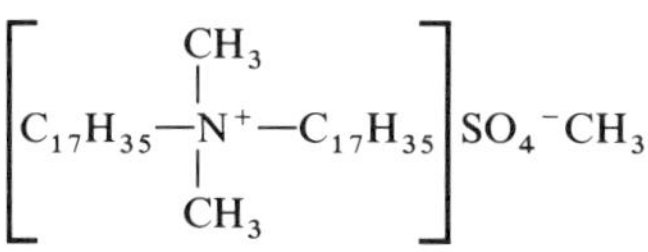

Figure 8.6 Dimethyldistearylammonium methosulphate.

ammonium ion, they show mild antistatic properties, and tend to be fairly hydrophobic as a consequence of their waxy nature.

The properties of these materials can be improved by modifying the structure, most commonly by replacing one or more of the methyl groups by ethoxylate chains, and/or by altering the nature of the fatty chains. Any such change will increase the cost of the material, but does give rise to a softener more suited to industrial application than the simple structures, which tend to be used for domestic softeners. The hydrophilicity of the softener on the fabric increases as a result of the presence of the ethoxylate groups, and the antistatic behaviour may also increase because of the molecule's ability to retain moisture, and so conduct static electrical charge. The use of 'hardened' fatty chains (that is not containing an unsaturated C=C group), gives rise to a less waxy, stiffer handle. In addition, the length of the fatty chain can be increased, although this tends to lead to a decrease in solubility (Figure 8.7). These types also show a moderate durability to domestic laundering. It is common to prepare an emulsion of these softener types, using a nonionic surfactant as emulsifier.

Lastly in this group are the imidazoline softeners (Figure 8.8), although these are less common than the above types.

Under the heading of cationic softeners are also grouped the wax emulsions based on the condensation of a fatty acid and an amine. The reaction between an acid and an amine results in an amide (Figure 8.9). Often a polyamide

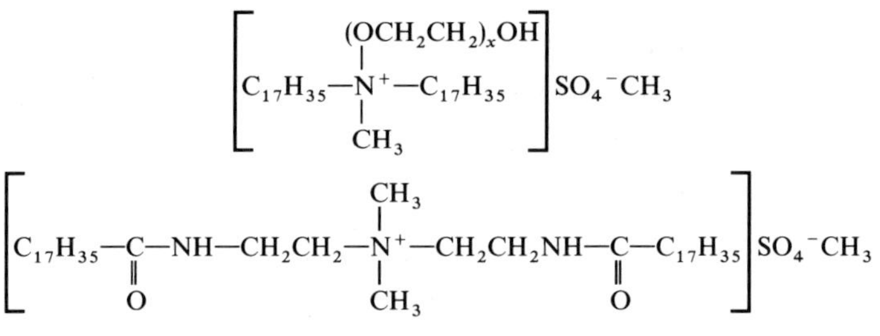

Figure 8.7 Increase in the length of the fatty chain of ethosulphates leads to a decrease in the solubility.

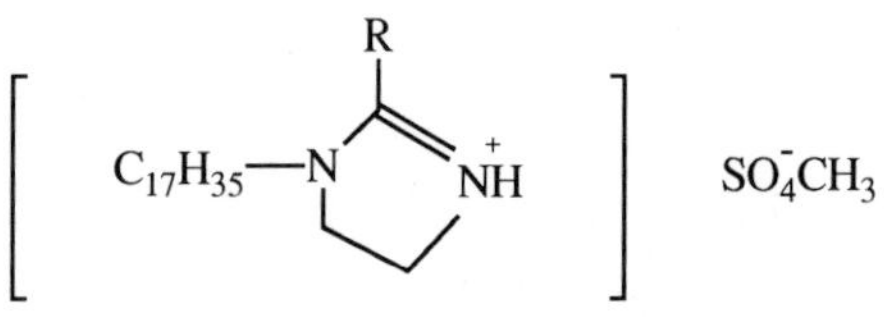

Figure 8.8 Imidazoline softener.

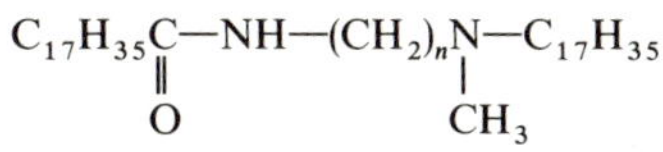

Figure 8.9 Amide softener.

such as diethylene triamine is used, resulting in a higher molecular weight. These materials are made cationic by virtue of their manufacture as acidic solutions, as which they are more stable. Wax emulsions tend to be hard waxes of relatively high melting point, which are dispersed in water with a suitable surfactant. They give a very lofty, waxy handle to many fibre types, and are the real workhorses of industrial softening. They are relatively hydrophobic and show no durability, being readily removed on laundering.

8.4.2.3 *Silicone softeners.* Probably the most widely used softener types in recent years are the silicones. The structures available are diverse, yet are all based on the same Si—O backbone. The simplest silicone softener is polydimethylsiloxane (PDMS, Figure 8.10), which is available as an oily material in the molecular weight range from about 1000 to 1 000 000 or more. These materials are emulsifiable in water and can then be applied to fabrics where they give a characteristically soft but greasy handle. They are also used as lubricants applied to sewing threads. The softeners have no real affinity to textile surfaces, and so must be applied by non-exhaustive techniques. Since they are not reactive materials, they have no durability to fabrics, and are readily removed by laundering. Better durability can be obtained by incorporating reactive groups on the simple molecules; most commonly, the silicone contains a terminal hydroxyl or other reactive group which reacts readily with a silane catalyst (Figure 8.11). The catalyst acts as a crosslinking agent between siloxane chains, so giving rise to an elastomeric structure, which might advantageously offer other benefits as well as softness. For example water and soil repellency are common features of these elastomeric finishes.

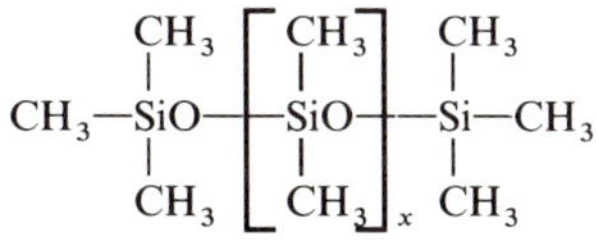

Figure 8.10 Polydimethylsiloxane (PDMS).

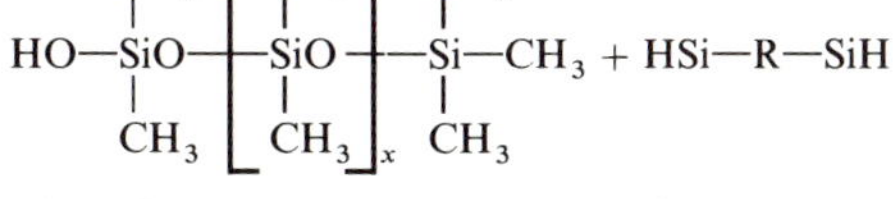

Figure 8.11 Incorporation of a reactive group on to a silicone softener using a silane catalyst.

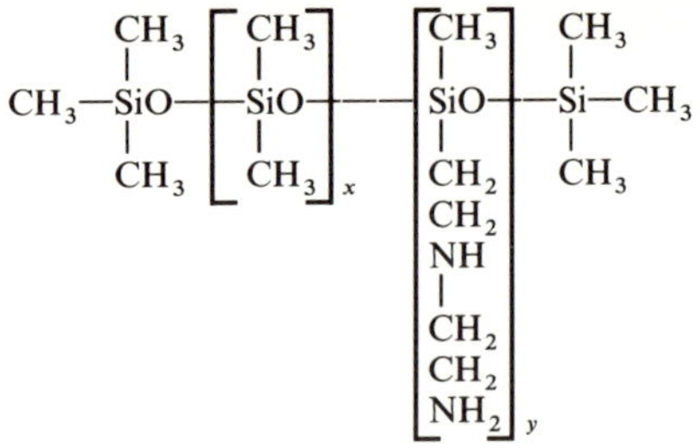

Figure 8.12 Aminofunctional softener.

Of the silicone softeners available, perhaps the most common in current industrial usage are the aminofunctional types (Figure 8.12). These materials offer a range of handles depending on the relative size of x and the ratio of x:y. They may be supplied as surfactant-stabilised emulsions in water, either mechanical or microemulsions. Mechanical emulsions contain large droplets which tend to coalesce on the fabric, giving surface effects. The microemulsions, of much smaller droplet size, will tend to migrate into the yarn and give an overall softness to the whole structure.

The aminosilicones may give a relatively dry handle where the x:y ratio is high, and a typically greasy handle where the x:y ratic is low. Because of the presence of the amino group, they tend to be of a mildly cationic nature (in acidic solutions for stability) and will therefore readily exhaust on to fibre surfaces. The amino group also results in a propensity to yellowing, particularly during curing or drying, and the likelihood of yellowing increases with increasing amino content.

A recent development is the introduction of related silicones containing not an amino group but an amide group (Figure 8.13). The benefits of these softeners are that they are essentially non-yellowing, and that the handle is very dry when compared to even the low-amine aminosilicones.

8.4.2.4 *Miscellaneous softeners.* The structures described in the previous sections form the majority of softener types in common usage in the textile

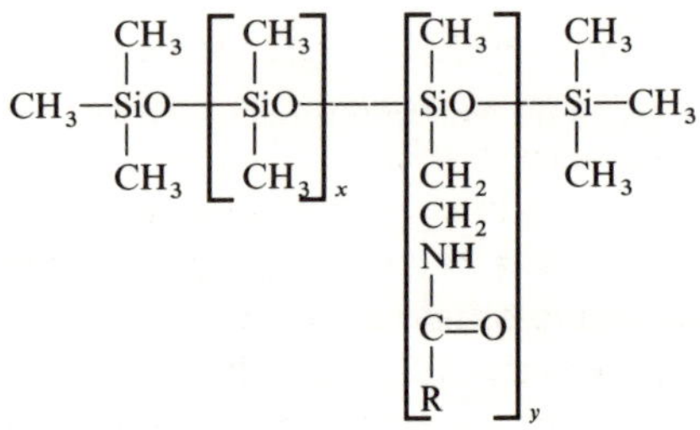

Figure 8.13 Silicone softener containing an amide group.

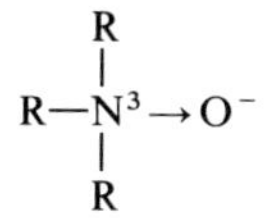

Figure 8.14 Long-chain amine oxide softener.

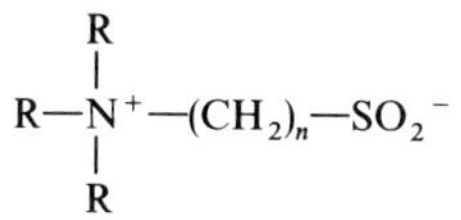

Figure 8.15 Sulphobetaine softener.

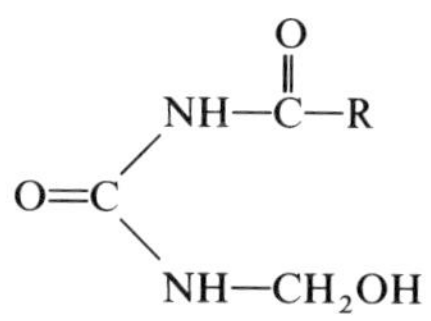

Figure 8.16 Softener for use with crease-resist resins.

field, but other types are also available. These other types tend to be more expensive, however, or are used only in specialised fields. Amongst these miscellaneous softeners can be included the amphoteric types, such as long-chain amine oxides (Figure 8.14) and sulphobetaines (Figure 8.15). There is also a class of reactive softeners useful in combination with crease-resist resins on cotton (Figure 8.16). These are essentially of similar chemistry to many crease-resist resins, in that they contain the *N*-methylol group, which requires an acid catalyst and baking to effect a crosslinking reaction. They are, however, conveniently applied with crease-resist resins via a padding operation. At higher application levels these reactive hydrophobes impart water repellency.

8.4.3 *Mechanical measurement of fabric properties*

The physical properties of a fabric are relatively easy to measure; the tensile strength, tear strength, bursting strength, etc. are all conveniently measured by various mechanical testing machines. However, while these methods have their value in determining fabric quality, they are all destructive methods, and represent extremes of wear. When evaluating fabric handle, we are largely dependent on the subjective opinions of the assessors, although the human fingers are remarkably sensitive to the small nuances of 'feel'.

Fabric handle may be considered to be related directly to factors such as the compressibility and frictional behaviour of a given fabric type. Other properties of the fabric are related to the tensile properties (extensibility) and shearing and bending properties (drape), for example. It would therefore be convenient to be able to compare such fabric characteristics by these readily measurable properties.

Extending this idea further, it would be useful to be able to measure the performance of a fabric under conditions more typical of actual wear. Such a system would measure the response of a fabric surface to small stresses, far lower than those imposed in traditional destructive testing, and more typical of the stresses normally encountered. The Kawabata KES-FB system (actually a series of four different instruments) measures six different fabric properties. From these, sixteen different parameters, relating to fabric processibility and wearability, may be measured or derived (Table 8.4). The biggest drawback with this system, however, is the extreme complexity and cost. Other systems are available, for example, the FAST (Fabric Assurance by Simple Testing) system developed by CSIRO, which is a simpler system than the Kawabata. A recent development is the use of an Instron universal testing machine, such as is commonly available in textile testing laboratories for measuring similar properties. These alternative systems provide similar information but by reducing the number and type of measurements, they are perhaps better suited to more routine use.

Table 8.4 Fabric properties and parameters measured by Kawabata

Property	Parameter
Bending	Bending rigidity Hysteresis of bending moment
Surface	Coefficient of friction Geometrical roughness Mean deviation of MIU
Tensile	Linearity of tensile curve Tensile energy Tensile resilience
Shearing	Shear stiffness Hysteresis at 0.5° Hysteresis at 5°
Compression	Linearity of compression curve Compressional energy Compressional resilience
Thickness	Fabric thickness
(Weight)	(Fabric weight)

Note: Fabric weight is measured on a balance, not on the Kawabata system.

With careful interpretation of the results obtained, all of these systems can yield useful information on the characteristics of a fabric, clearly defining differences resulting from yarn type (natural or synthetic, staple or single filament), construction (fabric weight, yarn count, weave) and the application of various finishes.

8.5 Mercerisation

In 1850 John Mercer discovered that cotton immersed in sodium hydroxide swelled and shrank. The alkali treatment converts the fibre's crystalline structure from cellulose I to the cellulose II form (see section 1.2.1.2). The commercial benefits of mercerisation are improved lustre through a more circular fibre cross-section; improved softness, strength and dyeability; and easy care properties. Typical processing conditions are 200 g/l sodium hydroxide, 10–20°C, 30–180 s immersion and tension to enhance lustre. Liquid ammonia can also be used to mercerise cotton.

References and bibliography

Alexander, P., Hudson, R F. and Earland, C. (1963). *Wool, Its Chemistry and Physics*, Chapman and Hall.
Burkitt, F.H. and Heap, S.A. (1971). *Review in Progress in Coloration*, Society of Dyers and Colourists.
Griffin, W.C. (1949). *J. Soc. Cosmet. Chem.*, **1**, 311.
Kawabata, S., Niwa, M., Ito, K. and Nitta, M. (1990). *Int. J. Clothing Sci. Technol.*, **2**, 18.
Mooney, W. (1980). *Textile Month*, October, 32.
Myers, D. (1992). *Surfactant Science and Technology*, 2nd edn, VCH Publishers Inc.
Olson, L.A. (1990). US patent 4,912,056.
Pan, N., Zeronian, S.H. and Ryu, H.-S. (1993). *Textile Res. J.*, **63**(1), 33.
Riva, A., Cegarra, J. and Prieto, R. (1993). *Journal of the Society of Dyers and Colourists*, **109**, 210.
Rossner, U. (1993). *Melliand Textilberichte*, **74**, 144.
Shore, J. (Ed.) (1990). *Colorants and Auxiliaries*, Vol. 2, *Society of Dyers and Colourists.*
Tyndall, R.M. (1990). *American Dyestuff Reporter*, May, 22.
Tyndall, R.M. (1992). *Textile Chemist and Colorist* **24**, 23.

9 The colouring of textiles

I.D. RATTEE

9.1 The development of a scientific approach

The dyeing, printing and decoration of textiles go back to the earliest times and, historically, have inspired craftspeople to remarkable levels of ingenuity as ways have been sought to exploit the colouring matters to be found in nature. The basis for most of the dyeing methods in use up until the nineteenth century was established by the ancient Egyptians, who developed the application of plant extracts often in association with mordanting (the pre-application of solutions of metal salts such as alum to the textile to give improved washing fastness and, in some cases, a specially desirable shade through the formation of complex compounds). For obvious reasons the plants and other materials used tended to be those indigenous to the region so that, for example, the production of scarlet shades from insects of the kermes family which originated in Persia did not feature in Egyptian dyeing practice. Similarly woad, later to be displaced by indigo, which was one of the most important colouring matters in use in ancient times, did not reach Egypt until about 300 BC. Another important ancient source of dyes was the molluscs used by the Phoenicians to produce the famous Tyrian Purple and related shades. It should not be forgotten that considerable experimentation and chemical expertise were required to make many of the developments which took place. The colouring matters derived from the indigo or woad plants and the molluscs used by the Phoenicians are chemically closely related, and to extract and use them requires several processes which could have come to be adopted only after much experimentation, which was carried out, let it be remembered, without any chemical knowledge that would be recognised today.

Prior to the sixteenth century dyeing was a secret and closely guarded technology. For example, the production in Turkey of the famous Adrianopolis or Turkey Red by the application of the extract of the madder root to cotton mordanted with alum was a secret which was retained for 250 years, in the pursuit of which many European industrial spies lost their lives. At a less dramatic level every dyer sought to retain secret methods. The fact that many of the 'secrets' were illusory and depended on some local factor, such as the presence of mineral traces in the water supply, or were based on totally spurious reasoning does not matter. However, with the expansion of trade and travel which accompanied the European renaissance and, more particularly, the development of the printing press, there was an explosion of knowledge and information which exposed much of the mystery to the light of day. To the chagrin of the European woad growers, Marco Polo returned from India with details of the use of the much more effective indigo as a dye (prior to this it had been used in Europe only as a pigment in painting). Many works were published in that period detailing dyeing and colouring procedures for different textiles, largely for domestic use. Of special importance, in that it was directed towards the industrial dyer, was *The Plictho* of Giovanventura Rosetti, published in 1548, which records, in addition to many dyeing recipes and methods of preparing plant and other extracts, procedures for the preparation of important chemicals such as hydrochloric acid (Rosetti, 1548).

With the expansion of knowledge and great increases in the volume of production that followed this period came the demand for more reliable quality control, and as a consequence the demand for greater understanding and the beginnings of scientific investigation into the phenomena involved. Some hundred years after publication of Rosetti's work, Colbert in France established regulations to control dyeing quality (Colbert, 1671), and this was followed sixty years later by the appointment of the first Inspector General of the Dyeing Industry, Dufay de Cisternay, who must be regarded as the founder of modern dyeing theory. Dufay established systematic fastness testing and quality classification in dyeing to replace the established and not very reliable *grand teint*, *bon teint* and *petit teint* qualities. As a result of his studies Dufay published what may be described as the first truly scientific account of dyeing processes based on physical chemical ideas (Dufay, 1737). He was the first to advance the concept of specific sites for dye adsorption based on the different compositions of the various textile fibres, a remarkable insight considering the fact that there existed no knowledge at that time of organic chemical structure or polymers. Following Dufay came a succession of distinguished chemists who developed such ideas as were possible, given the chemical knowledge available to them at the time, to account for the properties of mordants and, remarkably, the conception that dyeing was a physical binding process on a molecular scale (Berthollet, 1791).

The succession of Inspectors General of the Dyeing Industry in France came to an end with the French revolution. However, it was the blockade of Napoleonic France by England that prompted the search for substitutes for dyes and other compounds imported from the East and also the Americas. Napoleon's Minister of the Interior was the chemist and dyer, Jean Antoine Chaptal, who gathered around him the greatest chemists of the time and also established the first academic department for the teaching of chemical technology at the University of Montpellier. Chaptal advanced ideas for the development of technological education involving an interaction between industry and academic centres, which were to some extent subverted by the defeat of Napoleon. However, they were taken up to great effect in Germany, and formed the basis for the system of Technischen Hochschulen which enabled Germany to take advantage of, and play its own important part in, the emergence of the dyestuffs industry in the second half of the nineteenth century.

As early as 1740, experiments had shown the possibility of synthesising dyes. Naturally occurring indigo was treated with concentrated sulphuric acid by Barth to produce a product which was marketed as Saxe Blue. Later, in 1776, Scheele found that by treating uric acid with nitric acid a reddish compound was obtained which coloured the skin red. It was thus named purpuric acid. In 1818, Proust made the ammonium derivative and showed its use as a dye for tin-mordanted wool, mercury-mordanted silk and lead-mordanted cotton. Under the name murexide, the dye was manufactured in Mulhouse and in Manchester as bulk supplies of uric acid became available from South America. However, such discoveries could have no systematic basis. Firstly, it was generally believed that only living organisms or God could prepare organic materials and, secondly, there was insufficient chemical knowledge on which to base anything other than random excursions into chemical synthesis. However, a major change took place after Wöhler prepared urea from inorganic materials and also observed that urea and the inorganic ammonium thiocyanate were isomeric. The development of analytical methods by Liebig and the development of the coal gas industry, with its production of coal tar as a by-product, laid the basis for the next step.

The German chemist W.H. Hofmann was appointed to head the newly established Royal School of Chemistry in 1845. This became a centre for the analysis and isolation of the multiple components of coal tar produced in great quantity by the new and flourishing coal gas industry. W.H. Perkin, who became a student under Hofmann in 1853, was concerned with the reactions of aniline and, carrying out work in his own laboratory, attempted to prepare quinine by the oxidation of allyl toluidine using acidic potassium dichromate. There being no such concept as structural organic chemistry at that time, for Perkin quinine was $C_{20}H_{24}N_2O_2$, and the proposed synthesis was perfectly reasonable. In the event, the dark brown product achieved was extracted with water to yield a purple solution that was clearly not quinine. Perkin showed that the new product could be used to produce a novel purple shade on silk, and following favourable comment from Pullar's dyeworks, Perkin & Sons Ltd went into manufacture of Aniline Purple. Not only was it necessary to develop methods for the manufacture of aniline on an industrial scale and suitable reactors for the process, a not inconsiderable achievement in its own right, but Perkin had also to devise dyeing methods for the application of his product. This was because it did not behave like any naturally occurring dye in commercial use, being what would now be termed a basic dye, i.e. a salt with a coloured cation. In the event, bulk supplies came on to the market in the first instance from France where, taking advantage of Perkin's failure to secure adequate patent protection and the more advanced state of chemical technology in that country, several manufacturers were able rapidly to offer the dye. The name, Mauve, suggested by the French was more attractive commercially than Aniline Purple and eventually displaced Perkin's original name for the product. Following Perkin's discovery, an intensive period of experimentation began, to study many oxidative reactions of aniline and its derivatives to produce mostly new basic dyes. Another discovery which was to prove to be of major importance was made by another of Hofmann's assistants, Peter Griess, who, in 1858, found that by treating aniline and its derivatives with acidic sodium nitrite, compounds were formed which reacted with phenols and related compounds to produce brilliant yellows, reds, and other colours. This reaction proved to be one of the most important in colour chemistry. However, while the impact of these new colouring matters was considerable, progress was inevitably unsystematic. The development of structural organic chemistry following Kekulé's proposal for the ring structure of benzene seven years later in 1865 transformed research in the field from an empirical to a more systematic approach. Kekulé's ideas had little effect in England and France, but were taken up with enthusiasm in Germany, to which many of the emigrant chemists, including Hofmann, had returned in 1861. It should be said that despite their commercial success the dyes synthesised prior to 1865 were, in general, lacking in fastness to washing and to light. However, once it was realised that the important structural feature of organic compounds, particularly dyes, lay not in the ratio of the numbers of carbon, hydrogen, nitrogen and oxygen atoms but in their structural arrangement, it was possible to begin to look scientifically at dyes such as madder and indigo which, although derived from plants, were very stable. Commercial natural madder contained only 4% of the essential colouring matter, alizarin, but the establishment of the structure of alizarin and possible routes to its synthesis by Graebe and Liebermann in 1868 led rapidly to the availability of the highly concentrated synthetic product.

On the exploratory front it was possible to begin to relate the structure of dyes to dyeing behaviour and other properties. At that time it was possible to dye cotton only after the application of a mordant. By examining the dyeing behaviour of azo dyes of known structure formed using Griess's diazo reaction, Paul Böttiger found that linear dye molecules containing two azo groups (**1**) dyed cotton directly from a saline dyebath without mordants, and in 1884 he obtained a patent for the new dye, Congo Red (**1**). The C.I. Direct Red designation relates to the classification system in the Colour Index, discussed on p. 299. This was the first direct cotton dye, so called because of its direct applicability to cotton fibres. Congo Red is little more than an acid/base indicator, and the fact that it was regarded as a commercial proposition gives some indication of the prevailing standards of durability and fastness that obtained at that time. The opportunity presented by this development was seized upon by the German dyestuff manufacturers who never lost the impetus thus achieved.

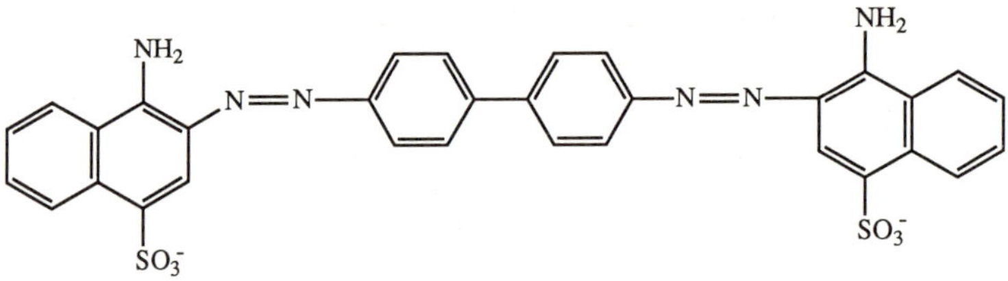

(1) Congo Red (C.I. Direct Red 28)

As more and more dyes came to be prepared and the structures of natural colouring matters were elucidated, the beginnings of a theoretical connection between colour and chemical constitution began to emerge. The earliest proposal was put forward by Graebe and Liebermann, the elucidators of the structures of alizarin, who suggested that colour was associated with unsaturated carbon atoms. A rather more effective extension of this idea was proposed by Witt (1876), who postulated that all coloured compounds contained an unsaturated group or groups which conferred the potentiality of colour. These Witt termed *chromophores*, e.g. $—NO_2$, —NO, —N=N—, $>C{=}O$, and $>C{=}C<$. This basic potentiality Witt suggested was modified by other substituents in the molecule which he called *auxochromes*, e.g. —OH, $—NH_2$, —NHR, and $—NR_2$. Auxochrome substituents did not in themselves give rise to strong colours but had a significant effect on the shades of chromophoric molecules. According to these ideas, there are present in the molecules of a coloured compound three sub-systems, a chromophore, an auxochrome and a system of conjugated double bonds linking the two. At the stage of development of organic chemistry in Witt's time his ideas were of little predictive value, but they provided a foundation upon which, with the development of quantum mechanics and molecular orbital theory, methods are available to the modern organic chemist that are far removed from the empiricism of the approach to dyestuff development characteristic of the latter part of the last century and the first half of this.

In contemporary terms chromophoric and auxochromic systems shift the principal absorption bands of the conjugated system to longer (visible) wavelengths. Griffiths (1976) has suggested a classification of chromophoric systems as follows:

(i) $n \rightarrow \pi^*$ chromogens
(ii) donor–acceptor chromogens
(iii) acyclic and polycyclic polyene chromogens
(iv) cyanine type chromogens.

Griffiths uses the term chromogen to describe any reasonably well defined unsaturated system which either is coloured or becomes coloured as a result of substitution by simple atoms or groups. He restricts his use of the term chromophore to any unsaturated group which is colourless. The donor–acceptor chromogens comprise the largest group and include the great majority of commercially useful dyes and pigments. Such systems contain an electron donor group possessing a lone pair of electrons which is connected directly to a conjugated system of π electrons. The whole thus contains a population of electrons which become excited by low energy radiation, i.e. visible light, and thus exhibit selective absorption and thus colour. The visible absorption band corresponds to the migration of electrons away from the donor into the general π system. When several atoms in the acceptor show an increase in electron density then the whole acceptor π system is best regarded as a complex acceptor. Commercial dyes and pigments for textile use generally fall into this category, while the simpler donor–acceptor systems have an important place in photoreprography and other areas where resistance to chemical attack is of less importance.

The effectiveness of a donor in the donor–acceptor chromogens is determined by the ease with which it releases its electrons. Thus the group $—N(CH_3)_2$ is highly effective, followed (in order of reduced donor power) by $—NH_2$, $—OCH_3$, $—NHCOCH_3$ and $—O \cdot COCH_3$ groups. The presence of a hetero (i.e. non-carbon) atom is an obvious essential to provide the lone electron pair, but the ease with which the electrons are donated will depend on the hetero-atom substituents. If these increase the electronegativity of the hetero-atom by electron withdrawal as, for example, in the case of the acetyl group, then the donating power is reduced, while on the other hand alkyl groups will reduce the electronegativity of the hetero-atom, thus increasing its donor power. The presence of several donor groups will produce an increasing bathochromic shift, raising the wavelength of the principal absorption band from the ultraviolet into visible wavelengths. In a simple donor–acceptor chromogen, the donor and acceptor groups are linked by a conjugated system which provides the path for the electron migration. However, the acceptor system can be itself conjugated and may offer a relatively delocalised destination for the electrons. Such systems, termed by Griffiths (1976) complex acceptors, when present in dye or pigment molecules generally offer the most intense colours and are characteristic of commercial dyes.

Space does not permit a full account of the range of chromogens to be found among commercial dyes and pigments today. However, some of the more important examples can be considered. The most important are the azo dyes, the first examples of which were prepared using Griess's reaction to form a diazonium salt, which was then coupled with an aromatic amine or a phenolic compound. The reaction is based on the formation of a diazonium compound by reaction of nitrous acid on an aromatic amine. This leads to the formation of a diazonium salt, which can act as an electrophilic reagent in the formation of an azo compound. In this the diazonium salt is called the 'azo component' and the electrophilic reagent is called the 'coupling component'. Because of its weakness as an electrophile it can only work successfully with reasonably strong nucleophiles, and consequently only aromatic compounds bearing —OH, $—NH_2$, —NHR and other electron-donating substituents. Typically, sodium nitrite is allowed to react under ice-cold conditions with the hydrochloride of the amine to form the diazonium salt, which is then added to an alkaline solution of the coupling component. A generalised reaction scheme is shown:

$$\underset{\text{diazotisation}}{Ar—NH_2 \rightarrow Ar—\overset{+}{N}{\equiv}N Cl^-}$$

$$\underset{\text{coupling}}{Ar—\overset{+}{N}{\equiv}N + Ar'O^- \rightarrow Ar—N{=}N—Ar'}$$

In 1861, Mêne reacted aniline with nitric acid to form Aniline Yellow (**2**). This compound is not soluble in water and cannot be used directly in dyeing. Kekulé's ideas being very new and the mechanism outlined above being as yet not understood, the full structural implications of the formation of Aniline Yellow were not clear, but its formation showed that the amino group of aniline and related compounds could be transformed into a colour-forming group. The first water-soluble azo dye, Bismark Brown (**3**), was prepared by Martius in 1862.

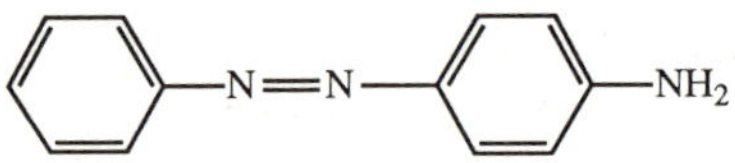

(2)

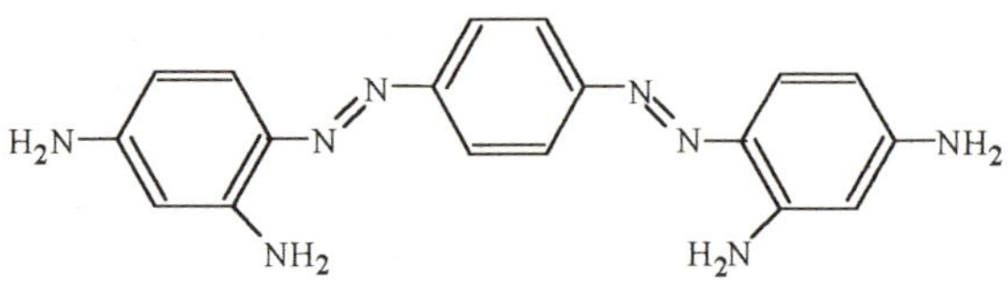

(3)

The easily performed diazo reaction offers the chemist almost limitless options in selecting dye structures for various end-uses. If the coupling component itself contains an amine group, the dye can then be diazotised again to form an even more complex system. Alternatively, there may be present in the original amine two diazotisable groups, so that a tetrazonium salt is formed, which can couple with two molecules of the same or different coupling systems. Examples (**4, 5** and **6**) of dyes of these three types are shown.

There is now available a variety of methods of forming diazonium compounds, and the azo chromogen has enormous flexibility. It is to be found in dyes ranging in shade from yellow through to black. An indication

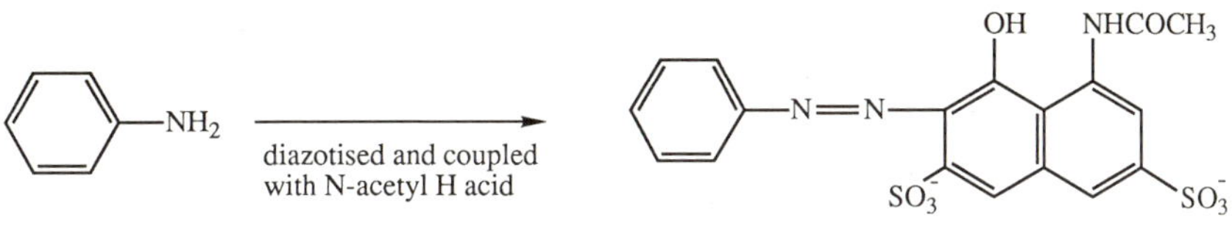

(4) (C.I. Acid Red 1)

of the scope offered to the chemist by this chromogen is provided by the fact that upwards of 3 million speculative examples of dyes have been prepared in the 135 years since it was first made. Another important feature is their potential for the preparation of metal complex dyes. By selection of coupling and diazo components, dyes may be prepared containing *o,o'*-dihydroxy or *o*-hydroxy–*o'*-amino groups relative to the azo group. Such a configuration readily complexes with copper or chromium to give colouring systems that are very stable to light. The formation of the metal complex involves a considerable broadening of the principal absorption band to produce shades which, while often stronger than those obtained without complex formation, are usually duller. Initially, dyers applied the metal-free dyes to fibres and then developed the metal complexes *in situ* by after-treatment with a suitable copper or chromium salt. Such methods are very much less used today for a variety of health and ecological reasons and preformed complexes are the norm. Typical examples (**7** and **8**) are shown.

The second most important group of commercial chromogens is the donor-substituted quinones, which provide a complex conjugated acceptor system. A typical example is anthraquinone (**9**), which absorbs light principally in the near ultraviolet. However, when it is substituted with simple donors intense colours ranging through orange to red and blue can be obtained. Examples (**10** and **11**) are shown.

Quinonoid dyes do not offer as wide a range of shades as do the azo dyes, but they provide the basis for bright fast blues and green shades, whereas the strength of the azo dyes lies at shorter wavelengths. They are used extensively in applications for which azo dyes are unsuitable and for textile substrates for which the small molecular dimensions of many quinone chromogens are especially suitable (see p. 318). Additionally they provide the majority of the class of vat dyes (see p. 321), which are temporarily solubilised during application by alkaline reduction of the quinone system. In this case, complex condensed quinone systems such as (**12**) are used.

The first of the vat dyes, the blue pigment indigo, contains quite a different chromogen (**13**). Alkaline reduction converts the carbonyl groups to ionised hydroxyl groups and the pigment to a water-soluble dye. Following its analysis, several substituted indigo analogues with different shades were manufactured.

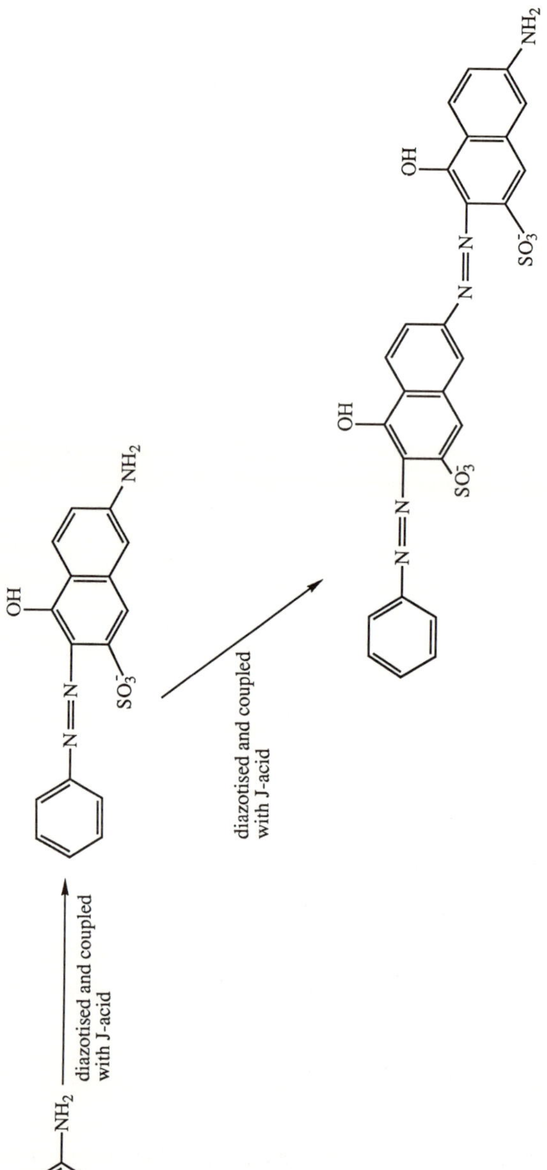

(5) (C.I. Direct Red 16)

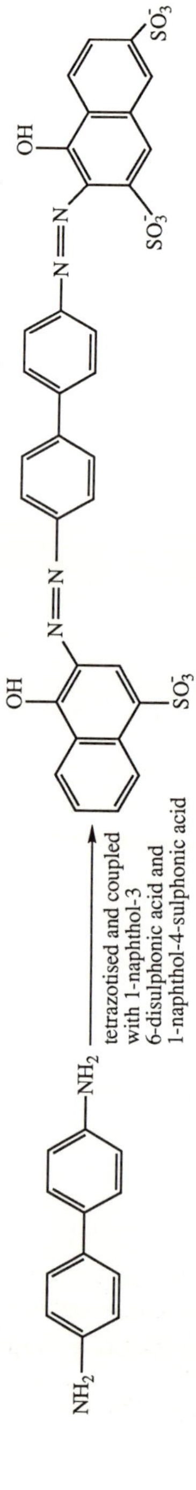

(6) (C.I. Direct Violet 3)

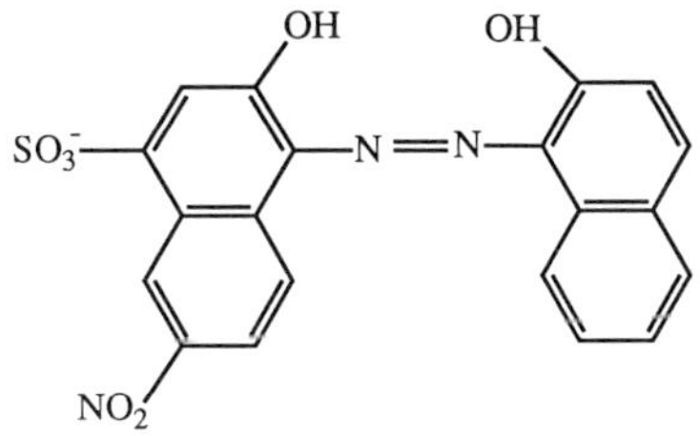

(7) (C.I. Mordant Black 1)

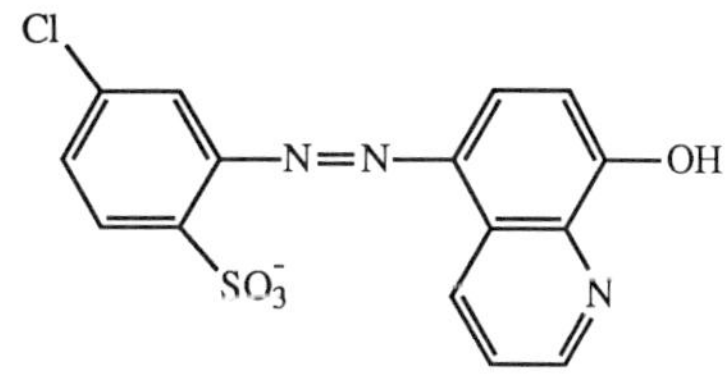

(8) (C.I. Mordant Orange 26)

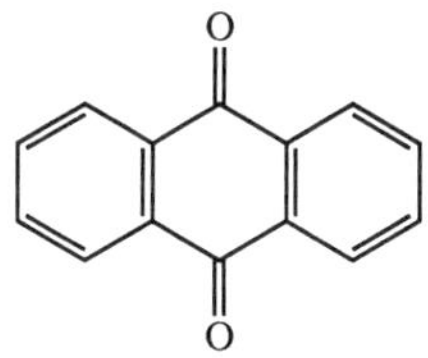

(9)

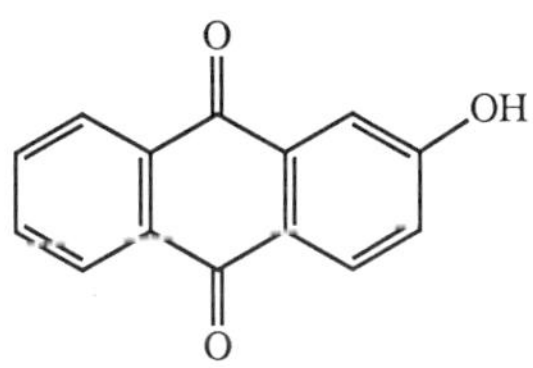

(10) $\lambda_{max} = 365$ nm

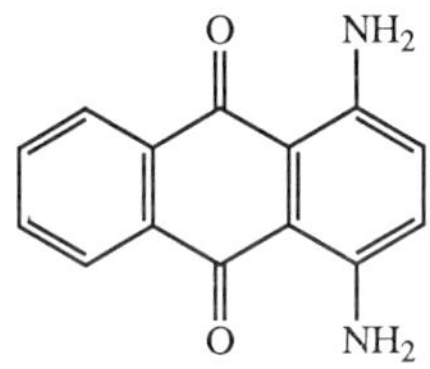

(11) $\lambda_{max} = 550$ nm

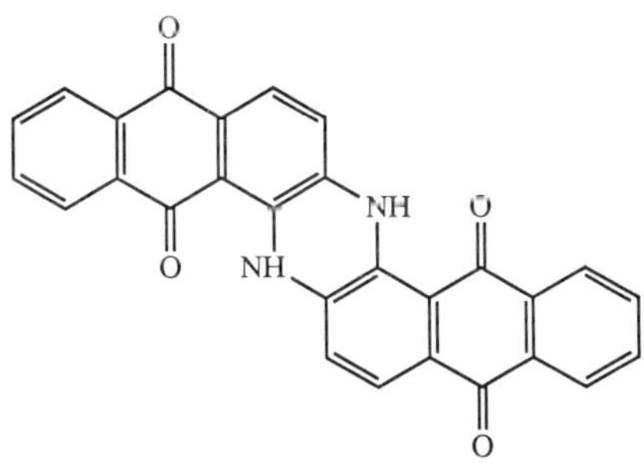

(12) (C.I. Pigment Blue 60)

Another extremely important but highly specialised chromogen which provides the basis of pigment blues and greens as well as soluble turquoise blue dyes for a variety of textile fibres is the phthalocyanine system (**14**). Substitution by sulphonic acid or quaternary onium groups provides solubility. In its unsubstituted form phthalocyanine provides very stable blue pigments.

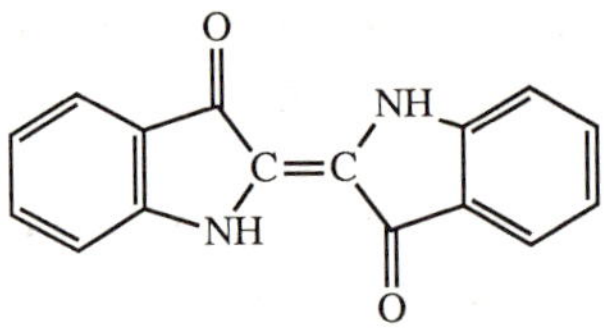

(13)

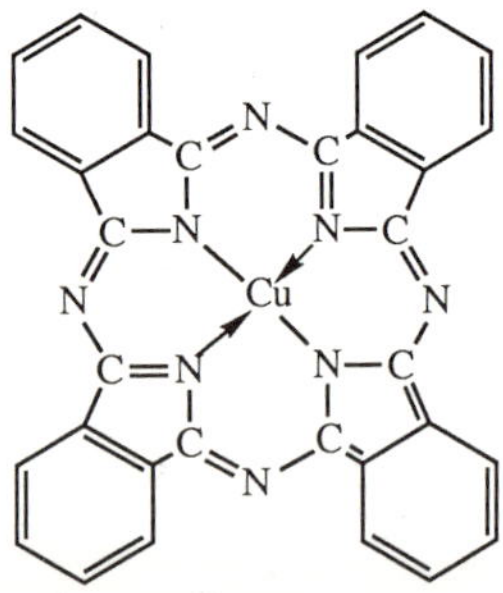

(14)

The modern dye chemist, unlike those of even only 50 years ago, is able to target with a good degree of precision the structures which will provide not only the shade required but also the dyeing properties which are needed to make the product fit in with the existing dyes alongside which it will be used. A good understanding of dyeing as a chemical process as well as a technical operation has made this possible. Accordingly, before discussing the characteristics of the various dye classes, consideration will be given to the physical chemistry of the dyeing process.

9.2 The physical chemistry of coloration

Investigators have speculated about the reasons why dyeing occurs for many centuries, but the scientific investigation of the phenomenon is relatively recent. This was due to the difficulty of measuring dye concentrations in solution or on the fibre without the availability of spectrophotometers. Today such instrumentation is commonplace, but this is a situation which has come about only during the last 40–50 years. The essential basis for any discussion of the physical chemistry of dye adsorption is a consideration of the energy of the system. When a textile fibre is immersed in a dye solution, there is a spontaneous transfer of dye from the solution to the fibre phase resulting in a reduction in the concentration in solution. This is because the movement of dye between the phases results in the system reaching a *state of lower energy* or higher stability. In order to reverse the process, it follows that energy must be put into the system, and consequently the adsorbed dye can be seen to behave as if it had formed bonds of some kind with the fibre or substrate that resist rupture. The adsorption does not normally lead to 100% uptake of dye and an apparent equilibrium is observed, which is characteristic of the formation of physical bonds. This is an essential feature of the adsorption of all dye classes on all textile fibres. In the case of acid, basic, direct and disperse dyes the whole dyeing process is essentially reversible. In the case of vat, reactive and azoic dyes, physical adsorption is an essential feature of the dyeing process, which is rendered irreversible at some stage by oxidation to form a vat pigment from the soluble leuco compound, reaction with the fibre or precipitation following reaction with a diazo component. Thus, from the physical chemical point of view, dyeing can be considered firstly as a single process covering all dye classes on all fibres, and secondly from the point of view of the special features of some of the particular methods employed to limit the reversibility of the adsorption process. The process can be analysed mechanistically with regard to the ways in which dye molecules become bound to substrates, or statistically with regard to the distribution of dye molecules in the dyebath and the kinetics of the dyeing process.

9.2.1 *Binding forces*

The two kinds of physical bonds which are involved are coulombic interactions between permanent (ionic) charge centres and dipolar or dispersion forces. It should be remembered that, since the adsorption generally occurs in the presence of water as a solvent, the interactions between all the components of the system have to be considered.

Coulombic interactions are involved wherever the adsorption of dye ions is involved. Thus, in all cases except that of disperse dyes, coulombic interactions are inevitably involved in some way in the dye adsorption process. Where the substrate is initially positively charged, dye anions will be attracted until the surface potential falls to zero. Further adsorption of any dye ions will lead to the development of a powerful electrical potential which will oppose further adsorption by ions of the same charge. Near to the charged surface there will be a distribution of ions such that oppositely charged ions are attracted to the surface and similarly charged ions repelled. Assuming a positively charged surface the concentrations C_+ and C_- for the anions and cations respectively are given by:

$$C_+ = C_0 \exp(-z^+ e\psi/kT) \quad \text{and} \quad C_- = C_0 \exp(+z^- e\psi/kT)$$

in which C_0 is the general concentration, z^+ and z^- are the charges on the two ions, e is the electronic charge, ψ is the surface potential, k is Boltzmann's constant and T is the temperature. Where more than one kind of anion is present in the above case, both will be attracted, leading to competition between them for the negative charge centres in the surface. Clearly, to some extent the presence of co-ions associated with the dye ion or due to the addition of another electrolyte will reduce the repulsive potential, but since the surface potentials involved are very high (several kV), little adsorption can take place when electrical saturation has been reached. It can be seen that a repulsive surface potential will act against the adsorption of dye ions by neutral or similarly charged surfaces.

Several theoretical models, notably those of Donnan, of Gouy and their numerous refinements, have been advanced to facilitate the calculation of the surface potentials of charged surfaces. The Donnan model has been used particularly effectively to calculate the pH inside fibres and to develop a quantitative theory of the physical chemistry of reactive dye fixation (Sumner and Taylor, 1967). However, coulombic interactions are quite nonspecific and are concerned only with charge centres. Consequently, it would not be expected that, if they were the only factor involved, dyes of the same charge would show much difference in behaviour or, for that matter, differ much from simple electrolytes. The fact that this is not the case shows that while coulombic interactions, both attractive and repulsive, may be of great importance influencing the way in which dye is distributed between the solution and fibre phases, they do not provide the strong bonding forces that are needed for a useful dye.

Unlike coulombic interactions, dipolar and dispersion forces are always attractive. Where two oppositely charged centres exist in a single molecule due to the presence of different permanent charge centres, as in a zwitterion, or a degree of charge separation exists due to the interaction of electrophilic groups, a molecule will operate as a dipole which is attracted by other dipoles or other molecules of the same kind. A dipolar molecule is capable also of inducing in a susceptible molecule a corresponding dipole and thus providing the basis for a useful binding force. This process of induction is carried further in dispersion forces, which are universal and very powerful.

Considering the simplest system of an atom consisting of an orbiting electron and a positive nucleus, it is clear that while it is not possible to be certain about the location and energy of the electron at any particular moment, an instantaneous dipole must exist. Since the relative positions of the electron and the nucleus are in constant flux even in their lowest energy states, the atom must exist as an oscillating dipole with a time-integrated dipole moment of zero. When two such atoms approach and the oscillations are in phase, then an input of energy will be needed if they are to be separated again. Consequently, it will appear as if an attractive force is operating to bring the two atoms together. The interaction between two equal permanent dipoles of moment d at a distance r is given by:

$$U = -2/3(d^4/r^6 kT)$$

That between a permanent dipole and an induced dipole is given by:

$$U = -(2\alpha d^2/r^6 kT)$$

where α is the polarisability or the ease of induction of the dipole. Considering two mutually induced dipoles such as operate in the case of dispersion forces:

$$U = -3/2\{I_a I_b/(I_a + I_b)\}\{\alpha_a \beta_b/r^6\}$$

where I_a and I_b are the ionisation potentials of the two species. In each case I can be replaced by hv_0 where h is Planck's constant and v_0 is the characteristic frequency. For simple substances I and hv_0 are almost identical, while for more complex substances such as dyes where the transitions are not in any sense ionisations, hv_0 is the proper term. The r^{-6} dependence of the potential *energy* becomes an r^{-7} dependence when considering the attractive *force*. The further development of this theory to take in asymmetric molecules brings in terms in higher negative powers of r. It might be thought that forces which diminish so rapidly with distance might be unimportant in dye adsorption, were it not for some decisive complicating factors. Firstly, with regard to dye adsorption, it is an assemblage of atoms in both the dye and the substrate that has to be considered. Casimir has shown that, due to co-operative effects, this reduces the distance dependence to r^{-3}. It also increases the time taken for the full force to develop, although this is still very short. Secondly, it has to be remembered that dye adsorption takes place in an aqueous environment. A water molecule existing surrounded by other water molecules will not experience an attractive force in any particular direction, since the surrounding water molecules will be all the same. However, when an alien molecule, e.g. a dye molecule, is present, the dispersion force interaction with an adjacent water molecule will change it sufficiently for it to interact with another water molecule in an abnormal way. Thus the initial dye–water interaction can be transmitted to an extent depending on the different susceptibilities of the dye and water molecules between the adsorbing surface and the dye or between two dye molecules over significant distances, leading to an above random probability of the two molecules coming together. As the distances between the interacting species grow less, then the normal dispersion interactions become increasingly important.

As an active dipolar molecule present in large excess, water is an active participant in the whole adsorption process. Water molecules readily form hydrogen bonds both with other water molecules and with appropriate groups in dye molecules or adsorbing surfaces. It is due to its powerful hydrogen bonding capacity and the resultant development of a complex structure in liquid water that it is relatively nonvolatile rather than a gas with a boiling point of −161°C. As a result of its structure and bonding behaviour, water plays a role in dye adsorption in two ways additional to those already discussed.

Due to their involvement in hydrogen bonding, water molecules compete with other bonding forces, leading to disruption of cohesive forces in textile fibres, with resultant swelling and opening of the physical structure to allow large dye molecules to penetrate the polymer matrix. They can also compete with dye molecules for adsorbing centres in the fibre. The latter effect is readily detected in the adsorption of nonionic disperse dyes by cellulose acetate or poly(ethylene terephthalate), since these dyes can be applied from the vapour phase in the absence of water as well as from aqueous dyebaths. Although both of these textile substrates are regarded generally as being hydrophobic and do not swell significantly when immersed in water, they do take it up to at least as great an extent in molecular terms as they do dye molecules (a 5% uptake of a dye with relative molecular mass 450 is equivalent to only a 0.2% water uptake in molar terms). It is found that, in the absence of water, these textile substrates adsorb far more dye than they do from an aqueous bath. Moreover, fibres dyed to heavy shades from the vapour phase desorb dye rapidly when they are exposed to an atmosphere containing water. It is not possible to make the same comparison with natural fibres, since they have to be dyed from aqueous baths, but there is no reason to suppose that the situation in these cases differs fundamentally from that operating with disperse dyes.

The second effect arises from the extensive structure of liquid water. When a large molecule is dissolved in water its presence results in extensive disruption of the structure with a consequent rupture of hydrogen bonds. Thus the free-energy gain which results from the dissolution is partially offset by the energy consumed in breaking up the water structure. The presence of ionic centres in the solute causes polarisation of the water molecules, thus favouring interactions between oppositely charged centres which can result in the release of water molecules. Of significance also is the effect of nonpolar molecules or substituents of molecules. These cause extensive structuring of the water to an extent which has led some writers to use the term 'iceberg' to describe the state of the water in the vicinity of the solute. The structuring arises from the combined effects of hydrogen bonding in the water and dispersion force interactions between the water and the nonpolar moiety of the solute, which lead to fully bonded water molecules similar to those in ice in a solvent envelope around the solute. The loss of entropy which results provides a driving force which can lead to a reorganisation of the system to minimise the amount of water disrupted, and this can be achieved by including two or more solute molecules within one 'envelope'. Thus, the system behaves as if the nonpolar moieties were bonding together. This phenomenon which is readily observed with detergents is called a hydrophobic interaction.

From this brief consideration of binding forces it should be clear that in devising a dyeing process the extent of assistance or impedance which can be expected from coulombic interactions has to be taken into account and the molecule designed to develop dispersion and dipolar binding forces to the degree required. This process is considerably assisted because a dye derives its colour from the fact that electrons in the molecular orbital are easily raised to an excited state by low energy radiation, namely visible light. Thus the same electrons are very readily perturbed in a dispersion force interaction. Thus, dyes with extended conjugation and high molecular mass would be expected to show high affinity for fibres as well as a tendency to form aggregates in solution due to self-bonding. The presence and the position of sulphonic acid groups in such molecules would also be expected to be important in modifying these bonding characteristics.

9.2.2 *Dyeing equilibria and kinetics*

After a material has been immersed in a dye solution, adsorption takes place until a steady state or equilibrium has been reached. At this point the percentage of the total dye present which has been adsorbed is termed the percentage exhaustion of the dyebath. The experimental values of the corresponding concentrations of the dye in the solution (D_s) expressed in appropriate mass units per litre and in the substrate (D_f) expressed in the same units per kilogram may be recorded in terms of an isotherm or an isostere. An isotherm expresses the relationship between D_f and D_s at a fixed temperature, while an isostere relates D_s to temperature at a fixed D_f. In practice, three isothermal relationships are observed, which can be expressed mathematically:

Partition isotherms	$D_f = k(D_s)$
Langmuir isotherms	$(D_f)^{-1} = k(D_s)^{-1} + k_1$
Freundlich isotherms	$D_f = k_1(D_s)^n$ (where $n < 1$)

Partition isotherms are observed with nonionic disperse dyes both on the relatively hydrophobic substrates to which they are normally applied, viz. polyesters, cellulose acetates and polyamides, and on natural fibres on which they are relatively weakly adsorbed. The partition isotherm is observed also in the case of ideal gases but this should not be taken to suggest that disperse dyes exist in ideal solutions, there being other more likely explanations.

The Langmuir isotherm was derived theoretically and shown to fit situations in which the adsorbate is taken up by a restricted number of fixed sites on the adsorbent. If the concentration of fixed sites is expressed as S with the adsorption and desorption processes treated as first order reactions:

$$\text{The rate of adsorption} = k_a(D_s)(S - D_f) \tag{9.1}$$

$$\text{The rate of desorption} = k_d(D_f) \tag{9.2}$$

In the steady state of equilibrium the two rates are equal so that:

$$k_a(D_s)(S - D_f) = k_d(D_f)$$

$$(D_f)^{-1} = (k_d/k_aS)(D_s)^{-1} + (S)^{-1} \tag{9.3}$$

The normal test for the operation of a Langmuir isotherm is a plot of $(D_f)^{-1}$ against $(D_s)^{-1}$. If a rectilinear relationship is observed, the line may be extrapolated to determine $(D_f)^{-1}$ at $(D_s)^{-1} = 0$, which gives the value of

$(S)^{-1}$, allowing the site concentration or saturation value to be calculated. It should be noted that reciprocal plots are not sensitive to actual curvature in the relationship and this, combined with the normal scatter of even good experimental results, may imply applicability of the isotherm when strictly speaking it does not exist. Consequently, the apparent applicability of the Langmuir isotherm should not be taken as proving that the adsorption is a simple case involving dye taken up by identifiable single sites. On the other hand, the reciprocal plot method can provide a good method of estimating saturation values. Acid and basic dyes of low relative molecular mass and showing little or no tendency to aggregation in solution, as well as simple inorganic acids, have been shown to follow the Langmuir isotherm relationship when adsorbed on amphoteric fibres such as wool or polyamides or, in the case of basic dyes, on acrylic fibres.

The Freundlich isotherm is an empirical relationship which has been found to apply in many adsorption situations. The normal test is to plot $\log(D_s)$ against $\log(D_f)$ giving values of k_f and n. Again, it should be remembered that log/log plots are very insensitive to curvature, and theoretical conclusions should be drawn with care. The main point regarding the Freundlich isotherm is that although the plot of (D_f) against (D_s) shows negative curvature, it does not predict any saturation value. Thus, this isotherm is generally taken to imply that the adsorbate has a tendency to aggregate on the adsorbing surface, building up multilayers. This situation is considered theoretically in another important isothermal relationship, the BET (Brunauer, Emmett and Teller) family of isotherms. However, due to the strong tendency of the direct dyes to aggregate in solution to a degree which is very difficult to determine, such refinements as applying the BET equations to such dyebath equilibria are not justifiable.

The observation of the applicability of the Langmuir isotherm to the uptake of simple acids by wool encouraged Rideal, Vickerstaff and others to apply classical thermodynamic concepts to dyebath equilibria. This leads to the following relationships,

$$\mu_s = \mu_s^0 + RT\ln a_s \qquad (9.4)$$

$$\mu_f = \mu_f^0 + RT\ln a_f \qquad (9.5)$$

where the μ terms are the chemical potentials of the dye in the solution and fibre phases respectively, the μ^0 terms are the corresponding standard-state chemical potentials and the a terms are the corresponding activities. At equilibrium the chemical potentials in the two phases become equal, so that:

$$\mu_s^0 + RT\ln a_s = \mu_f^0 + RT\ln a_f \qquad (9.6)$$

$$\mu_s^0 - \mu_f^0 = -\Delta\mu^0 = RT\ln a_f/a_s \qquad (9.7)$$

The term $-\Delta\mu^0$ is the change in standard chemical potential or standard molal free energy resulting from the transfer of dye from the solution to the fibre surface and is the *affinity* of the dye for the substrate under the relevant conditions. Equating concentrations with activities, it can be seen that a_f/a_s equals k_a/k_d in the Langmuir isotherm equation, so that

$$\Delta\mu^0 = RT \ln D_f(D_s(S - D_f))^{-1} \tag{9.8}$$

Further to this, from the standard thermodynamic relationship,

$$\Delta\mu = \Delta H - T\Delta S \tag{9.9}$$

$$\ln a_s = -(\Delta H/R)(1/T) + \Delta S/R + \ln a_f \tag{9.10}$$

Again, equating activities and concentrations, $\ln D_s$ may be plotted against $1/T$ at a constant D_f to give a value for the enthalpy of the adsorption. In fact, the underlying assumption is more justified in this case than in equation (9.8), since the error involved in the case of equation (9.10) appears as a constant log function with $\ln D_f$. The information needed to calculate a value for H is easily derived from a set of isotherms.

There are many reasons why the application of classical thermodynamics to dyebath equilibria may be regarded as being optimistic. The textile substrates are rarely sufficiently chemically stable to avoid the presence of dissolved degradation products in the dyebath, thus distorting the equilibrium. The dyes frequently aggregate both in solution and the adsorbed state. A combination of mechanisms of adsorption may well be involved, as has been discussed, and this may itself vary with adsorbed concentration. However, providing that mechanistic theories of adsorption are not based too much on reliance on thermodynamical analysis, these points are of less importance than the fact that by using the thermodynamic approach experimental data can be summarised extremely conveniently in two parameters, $\Delta\mu^0$ and ΔH, and the behaviour of different dyes can be compared systematically providing that their structures, in particular the extent of their conjugation, do not differ too much. It is clearly not possible to compare meaningfully affinities, etc., on different substrates.

There are several approaches to the subject of dye adsorption kinetics. The elementary processes involved are as follows:

(i) Mass transport of dye to the adsorbing surface
(ii) Material transport across the fibre/dyebath interface
(iii) Diffusion of dye molecules in the textile material
(iv) Interaction of the dye with binding centres.

Each of these four processes can, depending on conditions, become rate-determining. They are governed by different mechanisms. Process (i) is totally governed by the fluid mechanics of the system. Process (ii) is governed by a combination of fluid mechanics, the mechanisms of adsorption and diffusion, and the fabric or yarn structure. Processes (iii) and (iv) behave similarly to a chemical reaction in that they are governed by the temperature, concentrations, etc., as well as the polymer structure.

The approaches to stages (i) and (ii) are essentially phenomenological and are concerned more with the relationships between the environmental parameters of the system than with its chemistry. Thus, chemical engineering concepts, the thermodynamics of irreversible processes (TIP) and network thermodynamics have each been brought to bear on the problem with, in some cases, important results. However, these topics lie outside the scope of the present treatment, which has to be confined to stages (iii) and (iv). The diffusion of dyes in fibres is, in thermodynamic terms, an entropy-driven situation. It is governed by the same kind of linear relationship as are very many spontaneous processes, in this case Fick's equation,

$$J = -D \,\mathrm{grad}\, C \tag{9.11}$$

which relates the flux, J, to the gradient of concentration through a constant, the diffusion coefficient D. For the particular case of one-dimensional material transport, using cartesian coordinates,

$$J = -D(\mathrm{d}C/\mathrm{d}x) \tag{9.12}$$

which is usually termed Fick's Law. Since the dimensions of J are $\mathrm{m}^{-2}\,\mathrm{t}^{-1}\,\mathrm{m}^{-2}$ and those of the concentration gradient are ml^{-4}, it can be seen that the diffusion coefficient has the dimensions $\mathrm{m}^2\,\mathrm{t}^{-1}$.

In the experimental determination of diffusion coefficients in textile substrates two problems arise. Firstly, conditions have to be established such that dye molecules are supplied to the surface at a rate which exceeds the rate at which they diffuse into the polymer. This is less easy than it seems, because the rate of flow of a liquid past a fixed surface varies from zero to the general velocity with distance away from the surface. Thus, the situation readily arises in which the surface layers of solution become depleted. The layer in which depletion occurs is termed the diffusional boundary layer, and unless it is reduced to a minimum thickness the diffusion coefficients calculated from the experimental data are dependent on the circulation rate. This restricts investigations to studies on single fibres or sheets, as all other forms prevent effective circulation near the surface. The second problem arises from the fact that the diffusion coefficients of dye and other ions in fibres are concentration-dependent, due to potential and other effects, and so vary down the concentration gradient. Such effects can be offset by calculating a diffusion coefficient relative to some arbitrary standard state or by studying trace atom-labelled dye molecules in a system without a chemical potential gradient. Another factor which does not arise often, but has to be remembered, is the possibility that the adsorption process leads to physical changes in the substrate such as fibre swelling, with obvious implications for the observed diffusion coefficient. The mathematics of the solution of Fick's equation in the case of a concentration-dependent diffusion coefficient are complex, and generally analytical solutions are employed (Jost, 1952; Crank, 1956). Of these the simplest is that due to Boltzmann (1894), who derived an expression for the diffusion coefficient at concentration c, D_c,

$$D_c = \frac{1}{2t}\left[\frac{dx}{dc}\right]\int x\,dc \qquad (9.12)$$

which can be solved by numerical methods.

Tracer diffusion studies are based on the use of dyes labelled using, for example, ^{35}S or ^{14}C mixed with unlabelled dye. This allows the diffusion to be studied effectively without mass transfer, and with a potentially constant diffusion coefficient. Using this method it has been shown that for the chloride ion in nylon fibres the diffusion coefficient is a harmonic average of two coefficients reflecting site-bound chloride ions and chloride ions migrating in free entrained water (Chantrey and Rattee, 1969). It has also been shown that polyamide fibres have a previously unsuspected surface layer which is relatively resistant to the diffusion of dyes (Coates *et al.*, 1980). The thickness of this layer is dependent on spinning conditions and its presence affects dyeing behaviour. Similar effects have been shown to arise in wool dyeing.

Diffusion studies have their greatest value in relation to the dyeing properties of the polymer rather than the dye, and the migration of dye molecules through the polymer matrix provides information which it is not possible to obtain by other means. Below a certain characteristic temperature T_g, the glass transition temperature, amorphous polymers exist as a rigid glass, while above the T_g they become rubbery and behave like a highly viscous fluid. Useful textile polymers have a T_g below normally encountered temperatures, otherwise they would be unwearable under cold conditions. It has been shown that, below T_g, dyes can move only through solvent-filled defects in the polymer structure, while above T_g these do not exist and the dye must move through the *free volume* of the polymer (Murayama *et al.*, 1968). This represents that fraction of the total volume which is not occupied by the polymer itself. It exists in the form of voids appearing and disappearing in the polymer matrix. Clearly, only those voids which are big enough to accommodate the diffusing dye molecule will be involved in the diffusion process, and consequently a diffusional free volume must be defined which relates to the size of the dye in question and is in any case less than the polymer free volume.

Reactive dyes present a quite different picture from that of reversibly adsorbed dyes due to the need to take into account the influence of the reaction on the uptake and kinetics if dyeing. These matters are dealt with extensively elsewhere and only a brief introductory account is necessary here (Rattee, 1978). Considering first the reaction between a reactive dye and an alcohol in solution, three reactions are involved:

$A{-}OH \rightleftharpoons AO^- + H^+$	(ionisation of the alcohol)
$AO^- + D \rightarrow AOD$	(alcoholysis)
$HO^- + D \rightarrow HOD$	(hydrolysis)

The respective rates of alcoholysis and hydrolysis are given by:

$$df/dt = k_f[D]_t[AO^-]_t \quad \text{and} \quad dh/dt = k_h[D]_t[HO^-]_t \qquad (9.13)$$

where k_f and k_h are the bimolecular rate constants of fixation and hydrolysis, and time equals t.

Generally experimental conditions may be selected to give a large excess of the alcohol so that $[AO^-]$ and $[HO^-]$ are constants. This means that first-order kinetics apply and the total rate of disappearance of reactive dye is given by:

$$-dD/dt = df/dt + dh/dt = \{k_f[AO^-] + k_h[HO^-]\}[D]_t \qquad (9.14)$$

so that

$$k_{tot} = k_f + k_h = k_f[AO^-] + k_h[HO^-] = 1/t \ln[D]_0/[D]_t \qquad (9.15)$$

where k_f and k_h are the first-order rate constants of alcoholysis and hydrolysis. However, the ionisation of the alcohol is pH-dependent with an equilibrium constant K_a given by:

$$K_a = [AO^-]K_w/[AOH][HO^-] \qquad (9.16)$$

where K_w is the dissociation constant of water. Since alcohols are weak acids, the degree of ionisation is small at any pH used for reactive dye fixation, so that effectively [AOH] is independent of $[AO^-]$ and $[H^+]$, and the value of [AOH] used may be the gross concentration. Therefore:

$$k_{tot} = k_f[K_a/K_w][AOH][HO^-] + k_h[HO^-] = 1/t \ln[D]_0/[D]_t \qquad (9.17)$$

Consequently, the reaction behaves as a first-order reaction if $[OH^-]$ is kept constant, and there is an apparent bimolecular rate constant given by:

$$K_{tot}/[HO^-] = k_f[K_a/K_w][AOH] + k_h \qquad (9.18)$$

Equation (9.18) may be simplified by combining the dissociation constants of the alcohol and water so that:

$$K_a[AOH]/K_w = K_0 \qquad (9.19)$$

$$k_{tot}/[HO^-] = k_f K_0 + k_h \qquad (9.20)$$

and the ratio of reactivities is

$$k_f/k_h = K_0\{k_{tot}/(k_h[HO^-]) - 1\} = R \qquad (9.21)$$

It can be seen from equation (9.21) taken with equation (9.18) that the reactivity ratio is a pH-independent constant varying with k_h and the properties and the concentration of the alcohol.

The efficiency of the reaction with the alcohol is given by the ratio of the reaction rates d_f/d_h, i.e.:

$$d_f/d_h = k_f[AO^-]/k_h[HO^-] = k_f K_0/k_h \qquad (9.22)$$

when the ionisation of the alcohol is proportionally small. It can be seen that the properties of the alcohol are important, and in comparing reactivity towards different alcohols this must not be neglected.

Many of the reactive dyes based on heterocyclic systems, particularly the *s*-triazinyl and 2,4-diazinyl systems, exhibit reaction rate constants which vary with pH. Horrabin (1963) has suggested this is due to an acid–base equilibrium:

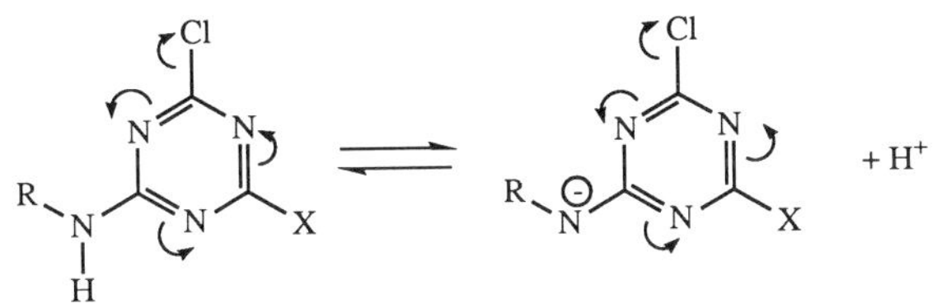

in which the deprotonated molecule is less reactive than the normal form. Some doubt has been cast on this hypothesis, since it has been shown that the way in which the reactivity varied with pH and temperature together did not conform to what would be expected of an acid–base equilibrium (Murthy and Rattee, 1969). It was considered that changes in the solution state may be the cause of the effect. Further to this, Rys and Zollinger (1966) have shown the existence of tautomeric forms of the dye with different reactivities due to changes in ring conjugation.

Dyeing cotton with reactive dyes involves a heterogeneous system requiring consideration of dye diffusion with simultaneous reaction. The rules governing the reaction are the same as those applying to alcohols in solution, but allowance must be made for the rate at which the dye is delivered to the reaction sites. The diffusion equations have been solved for a variety of conditions including chemical reaction, and it has been shown (Sumner and Weston, 1963) that:

$$Q_t = [\mathrm{D_f}](t + 1/2k_\mathrm{f})(Dk_\mathrm{f})^{\frac{1}{2}} \qquad (9.23)$$

where Q is the quantity of dye reacting in time t and D is the diffusion coefficient. The rate of fixation is given by:

$$\mathrm{d}Q/\mathrm{d}t = [D_\mathrm{f}](Dk_\mathrm{f})^{\frac{1}{2}} \qquad (9.24)$$

and the efficiency of fixation, E, after allowing for the negative surface potential of the cellulose and the consequent differences between internal and external phases is given by:

$$E = \frac{[D_\mathrm{f}]}{[D_\mathrm{s}]}\left[\frac{DRK}{k_\mathrm{h}[\mathrm{HO_s^-}]}\,\frac{\exp(e\psi/kT)}{(1 + \exp(2e\psi/kT)^2)}\right] \qquad (9.25)$$

in which ψ is the surface potential of the cotton due to the ionisation of hydroxyl groups. Thus, the main factors governing the efficiency of fixation can be seen to be the substantivity ratio, $[D_f]/[D_s]$, and the surface potential factor. The value of the electrical potential function falls with ψ when it is more negative than 14 mV, as it always is with cellulose. Thus, factors which minimise the influence of the substantivity ratio or optimise its value are of primary practical importance. Fixation at heavy shade depths will be less than at paler depths due to the higher value of ψ which is brought about by the higher concentration of negative charges. Efficiency will also decrease with increasing pH due to the reduction of $[D_f]$ and a fall in the substantivity ratio as a consequence of the increased negative potential of the surface.

The behaviour of dyes in solution is of great importance. A dye molecule capable of participating in hydrophobic interactions with a textile fibre or which has been designed to develop strong perturbational effects must be capable of participating in analogous dye–dye interactions. At the theoretical level, interactions which lead to thermodynamically nonideal behaviour clearly have to be taken into account. At the practical level, grosser effects leading to association in solution or the formation of aggregates can affect dyeing processes. The determination of the activity coefficients of dyes in solution is difficult, partly because of the very dilute nature of dye solutions used in practice (10^{-3}–10^{-6} molar) but also because of their strong tendency to interact with any available surface, including those of the apparatus being used to contain or examine the solution. For example, even the simple matter of determining the pH of an aqueous solution of a dye acid can in many cases be complicated by dye adsorption on to glass electrodes. Nevertheless, studies using vapour pressure measurements (Chadwick and Neale, 1958; Friedrich and Stammbach, 1968) have demonstrated that dye–dye association is common even with very simple dyes.

The aggregation of dyes is more easily studied, although the results frequently serve only to show how complex a dye solution can be. Two methods have been used. The first is based on the measurement of the diffusion coefficient of a dye in solution through a glass sinter. The diffusion coefficient can be related to the aggregate radius via the Stokes–Einstein relation:

$$D = RT(6\pi\eta Nr)^{-1} \qquad (9.26)$$

in which η is the viscosity, N is Avogadro's number and r is the radius of the diffusing species. With dye ions there is a problem that the co-ions, e.g. Na^+, diffuse more rapidly than do the dye ions, and they set up a diffusion potential which accelerates their flux across the interface. Theoretical treatments allowing for corrections to be applied using ion-mobility data from conductivity measurements may be applied, and by using this technique the very marked aggregation of some direct cotton dyes has been demonstrated (Robinson and Moilliet, 1934). Diffusion coefficients in solution have been measured also by polarography (Hillson and McKay, 1965) to provide similar results. The second approach to the study of aggregation is to utilise the fact that the dye–substrate interaction involves the same electronic orbitals as absorb light and create the colour. Thus, aggregation produces a shift in the absorption spectrum of dye solutions, which can be used to calculate association constants, subject to the assumption that only dimers are formed. The temperature jump-relaxation technique has also been used to measure extremely small spectral shifts. C.I. Acid Red 66 has been shown by this technique to be 66% dimerised at 20°C even at a concentration as low as 3×10^{-5} molar (Hague *et al.*, 1971).

Dyes in solution are, of course, able to interact with other solutes which may be present, and this has been put to use in the development of surfactants which, added to dyebaths, complex with the dye present to modify dyeing behaviour. Such interactions have been studied extensively (Rattee, 1953; Craven and Datyner, 1967). The practical application of this approach is discussed later in connection with particular situations.

9.3 Classes of colorants and their applications

Dyes and pigments may be classified by their chromogenic systems, and this has been achieved in the Colour Index (C.I.) published jointly by the Society of Dyers and Colourists and the American Association of Textile Chemists and Colorists. The Colour Index is the most comprehensive and authoritative reference work in the field. Technical information relating to fastness properties and methods of synthesis is provided. The classification in which dyes are identified by C.I. structure numbers according to chromogen cross-refers to a further classification in which the dyes are collected into groups according to use. For convenience the treatment which follows is based on the latter.

There are four main methods of colouring textiles.

(a) The direct adsorption and attachment of water-soluble dyes. This category includes:
  (i) Acid dyes
  (ii) Direct cotton dyes
  (iii) Basic dyes
  (iv) Reactive dyes
  (v) Disperse dyes
(b) The adsorption of temporarily solubilised dyes converted to an insoluble form by an after-treatment. This category covers:
  (vi) Vat dyes
  (vii) Sulphur dyes
(c) The adsorption by the fibre of components which react to form an insoluble pigment within the fibre. The main dye class behaving in this

way is:
(viii) Azoic dyes

(d) The pigmentation of fibres by either mass pigmentation or the attachment of pigmented resins. These processes do not involve dyes and necessitate consideration of a ninth class of colouring matter, i.e.:
(ix) Pigments

These nine groups are now considered in detail.

9.3.1 *Acid dyes*

Substituting the chromogen with a sulphonic acid group produces an anionic dye. The adsorption of such dyes by a textile fibre introduces on to the adsorbing surface a strong negative potential which can prevent further dye adsorption. However, wool, silk and nylon fibres are amphoteric in character, containing carboxyl and protonated amino groups. Under acid conditions the carboxyl groups can be back-titrated, leaving a net positive charge which enables dye anions to be adsorbed.

Wool is a polypeptide of complex physical structure consisting of some 18 condensed α-amino acids in polypeptide chains. While the characteristics of wool fibres depend on the presence and arrangement of all the amino acids, of major importance are the amino acid cystine, which provides disulphide cross links between polypetide chains, the amino acids lysine and arginine, which supply side-chain amino groups, and the amino acids aspartic acid and glutamic acid, which provide carboxylic acid side groups. Silk contains most of the same amino acids, albeit in a different arrangement, but contains no cystine. Polyamide fibres are prepared by the condensation of a dicarboxylic acid and a diamine or by the self-condensation of an ε-amino acid to give a synthetic polymer with terminal amino and carboxylic acid groups. For example, nylon 6.6 is prepared by the condensation of hexamethylene diamine and adipic acid, while nylon 6 is prepared by the self-condensation of ε-amino caproic acid. The adsorption of acid dyes by amphoteric fibres has sometimes been discussed as if the charge centres were the driving force for the adsorption. However, if this were all, acid dyes would have much the same behaviour as, and would differ little from, simple anions. It is now understood that the role of the positive surface potential which exists under acid dyebath conditions is to enable adsorption to proceed without building up a potential barrier against the dye molecules. The driving force for the adsorption comes from other forces, as already discussed. Thus, the dyeing process is governed by the relative strengths of the two influences, coulombic forces which assist or oppose adsorption depending on pH conditions, and perturbational forces which are always attractive. With a simple acid *levelling* dye, e.g. (**15**), the coulombic repulsion can be dominant, so that a high concentration of positive charges has to be built up by dyeing at a fairly low pH of 2.5–3.0. To control the rate of adsorption and assist uniform dyeing, an inorganic salt is usually added to provide colourless competing anions which increase the frequency of the adsorption–desorption process. Sulphate ions have relatively little affinity compared with acid dyes, but their ability to reduce the positive surface potential so that they have to be displaced by the dye anions creates a kind of competition.

By increasing the number of dipolar substituents or by incorporating large nonpolar substituents it is possible to make the dye less susceptible to electrostatic repulsion. More complex dyes such as (**16**) and (**17**), for example, show good exhaustion at a dyebath pH as high as 5–6, when the back-titration of carboxyl groups present is very much less. In these cases strong bonding forces supply the energy needed to overcome the potential barrier, with the process being assisted by the uptake of sodium ions present as co-ions. In the case of (**17**), the affinity is enough to allow wool to be dyed from neutral or even alkaline dyebaths, providing that a fairly high salt concentration is used. The increased molecular complexity of dyes such as (**16**) and (**17**) also reduces the ease with which the dye molecules can diffuse through the polymer structure. Consequently, the dye is more resistant to desorption processes such as washing or other wet treatments, particularly alkaline treatments, and acid dyes of this type are termed *neutral dyeing acid* or *acid milling dyes*. (Milling is a severe alkaline finishing process applied to certain types of wool cloth.) An interesting feature of (**17**) and similar structures is their marked tendency to develop hydophobic interactions with soap or surfactants. This means that if any dye is desorbed during alkaline washing it complexes with the detergent and is prevented from dyeing back on to adjacent white material. The same tendency can be employed usefully to control the adsorption process in the dyeing operation, and this can be important, since the relative lack of influence of coulombic forces means that small anions cannot be used to control adsorption, as is the case with simple acid dyes.

Using selected dye intermediates it is easy to produce acid dyes containing *o,o′*-dihydroxyazo functionalities or similar configurations which are able to form complexes with chromium. Such dyes are normally classified separately as *mordant dyes*, but they may be considered as a subgroup of the acid dyes. Dyes of this kind were first applied to wool previously mordanted by boiling the material in a dilute potassium dichromate solution. The latter reacted with reducing groups in the wool resulting from the hydrolysis of disulphide bonds, leaving the fibre containing chromic ions. Application of the appropriate dye to the mordanted fibre produces a 1:1 chromium-dye complex which is extremely fast to washing due to the chelation of the metal atom with both dye and fibre. This method fell into disuse when it was shown that a more reliable and rapid procedure was to apply the dye as a regular acid dye and then after-treat the dyed fibre with the dichromate in the after-chrome process. This method gives dyeings of very high washing-fastness combined with high

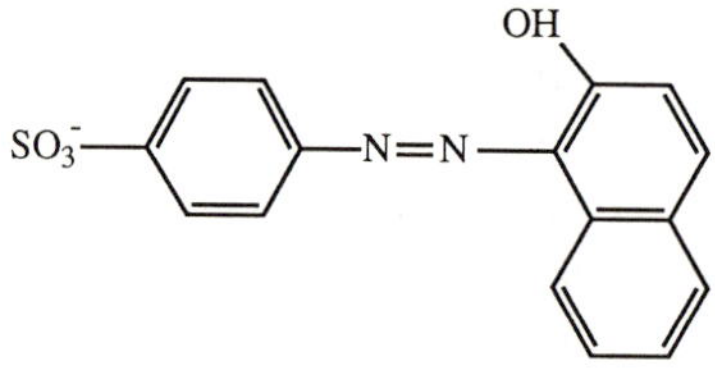

(15) (C.I. Acid Orange 7)

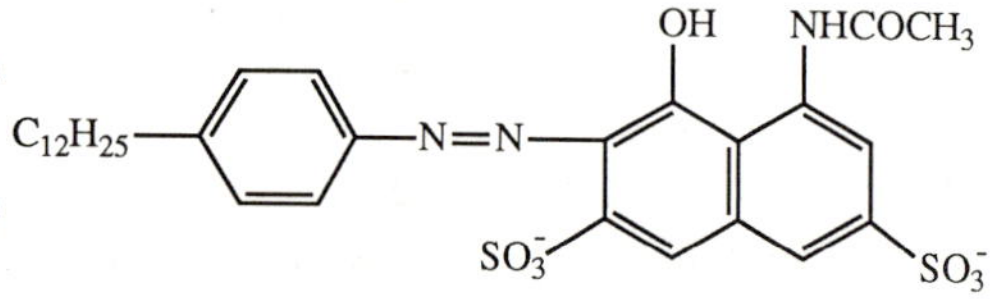

(17) (C.I. Acid Red 138)

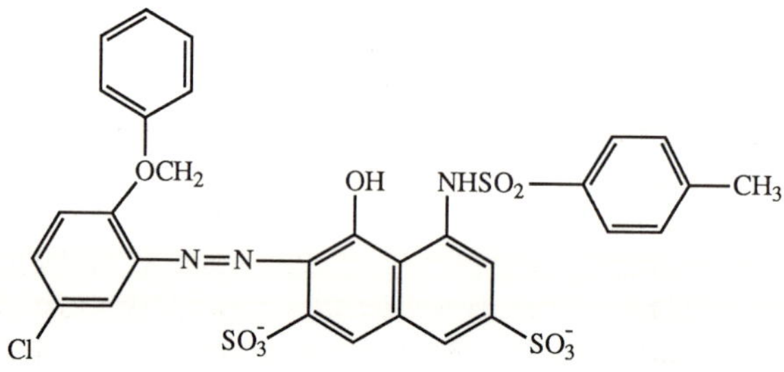

(16) (C.I. Acid Red 172)

lightfastness, resulting from the stabilisation of the azo group to light through chelation, without the harshness of handle produced by premordanting the wool. Another version of the procedure, the metachrome process, which as the name implies involves simultaneous application of the dye and the dichromate, was introduced in 1900. Dyes suitable for metachrome application had to be stable to the oxidative action of the dichromate in the dyebath, and the method was more restricted than the after-chrome method. Both the after-chrome and the metachrome methods continued to be used extensively until relatively recently, when environmental legislation seriously discouraged dyers from discharging spent dye liquors containing chromium. However, in 1920 the preformed 1:1 chromium complex dyes appeared, e.g. (**18**).

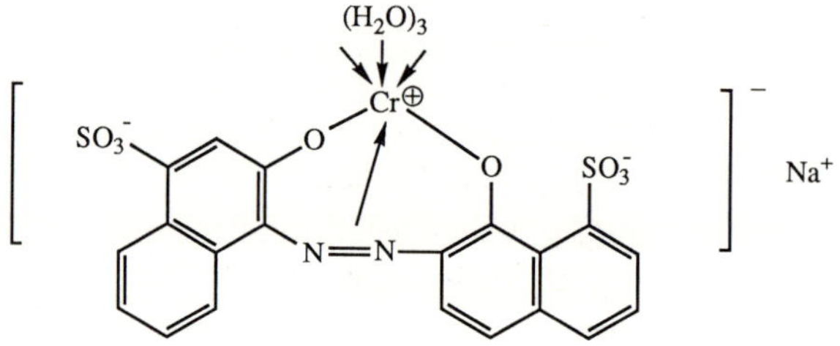

(18) (C.I. Acid Blue 158)

Because they were preformed complexes and no treatment of the wool with chromium salts was necessary, reliable production of shades was facilitated. However, it was not easy to produce level results because the dyes were very strongly adsorbed at normal acid dyebath pH values. By lowering the dyebath pH to 1.2–1.5 the tendency of the chromium atoms in the dye to chelate with the fibre was reduced, and this enabled dye adsorption to be controlled, albeit at the expense of fabric handle. Better methods were later developed which allowed a higher dyebath pH to be used, utilising the principle of temporarily complexing the dye molecules in solution with polyethanoxy surfactants (Valko, 1972). A much more satisfactory solution to the problem presented itself when it was found that the 1:1 chrome complexes could be reacted with a second complexing dye molecule under alkaline or weakly acidic conditions to form a 1:2 chrome complex, exemplified by (**19**). Such complexes formed from sulphonated dye intermediates were found to be very difficult to apply evenly, but a detailed study of the relationship of the ionic charge on the dye molecule and dyeing behaviour showed that sufficient solubility was conferred on dyes using even such weakly acidic groups as *N*-substituted or unsubstituted sulphonamido groups. Such dyes gave readily controlled dyeing rates in weakly acidic dyebaths. The development of level dyeing assistants that were either weakly cationic or nonionic surfactants capable of complexing with the dye and forming coacervates in the dyebath resulted in the very successful introduction of the Irgalan dyes by Geigy in 1949 (Schetty, 1955). In addition to the 1:2 chrome complexes which comprise the overwhelming majority of dyes of this type, some 1:2 cobalt complexes were also introduced to enhance the shade gamut. Due to the fact that in the 1:2 metal complex dyes all the coordination valencies of the metal atom are occupied within the dye molecule, no further complexing with groups in the fibre is possible. Consequently, the 1:2 metal complex dyes do not exhibit the very high fastness to alkaline treatments that is shown by the after-chrome dyes. Their advantage lies rather in their high lightfastness, their excellent compatibility of dyeing behaviour and the excellent appearance of the dyeings they produce.

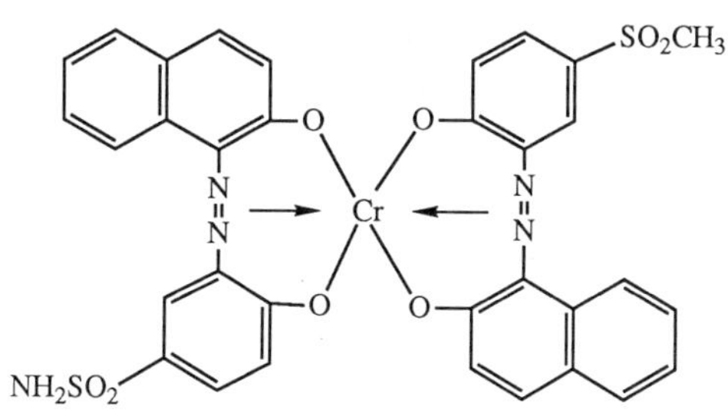

(**19**) (C.I. Acid Violet 78)

9.3.2 *Direct cotton dyes*

Cotton and other cellulosic fibres such as the various forms of viscose rayon consist of a polymer of 1,4–anhydroglucoside units (**20**).

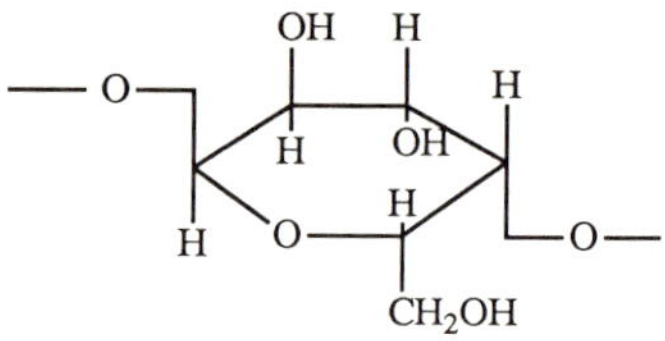

(20)

As a result of natural oxidation occurring during growth or subsequent processing, some of the pendant methylol groups are oxidised to carboxylic groups. There are no crosslinks between polymer chains and the polymer structure is strongly hydrophilic. Due to the high chain length and extensive hydrogen bonding between the chains, cotton cellulose is not water-soluble as is the case with many similar but smaller molecules. Due to the presence of the small number of carboxylic groups along the polymer chain, cellulose has a weak negative charge in dyebaths with a pH higher than 4. If the dyebath pH is raised above 8, some of the hydroxyl groups present on the methylol side chains ionise, increasing the negative charge significantly. The charge repulsion between polymer chains which results causes fibre swelling to occur. If the pH is raised further some of the other hydroxyl groups begin to ionise, and extensive swelling with polymer chain rearrangement can occur. Such extreme conditions are not normally encountered in dyeing situations.

The negative charge on the adsorbing surface of cellulose clearly creates a very unfavourable situation for the adsorption of anions. Although empirically based, the work of Böttiger and of Griess leading to the introduction of the direct cotton dyes pursued exactly the line of thought which would be suggested by our present understanding. By building up the molecular complexity of the acid dye and incorporating dipolar and similar substituents, sufficient dye–fibre bonding capacity can be introduced to offset the charge repulsion effect. This meant extending the length of the molecule by preparing bis- and tris-azo dyes. Furthermore, by applying the dye in the presence of an electrolyte such as sodium chloride, a high concentration of cations is made available to accompany the adsorbed dye molecules and nullify the negative surface charge. In essence, this is the way in which direct cotton dyes are applied.

Apart from the azo groups present which make a major contribution to dye affinity and other substituents normally present in acid dyes to modify the colour the main substituents employed to enhance direct dyeing behaviour are:

Benzoylamino groups 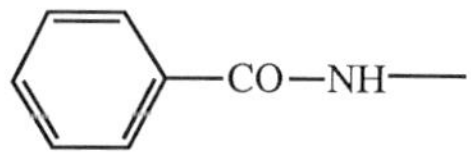

Ureido groups —NH—CO—NH—

Triazinyl systems 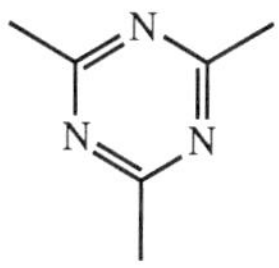

The second group is found in direct cotton dyes as a link between two chromogens, and the dyes can be produced by condensation of two molecules containing free amino groups with phosgene, e.g. (**21**).

The triazinyl systems, which are especially effective in conferring affinity on the dyes, are used in a similar way, but because it is possible to build the bridge in a stepwise manner, dyes can be made by bridging different chromogens, e.g. (**22**).

The way in which the substituents are introduced into the dye molecule is important. The normal configurations of an unsubstituted polyazo system is planar, and this allows optimum development of the molecular orbital. If a substituent is introduced that distorts the planarity of the molecule, the colour strength and the dye affinity, both of which depend upon the delocalised electron system in the molecule, are reduced significantly.

The direct cotton dyes have been subclassified in accordance with their relative needs for salt in the dyebath. Clearly, as the dye complexity is built up, the need for the addition of salt becomes less. The simplest of the direct cotton dyes require high concentrations of salt (Class A dyes), and the most complex exhaust with relatively little salt requirement (Class C dyes), while between the two are found the Class B dyes. Normally, dyes from the different subclasses would not be applied together.

Since the adsorption of the direct cotton dyes from the dyebath is promoted by the presence of salt, it follows that, after the dyeing is completed and the dyed goods are washed off, a major factor causing the dye to remain adsorbed has been removed. Therefore, it is not surprising that direct cotton dyeings do not possess good fastness to washing and other wet treatments. However, most washing treatments are of short duration, and providing that they are carried out at a low temperature at which desorption is slow the amount of dye lost is acceptable for certain end-uses. By carrying out dyeing at (typically) 85°C for 45 min, a satisfactory dyeing rate is achieved, while washing instructions would normally specify hand washing at 40°C. More serious than the loss of colour strength during wet treatment is the tendency of

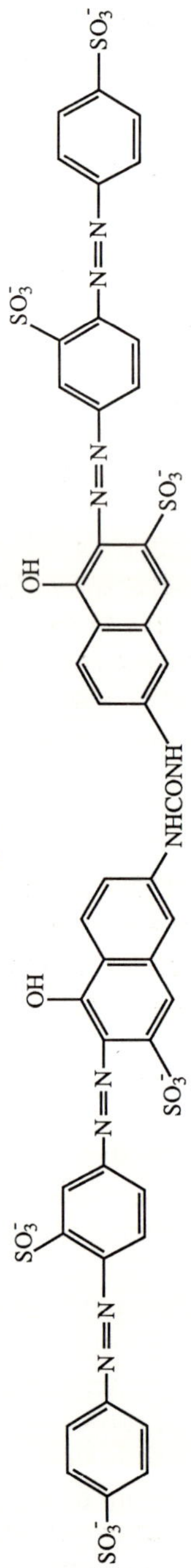

(21) (C.I. Direct Red 80)

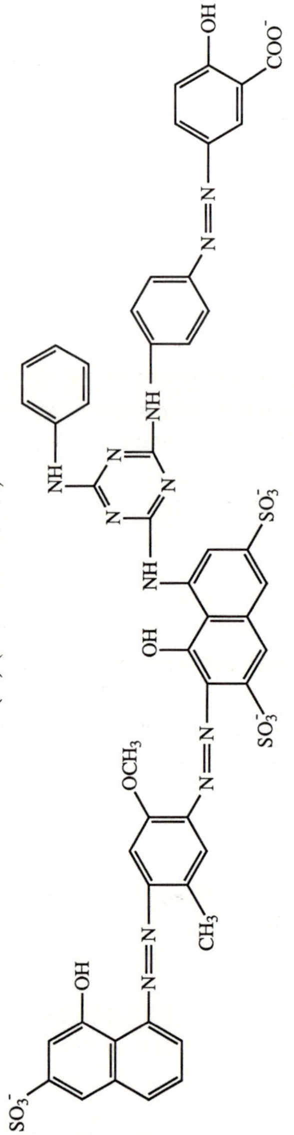

(22) (C.I. Direct Green 26)

desorbed dye to stain adjacent cellulosic materials. So far as can be determined, no attempt has been made to exploit dye–detergent interactions in a manner comparable with that described for acid milling and 1:2 chrome complex dyes. There have been many attempts to devise methods of after-treating direct cotton dyeings to produce a final dyeing of good fastness. Technically the most successful, although not all that widely adopted commercially, has been the use of direct cotton dyes carrying substituents with replaceable hydrogen atoms, for example amino, alkylamino, etc., which can be reacted with a so-called bridging compound to bond the dye chemically to the fibre. Such treatments are inevitably accompanied by a shade change, and any small proportion of the dye molecules escaping modification are available to produce a stain on adjacent material. Another form of after-treatment which has been used is to after-treat the dyeing with a high molecular weight cation, e.g. cetyl pyridinium bromide, to produce an insoluble salt. This is quite effective in improving fastness to perspiration and does assist washing-fastness. However, in a great many cases the lightfastness is significantly reduced by such treatments. With direct cotton dyes that have the capability of forming a stable copper complex due to the presence of an *o,o′*-hydroxyazo configuration it is possible to combine after-treatment with a cationic salt, and after coppering to obtain good lightfastness with improved washing fastness.

9.3.3 *Basic dyes*

An alternative name for this application class is the cationic dyes. This indicates clearly that they mirror the acid dyes in that they are water-soluble with a coloured cation. In a sense the basic dyes represent, as far as synthetic dyes are concerned, the oldest dye class, as Perkin's Mauveine (**23**) was a basic dye.

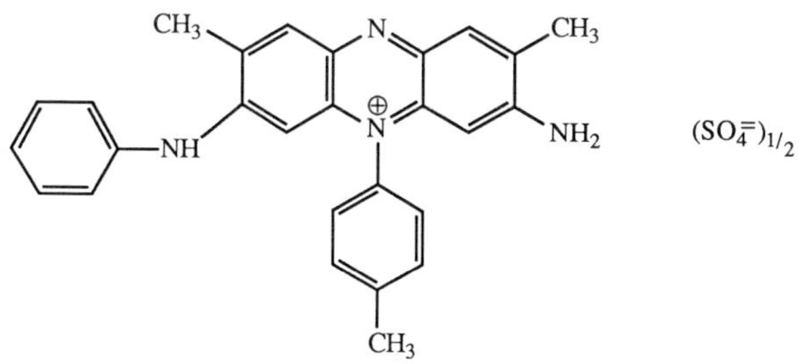

(23)

The same rules apply to basic as to acid dyes but in the opposite sense. Their adsorption on to wool and silk is promoted by using alkaline dyebath conditions, which develop a negative charge on the fibre, and is reduced in acid dyebaths. They are weakly adsorbed on to cellulosic fibres, and when first introduced were applied to mordanted material.

With the acid and direct cotton dyes, the sulphonic acid groups providing solubility are substituents and not an intrinsic part of the chromogen. Therefore, it is possible to modify the structure and the dyeing behaviour relatively independently of the dye charge. However, most basic dyes resemble Perkin's Mauve in that the charge on the cation which renders the dye water-soluble is an integral part of the chromogen. Modifying the dyeing behaviour is consequently less easy than it is with acid dyes.

The brilliant shades offered by many basic dyes makes them attractive as textile dyes from the point of view of the consumer and the designer, but their poor fastness to washing and to light rapidly limited their use and eventually led to their decline as far as textiles were concerned. For many years basic dyes were encountered mainly in nontextile applications. This situstion was changed dramatically with the introduction of fibres derived from polyacrylonitrile in 1948. These are straight polymers with the structure:

$$\left[-CH_2-\underset{\displaystyle CN}{\underset{|}{CH}}-\right]_n$$

where n has a value between 600 and 2000. Various modified or modacrylic fibres are also available, which are made by copolymerising acrylonitrile with other monomers. As a group the acrylic fibres have aesthetic qualities similar to wool. In order to catalyse the polymerisation, redox catalysts are used (sodium bisulphite/potassium persulphate) and this results in the presence of terminal sulphonato- and sulphato-groups. In addition, carboxylic groups may be present in copolymerised systems. Consequently, the acrylic fibres readily adsorb basic dyes. The technical importance of this arises from the fact that many basic dyes have considerably greater lightfastness on acrylic fibres than they do on any other textile substrate. Thus, the introduction of these fibres led to a resurgence of interest in basic dyes for textiles and an extension of the chromogens employed to produce coloured cations to include azo and anthraquinonoid systems. Many examples are now available in which the positive charge centre is isolated from the chromogen by a saturated carbon link.

9.3.4 *Reactive dyes*

In the discussion of acid, basic and direct cotton dyes it has been made clear that control over both colour and dyeing behaviour normally involves modifications to the chromogen. Since the more complex the chromogen the broader, in general, is the principal absorption band of the dye, it follows that the production of a direct cotton dye involves using chromogen systems that offer duller shades than are achievable with the simpler acid dyes. It was realised at an early stage that a valuable solution to this problem would be to develop dyes that achieved fastness by the formation of a covalent link between chromogen and fibre. However, the principal realisation of this idea, the fibre reactive dyes, took over sixty years. Chemical reaction as a means of producing fast dyeings has attractions also for the dyeing of wool, silk and polyamide fibres, although the problem here is not the production of fast bright shades so much as the relative difficulty in devising reliable level dyeing methods with the structures which confer fastness by virtue of their high molecular weight and slow diffusion in the fibre.

Early investigations of the chemistry of cellulose demonstrated that cellulose treated with strong sodium hydroxide solution to ionise the primary and some of the other hydroxyl groups may be benzoylated (Cross and Bevan, 1895). Starting with *p*-nitrobenzoyl chloride they prepared a cellulose derivative bearing a pendant nitro group, which they then reduced to amino. This was then diazotised and coupled to a phenol to give a coloured cellulose derivative in which the dye was covalently bonded to the fibre. While this early exercise showed that this kind of dye–fibre was possible, the means of its achievement did not provide a practical dyeing system. This had to await the work of Stephen and Rattee almost 60 years later, which resulted in the efficient reaction of chlorotriazinyl amino dyes to cellulosic fibres under mild conditions of alkalinity and temperature in practical dyeing times (Zeneca, 1954; Vicherstaff, 1957).

The first commercial reactive dyes for cellulose were effectively simple acid dyes bearing a chlorotriazinyl amino group, exemplified by (24). The successful

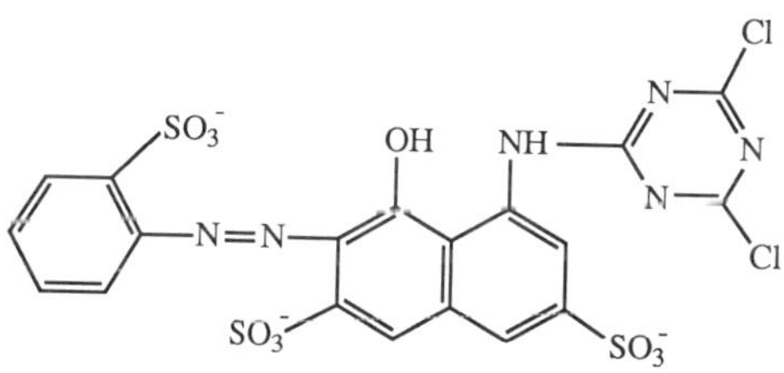

(24) (C.I. Reactive Red 1)

production and use of this type of dye depended on two observations. Firstly, it was shown that these very reactive compounds can be isolated and marketed as stable dye powders by preparing the final press cake (i.e. the filtered precipitated dye prior to drying and standardisation) at a buffered pH of 6.5, under which conditions the hydrolysis rate is at a minimum. Additionally, by incorporating in the dye powder suitable buffers, it was possible to prevent the pH from falling if any of the dye powder hydrolysed in the atmosphere to generate hydrochloric acid. Secondly, it was demonstrated that these dyes, while reacting with cellulose and with water at much the same rate at a pH of around 10, reacted preferentially with the cellulose if dyeing conditions were selected that caused the dye to be concentrated in the cellulose rather than the water phase. The physical chemical aspects of this situation have been discussed (section 9.2). The mechanism of the formation of the cellulose derivative in the case of the chlorotriazinyl dyes involves the nucleophilic substitution of the chlorotriazinyl ring by the ionised group in the cellulose as shown in Scheme 9.1.

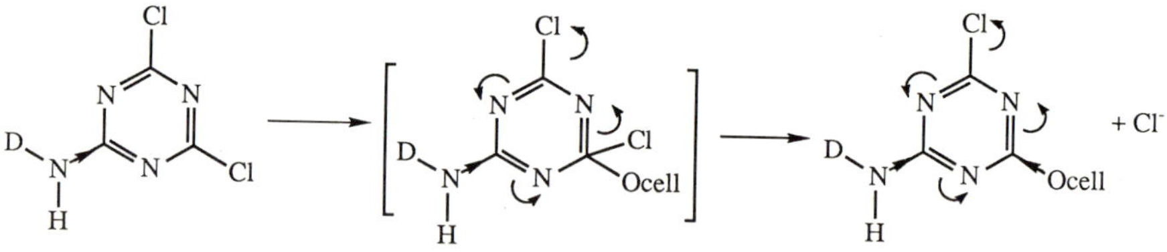

Scheme 9.1

Following the introduction of the dichlorotriazinylamino dyes a number of other halogenoheterocyclic systems were introduced for application by the same general methods, such as (**25**)–(**30**).

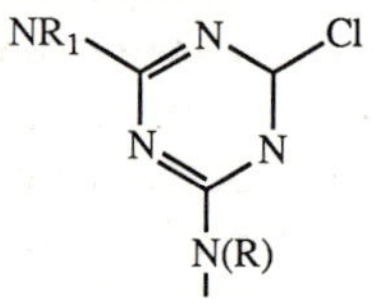

(**25**) monochlorotriazinyl

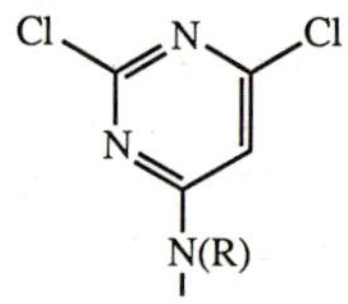

(**26**) dichloropyrimidinyl

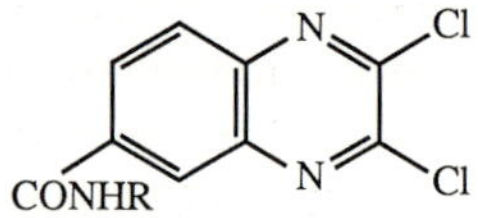

(**27**) dichloroquinoxalinyl

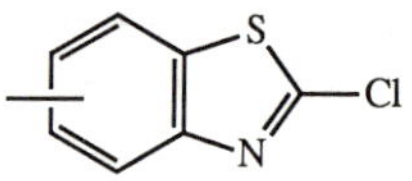

(**28**) chlorobenzothiazole

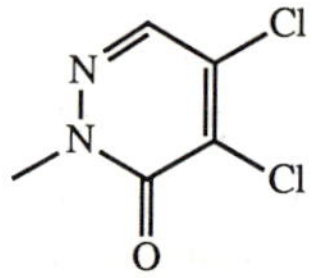

(**29**) dichloropyridazone

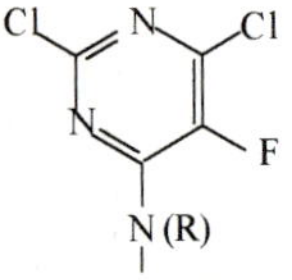

(**30**) dichlorofluoropyrimidinyl

An alternative approach was taken in the development of the chloroethylsulphonyl reactive system. The mechanism of reaction with this system involves nucleophilic *addition* following a general base catalysed elimination of HC1:

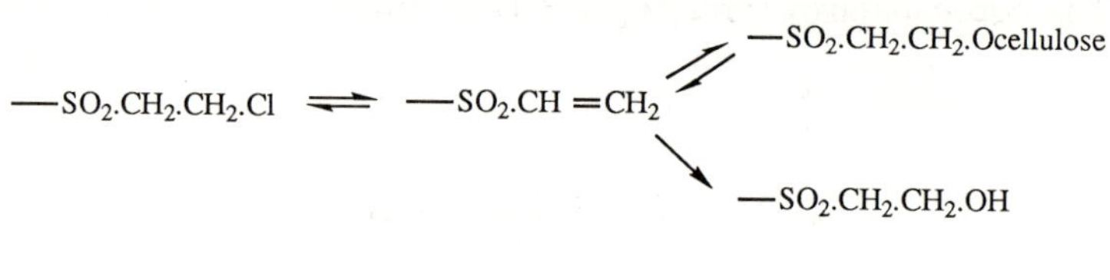

The reactive dyes established themselves very rapidly as a major dye class for cellulosic fibres due to their ease of application, broader shade gamut than could be obtained with other dyes from cellulose, and excellent fastness properties. With their introduction came new and simple application methods which offered continuous and semicontinuous dyeing methods at low capital initiation costs such as pad–dry–wash and pad–batch–wash. Very reliable printing methods also came to be established. A summary of the principle application methods is given.

(i) Exhaustion from an aqueous dyebath by salt addition followed by the addition of alkali to promote dye fixation. Using the most highly reactive dyes, the initial exhaustion stage is completed before the alkali is added in order to minimise hydrolysis of the dye in the dyebath. When dyes of lower reactivity and higher affinity are being applied the salt and alkali may be added together. Since the addition of alkali causes some ionisation of cellulose hydroxyl groups and the negative charge repels the dye anions, it is necessary to keep the dyebath pH as low as possible consistent with completing the reactions within a practical time. In practice this means a dyebath pH of 9–10 and the use of sodium carbonate rather than stronger alkalis. Depending on the properties of the fabric to be dyed, the reactivity of the dyes used and their affinity for the fibre, the dyeing temperature may lie between ambient temperatures and 85–90°C. At the end of the dyeing process the goods are rinsed and finally washed with successsive treatments until all unfixed exhausted dye has been removed.

(ii) Semi-continuous application (pad–batch) by impregnating the fabric with an alkaline dye solution using a padding mangle and storing or batching the wet material for the reactions to proceed at the ambient temperature. Under these conditions fabric contains no more than 70% of its own weight of dye liquor, whereas in exhaustion dyeing there is 5–10 times as much solution as fabric. These so-called 'short' liquor to goods ratios lead to high levels of dyebath exhaustion, and salt addition is not needed to achieve efficient fixation, because the fixation efficiency shows very little dependence on the substantivity ratio, $S = D_f/D_s$ (cf. equation 9.25). The reason for this is as follows. The total amount of dye present, D is given by

$$D = D_f W + D_s V \tag{9.27}$$

The liquor to fabric ratio, L, is V/W so that

$$D = SD_s VL + D_s V \tag{9.28}$$

$$D = D_s V(S + L)/L \tag{9.29}$$

Therefore

$$D_s V/D = L/(S + L) \quad (9.30)$$

Since S is generally more than 20 when L is small, the proportion of the total dye which is in the liquor is small and its influence on the overall efficiency is minimised. Dyes of higher reactivity are not stable in alkaline solution and special methods have been devised to produce continuous mixing of dye and alkali and minimise the time available for hydrolysis. The batching time depends on the pH of the padding liquor and the reactivity of the dye used. At pH 8 ($NaHCO_3$ used as alkali) with a dichlorotriazinyl dye, 24 h is required to complete fixation; at pH 12 (NaOH used as alkali) fixation is complete in 2–3 h with a dichlorotriazinyl dye or 48 h with a dichloropyrimidinyl dye. At the end of the batching time unfixed dye is removed by circulating hot water through the batch or by using a washing range. The method is very convenient for woven and knitted fabrics, as the equipment required, apart from the padding mangle, is cheap and simple. It is also very effective when dyes of lower affinity have to be applied, and is widely used for the application of the chloroethylsulphonyl dyes.

(iii) Continuous applications, all of which involve padding fabrics where the fabric passes in a continuous web through the full sequence of impregnation, fixation, washing off and drying. The first continuous reactive dyeing sequence was pad(dye)–dry–pad (NaOH + salt)–steam–wash–dry. This complex sequence required expensive and elaborate machinery and is little used now. Later, simpler variations such as pad(dye + $NaHCO_3$ or Na_2CO_3)–dry–steam–wash–dry were more adopted, although current market conditions lead to smaller batches which are not appropriate for continuous applications. One very interesting variant of this application sequence arose from the observation that, when the dye carried a dichlorotriazinyl system, the decomposition of $NaHCO_3$ to Na_2CO_3 during drying caused a sufficiently increased pH to bring about fixation of pale to medium shade depths without the need to steam. A corresponding method for less reactive dyes, e.g. monochlorotriazinyl dyes, was developed which used Na_2CO_3 and a pad liquor containing 20% urea followed by drying and baking. The urea was believed to act a secondary solvent, i.e. fixation was thought to take place in the molten urea. In view of the chemistry of urea there is some doubt about this idea, as urea decomposes quite rapidly at 110–115°C to biuret which, in turn, melts at 190°C and is unlikely to provide a molten medium for reaction. No research has been carried out in this area and the subject is open for speculation.

With the passage of time two reactive systems have come to dominate the reactive dye market. The chloroethylsulphonyl dyes are the cheapest and simplest to manufacture of the reactive dyes for cellulose. The reactive system (usually called the vinyl sulphone system) contributes little to the affinity of the dyes for the fibre, and consequently their main outlets have been in those areas where dye affinity is of lesser importance, i.e. textile printing and semicontinuous dyeing processes such as pad batch–wash. They have established themselves as predominant in the production of goods for discharge printing, which is discussed in detail later. The preparation of vinyl sulphonyl dyes with greater affinity for cellulose is governed by the same principles as apply to the preparation of direct cotton dyes, and with the same limitations restricting the brightness of shade of the more complex dye structures. Nevertheless, the problems have been overcome to enable black and dark blue dyes, which are a vital feature of any commercial range of dyes, to be produced.

The heterocyclic reactive systems offer the chemist greater flexibility, especially when the reactive component is multifunctional e.g. 1,3,5-trichloro-2,4,6-triazine (**31**). The important characteristic of (**31**), which has been a

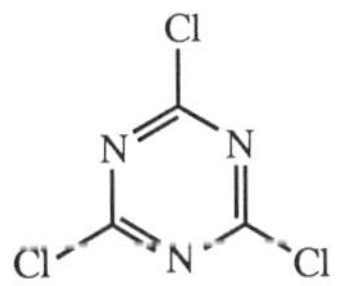

(31)

major factor in making it the most widely used of the heterocyclic reactive intermediates, is the fact that the chlorine atoms are mutually activating. Thus, replacement of one chlorine atom leads to a deactivation of the remaining two, and the replacement of the second chlorine leads to a further deactivation of the third. It is possible, therefore, to replace the chlorine atoms in an exactly stepwise fashion. This feature was used in the direct cotton dyes incorporating a triazinyl bridge (see **22**), and some of the dyes which were marketed actually contained a residual chlorine atom which could be reacted with cotton under the right conditions. However, in preparing reactive dyes the retention of unreacted chlorines is deliberate and essential. The most reactive triazinyl dyes are those containing two residual chlorine atoms, and these are prepared by condensation of the trichlorotriazine with either an intermediate containing a primary or secondary amino group or the dye molecule similarly substituted. The dye is then isolated by precipitation from a buffered solution using salt, filtration, drying and final standardisation using an appropriate buffer to optimise storage stability. The main difference between dyes prepared using primary and secondary amines lies in their solubility and affinity for cellulose. The triazine ring is the most powerful physical bonding system in the molecule and an *N*-alkyl group sterically hinders the ability of the ring to participate both in the formation of aggregates and in physical bonding with the cellulose. This option is of great value when the dye formed using a primary amine has inadequate solubility or too great a tendency to stain adjacent materials in washing. However, while very widely applicable and providing the basis for new dyeing methods, the dichlorotriazinylamino dyes proved too unstable for some applications, notably printing, which requires stable print pastes that can be stored for some weeks, and also exhaustion methods employing higher temperatures, which are frequently needed to ensure levelness and fabric penetration. For such applications, the second chlorine atom can be substituted to give a less reactive dye suitable for the more severe conditions of application and offering solution stability when required. By selecting the second nonreactive substituent with care it is possible to fine-tune affinity, solubility and even reactivity to provide dyes designed for particular application methods with a precision previously unattainable. In addition, it was found possible to produce dyes designed for textile printing which contained more than one monochlorotriazinyl group to optimise reaction potential and at the same time to provide extra solubilisation to assist print paste preparation and the removal of any unfixed dye at the end of the printing process, e.g. (**32**) and (**33**). Simple monochlorotriazinyl dyes with structures such as (**32**) are valuable for exhaustion dyeing processes from warm dyebaths. The inert substituent may be selected to enhance affinity or solubility, the rules being the same as those applying to the design of direct cotton dyes. Dyes with a binary reactive system exemplified by (**33**) offer many opportunities for modifying the solubility and reactivity, including, as will be discussed below, the use of two reactive systems of different kinds.

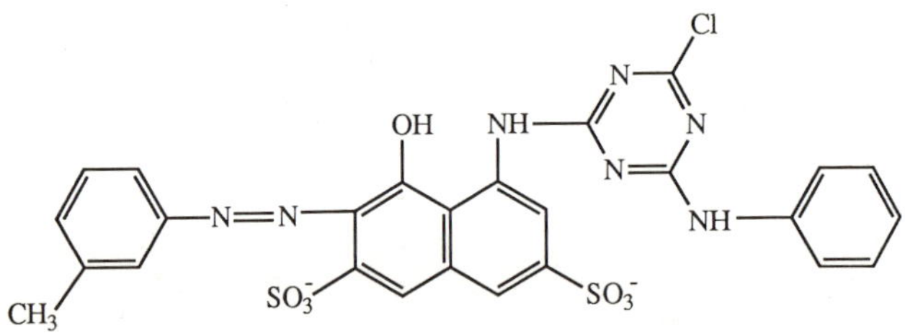

(**32**)

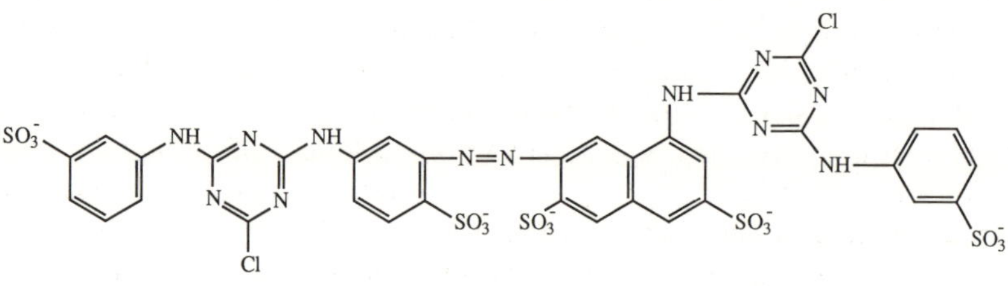

(**33**)

Dyes designed for high-temperature exhaustion applications have a structure exemplified by **(33)**. The similarity to direct cotton dyes incorporating a triazinyl bridge, e.g. **(22)**, will be noted. The difference in practice lies in the knowledge the modern dye chemist has when it comes to the design of the dye molecules for specific end-uses, which allows a high fixation efficiency to be achieved.

Other heterocyclic systems, e.g. **(26)**–**(30)**, are not dissimilar but do not offer the same degree of flexibility as the triazinyl system. The most extensively explored of the alternative heterocyclic ring systems has been that based on pyrimidine **(26)** or 1,3-diazine. Due to the fact that the ring contains two instead of three nitrogen atoms as in the triazine ring, there is less ring activation, and although there is some mutual activation by the chlorine atoms in the dichloropyrimidinyl dyes the two chlorines are not equivalent due to the asymmetry of the system, the chlorine atom in the 2-position being more reactive than that in the 4-position. In practice, only the more reactive of the two can be used to form a bond with the fibre. In order to complete a reaction in a reasonable time the application pH has to be fairly high, to provide a high concentration of ionised sites in the cellulose for the nucleophilic substitution. This makes it more difficult to achieve exhaustion, due to the charge repulsion of the dye anions. Since the pyrimidine ring, in any case, does not contribute very much affinity to the dye, the net result is a limitation on the value of the system in dyeing processes depending on exhaustion. Compensation for this can be obtained by careful selection of chromogens with good affinity, but this approach imposes limitations which do not apply as strongly with a more reactive system. It is possible to increase reactivity in the diazine ring by suitable substitution in the 5-position. A 5-chloro substituent increases reactivity slightly but usefully, and this has been used commercially. A 5-fluoro substituent **(30)** has a greater activating effect and, although relatively expensive, has been used. An instructive but technically useless example of activation is provided by the 2,4-dichloro-5-nitropyrimidinyl dyes. The nitro group has a very powerful activating effect and the dyes can be reacted with cellulose under quite mild conditions. However, the nitro group remains on the reacted dye molecules and continues to activate the dye–cellulose condensation product, which is only marginally less readily hydrolysed than the original reactive dye. Consequently, the dye–fibre bonds are broken down by the alkali present to induce the fixation at a rate which is just a little slower than the fixation itself. This example provides a valuable reminder that reactive systems used in dyes generally possess both permanent and temporary activation, but that the most effective reactive systems are those which depend on the latter rather than the former.

The dichloroquinoxalinyl system **(27)** uses the 1,4-diazinyl ring in a condensed ring system. The chlorine atoms in the 2- and 3-positions are mutually activating, and if one of them is lost the system is relatively inactive. Consequently, in order to attach the group to a chromogenic system a —COCl group is introduced in the 6-position, which can be reacted readily with a chromogen bearing a free amino or methylamino group. The dyes are very effective as reactive dyes for cellulose, but the amide bridge linking the reactive system to the chromogen is less stable to acid hydrolysis than the imido bridge found with other reactive systems, and this leads to a lack of fastness to acid conditions.

A number of reactive dyes for cellulose have been introduced which incorporate both a halogenoheterocyclic and a vinyl sulphonyl reactive system. The underlying concept is that not only is the probability of fixation rather than hydrolysis increased by having two systems, but also that the two systems, having different reactivity characteristics, can be brought into play separately during the dyeing process, giving the dyer greater control.

Wool, silk and polyamide fibres all have available amino groups with which reaction under mildly acid conditions is easily achieved. However, it was not until the mid 1930s that dyes were marketed that depended to some degree on dye–fibre reaction for their fastness. These were the chloroacetylamino dyes, the chlorine atom of which is readily substituted by an amino or thiol group, both of which are available in wool. A typical structure (**34**) is shown.

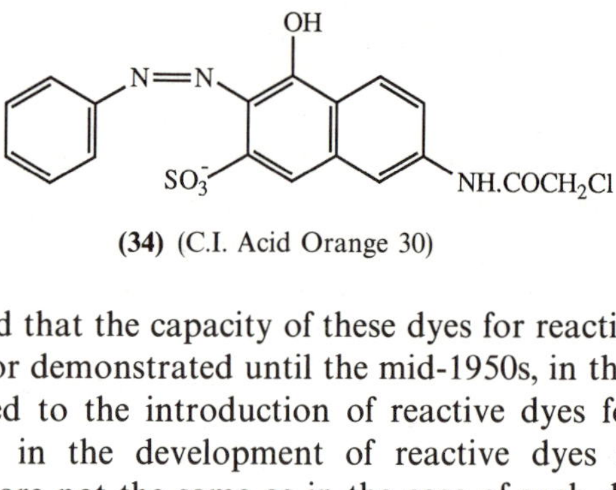

(**34**) (C.I. Acid Orange 30)

It should be noted that the capacity of these dyes for reaction with the fibre was not realised or demonstrated until the mid-1950s, in the course of work that ultimately led to the introduction of reactive dyes for cellulose. The problems arising in the development of reactive dyes for protein and polyamide fibres are not the same as in the case of such dyes for cellulose. With the latter, unreacted or hydrolysed dye does not exhaust strongly on to the fibre and is, in any case, fairly readily removed. In addition, cellulosic fibres are relatively pure and take up dyes uniformly under fairly mild conditions. However, dyes diffuse relatively slowly in wool, silk and polyamide fibres, and temperatures at or approaching the boil are generally needed to complete dyeing processes within acceptable times. Further to this, it is found that there are variations between one wool fibre and the next, which gives rise to non-uniform dyeing characteristics, and uniformity of shade requires either prolonged dyeing at high temperatures to equalise dye distribution or specially selected dye structures and assistants to obtain uniform initial 'strike'. These considerations mean that the conditions that exhaust an acid dye on to wool and the other similar fibres also cause strong exhaustion of nonreacting dye. This is then removed from the fibre only with difficulty, and by methods which easily lead to fibre damage. In addition, wool itself hydrolyses during the dyeing process to produce dissolved protein in the dyebath, and this is able to react with some of the dye to give a product which then exhausts back on to the fibre. Consequently, an ideal reactive dye for these fibres needs to have a reactive substituent that is resistant to hydrolysis under fairly severe aqueous dyeing conditions and a hydrolysed form that has low affinity for the fibre. No reactive system for these fibres has been developed which fully meets these criteria. However, two systems have been developed that possess a balance of properties which provides a sufficiently good level of behaviour for practical purposes. These are (**35**) and (**36**). Dyes of both types are applied to the fibre under mildly acidic conditions

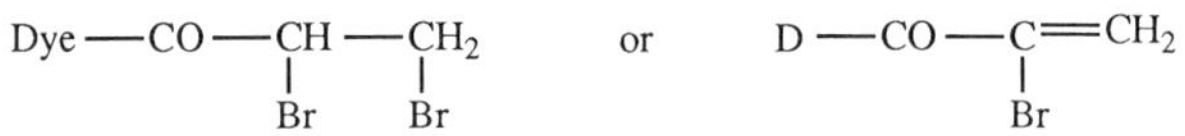

(35)

$$D{-}SO_2{-}CH_2{-}CH_2{-}O.SO_3Na$$

(36)

with appropriate level dyeing assistants. Under these conditions reaction with the fibre is not significant until the dyeing temperature approaches the boil, when a condensation reaction involving nucleophilic addition begins with available amino groups. In wool an additional side reaction takes place with thiol groups present due to hydrolysis of the disulphide bond of cystine. The product of this reaction is not very stable to alkali and can give rise to washing-fastness problems. The amount of hydrolysed dye adsorbed by the fibre is small if dyeing conditions are correctly chosen, but is sufficient to cause problems if the highest standards of fastness are demanded. Consequently, dyeings may be after-treated in a mildly alkaline bath to give the fibre a weakly negative charge leading to the ready desorption of unwanted dye anions.

9.3.5 *Disperse dyes*

The dyes so far discussed have all been water-soluble salts in which the coloured moiety is an anion or a cation. It was found in the 1920s, when cellulose acetate was first introduced as a textile fibre, 'artificial silk', that dyes of this type had very little affinity for the new fibre and that a new type of dye would have to be developed if cellulose acetate fibres were to become economically viable as a textile. The problem is that cellulose acetate is relatively hydrophobic and takes up very little water as compared with natural fibres such as cotton or wool. This greatly increases the electrical repulsion effects which are a very important factor when dye ions are adsorbed, as has been discussed already. In addition, the low water uptake reflects low swelling of the fibres and consequently they are permeable with difficulty and by only small dye molecules. The solution to the problem was found in the use of small dye molecules of low polarity and consequently very sparing aqueous solubility, which are present in the dyebath initially as a dispersion. In the early stages of development of this class of dyes various ideas were advanced to explain their mode of action, which were based mostly on the somewhat dubious concept of a process of direct transfer from the dye in the solid state to a 'solid solution' in the fibre which was regarded as a 'solid colloid'. This concept received apparent support from the fact that cellulose acetate could take up dye by direct contact with the warmed dye powder and also by transfer from the vapour phase under certain circumstances. It was not until the 1950s as a result of a series of studies led by C. L. Bird (Bird, 1954; Bird *et al.*, 1954; Bird and Harris, 1957; Bird and Firth, 1960; Bird and Rhyner, 1961), that a true picture began to emerge of the mechanism whereby these dyes were being taken up.

The lack of an accurate theory of the mechanism of the adsorption of the disperse dyes did not, however, seriously impede the development of the dyes in the early years, since an empirical approach was adequate to deal with such problems as arose and, in any case, cellulose acetate remained the only synthetic or man-made textile fibre until the much later appearance of the polyamide and polyester fibres. The first dyes developed were based on azobenzene, which was substituted with a variety of substituents in the 4,4′-positions, and on diphenylamine, as exemplified by (**37**) and (**38**). Such dyes gave a shade gamut largely restricted to yellow, orange and red shades, although in recent years this has been considerably extended. The overwhelming majority of early blue disperse dyes were based on simple substituted anthraquinones, which provided excellent bluish reds, violets and blues to give a full shade coverage, e.g. (**39**), (**40**) and (**41**).

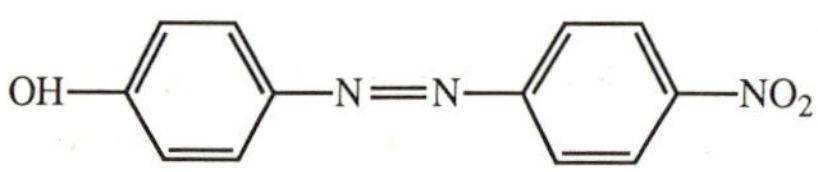

(**37**)

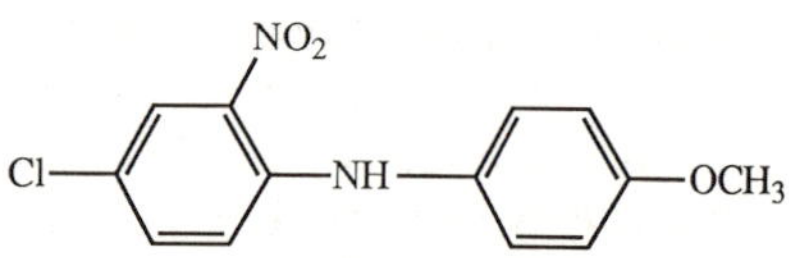

(**38**)

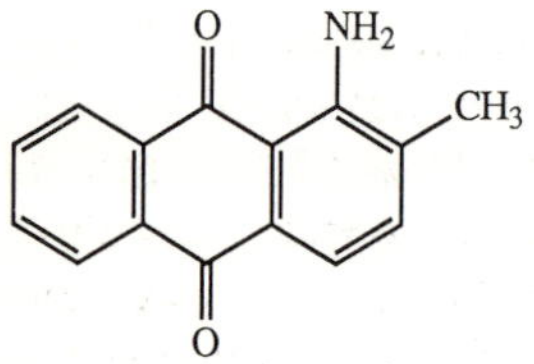

(**39**) (C.I. Disperse Orange 11)

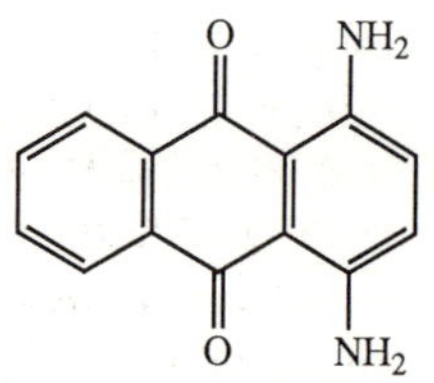

(**40**) (C.I. Disperse Violet 1)

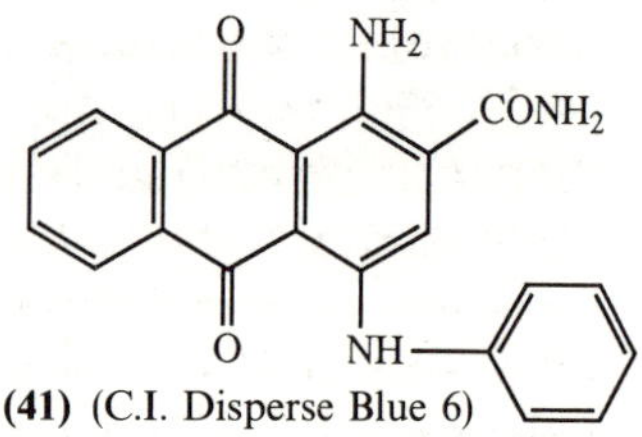

(**41**) (C.I. Disperse Blue 6)

With the appearance of polyamide and polyester fibres the importance of the man-made textile fibres extended rapidly, seriously eroding the dominant position of cotton in the world textile market. The demand for disperse dyes grew considerably and supported extensive research leading to the development of dyes of higher fastness. Of particular importance have been advances in the chemistry of azo dyes, which have led to an extension of their colour gamut to cover most of the visible range, through the development of 2-cyano-4-nitroaniline as a diazo component, and bluish reds, e.g. (**42**), and bright blues, e.g. (**43**), are possible.

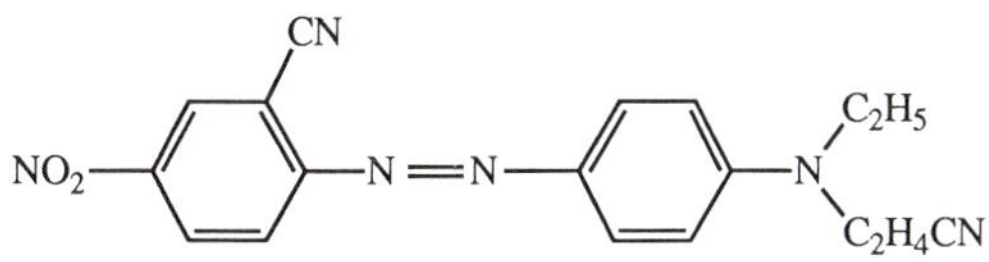

(42)

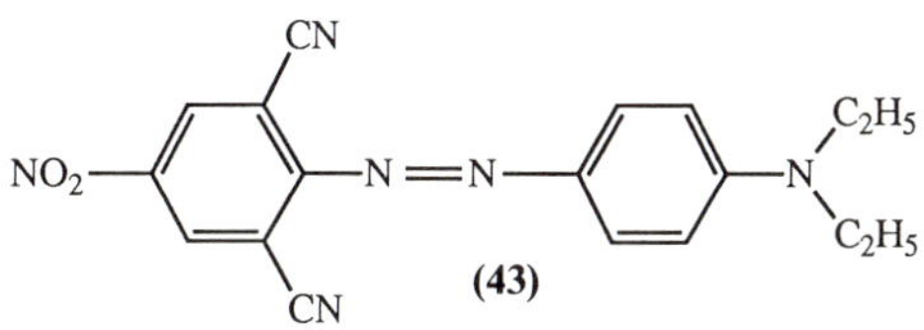

(43)

An important feature of the post-polyester developments has been the growth in the number of disperse dyes of higher molecular weight and lower vapour pressure. These have higher fastness, particularly to heat processing, than the earlier disperse dyes, but the effect of these factors is to reduce the aqueous solubility at normal dyeing temperatures to very low levels and to reduce the rate of diffusion of these larger molecules through the fibre, so that the higher dyeing temperatures are needed to complete the dyeing process in commercially acceptable times. Thus were developed high-temperature dyeing procedures carried out under pressure at temperatures of 130–140°C. It is interesting to note that at these temperatures and at dyebath concentrations the dyes are in solution for most of the dyeing time and the term disperse dyes, in a sense, no longer strictly applies.

It is a characteristic of textile coloration developments that a serious disadvantage in relation to one application can be turned to a positive advantage in another. In the case of the disperse dyes, the relatively high vapour pressure of many of these dyes has been put to good use in the development of sublimation transfer printing for polyester fabrics (de Plasse, 1972). This process had its beginnings in the early development of disperse dyes by Ellis and his colleagues in the 1930s, who observed that certain disperse dyes could be transferred by heat from a printed paper to cellulose acetate. This observation was of little practical value at the time, but particularly with the widespread introduction of knitted polyester fabrics the basis was created for the development of the early results into a fully-fledged production process. In essence, a transfer printing paper is produced by printing a design on a suitable selected paper using a specially formulated ink, in which the colouring matter is a disperse dye selected for its volatility under high-temperature conditions. The paper is then brought into close contact with a polyester fabric under light pressure at 200–220°C for 20–30 s in a specially designed calender unit. When the paper is peeled off the fabric, the bulk of the dye is found to have transferred on to the fabric, which is thus decorated with the initial design in reverse. The available disperse dyes have been classified according to their degree of volatility as Class A, B, C and D dyes, and it is the dyes of Classes B and C which are useful for this application. Dyes of Class A are so volatile that it is difficult to preserve the integrity of the design during the transfer, while dyes of Class D have insufficient volatility for significant transfer. Although its use has declined since the polyester 'boom' of the 1960s, transfer printing remains a significant coloration process for polyester fabrics. While a limited number of new disperse dyes has been developed to fill in gaps in the range of dyes available for transfer printing, the great majority of new disperse dye developments have fallen into Class D.

9.3.6 *Vat dyes*

The vat dyes owe their name to their ancient origins in the application of indigo. It was discovered in the Bronze Age that by grinding the leaves of the woad or indigo plants in water and then leaving the paste to ferment, and adding an alkali (at that time derived from fermented urine, lime water or wood ash), a weakly coloured solution was obtained which could be used to dye textile materials. When the dyed fabric is exposed to air a durable blue colour develops. This is not the place to speculate by what ingenious chain of thought this complex path to a dyed fabric was discovered. Later, the blue colour, indigo, was isolated and became a marketable product after it was shown that the blue insoluble pigment could be solubilised by treating it in an alkaline reducing bath or vat to convert it to its soluble form, called the leuco form due to its pale colour (Greek leucos = white). Indigo remained the sole member of the class of vat dyes until its structure (**13**) had been elucidated. Following this, numerous indigo derivatives were prepared by halogenation, for example (**44**) and (**45**).

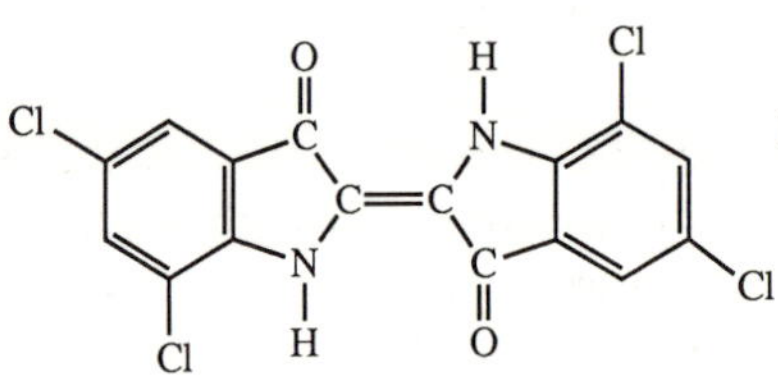

(**44**) (C.I. Vat Blue 41)

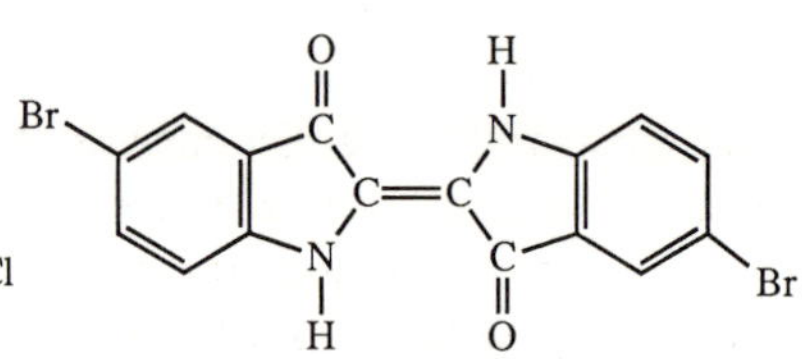

(**45**) (C.I. Vat Blue 35)

Later the thioindigoid vat dyes were prepared, in which the two —NH— groups are replaced by sulphur atoms, for example (**46**).

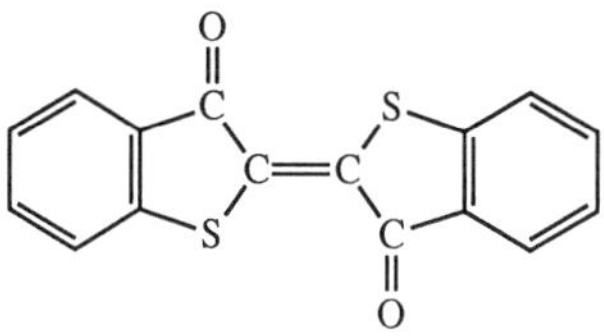

(**46**) (C.I. Vat Red 41)

However, after the production of synthetic indigo, the most important development in the vat dyes was the synthesis of the anthraquinonoid vat dyes in 1901 by R. Bohn. As with many important dyestuff discoveries, their first synthesis resulted from a failed attempt to produce something else. In this case it was Bohn's attempt to produce an analogue of indigo by the caustic fusion of 2-anthraquinonylglycine. Instead of indigo, what was formed was the blue vat dye with the structure (**47**). Over the next decades this important dye, which demonstrated the superiority of the polycyclic quinones over the indigoid dyes, was joined by anthraquinone derivatives, e.g. (**48**), anthrimides, e.g. (**49**), anthanthrones, e.g. (**50**), pyranthrones, e.g. (**51**), and dibenzanthrones, e.g. (**52**), as well as other condensed aromatic ring systems.

These chemical developments would have been largely academic if there had not been also great advances, made during the last thirty years of the nineteenth century, in the chemistry of vatting. Up until that time, vat dye reductions had been obtained by treating the pigment by taking advantage of the reducing action provoked by the micro-organisms on fermentable sugars, fruit, etc., or latterly the reducing action of zinc in aqueous alkaline solution. In 1869, hydrosulphurous acid solution was first prepared and shown to be highly effective in reducing indigo when made alkaline. Dyeworks wishing to take advantage of this discovery had to prepare the substance themselves, but in 1898 solid sodium hydrosulphite, more correctly termed today sodium dithionite, $Na_2S_2O_4$, was manufactured. This reacts readily under alkaline conditions with the keto groups of vat dyes to form the leuco compound:

$$S_2O_4^{2-} + 4OH^- + O{=}C\langle \rightarrow 2SO_3^{2-} + 2H_2O + {}^-O{-}C{-}$$

The vatting reaction is heterogeneous, taking place at the surface of the vat dye pigment particles, and has not been fully elucidated. It is thought that the electron donor in the reduction is the SO_2 radical. The vat dyeing process involves adding sodium dithionite to a suspension of the vat dye pigment and allowing the reduction to proceed before adding the leuco solution to

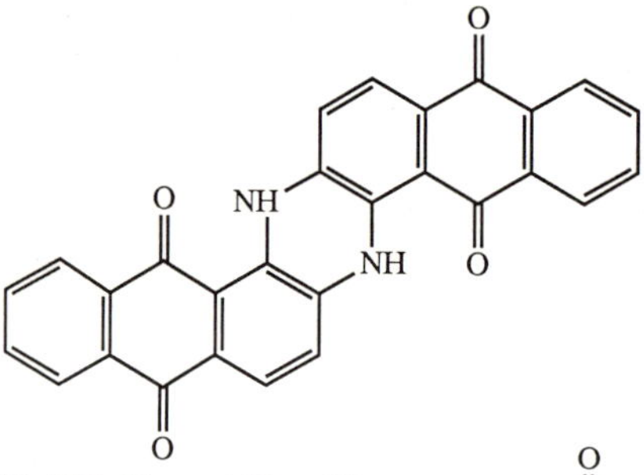

(47) (C.I. Pigment Blue 60)

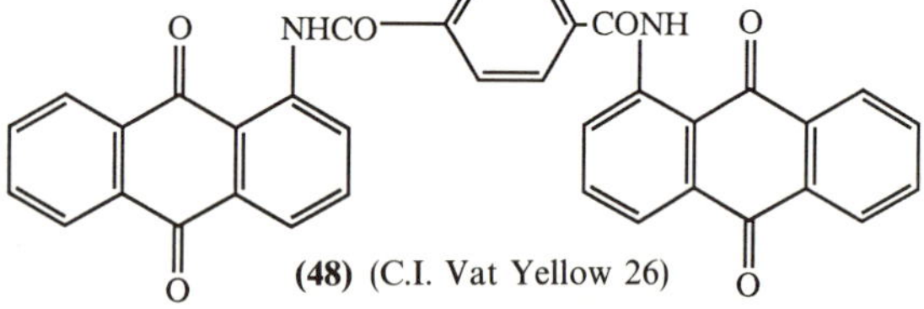

(48) (C.I. Vat Yellow 26)

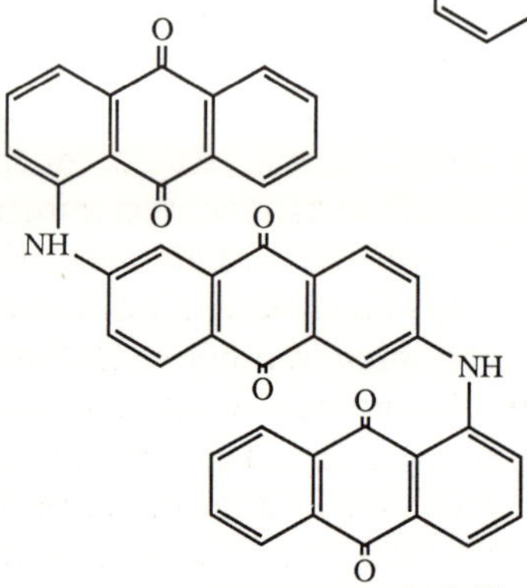

(49) (C.I. Vat Red 48)

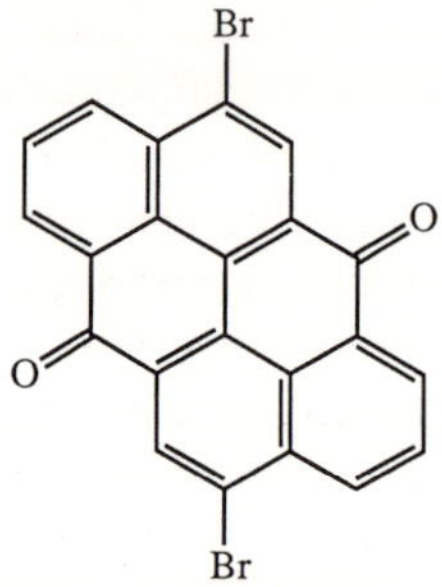

(50) (C.I. Vat Orange 3)

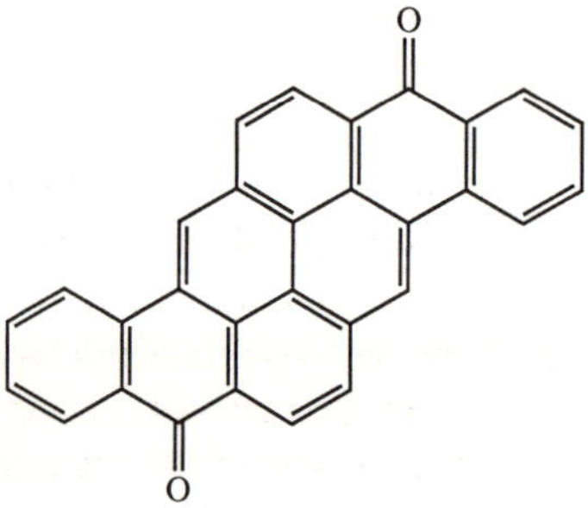

(51) (C.I. Vat Orange 9)

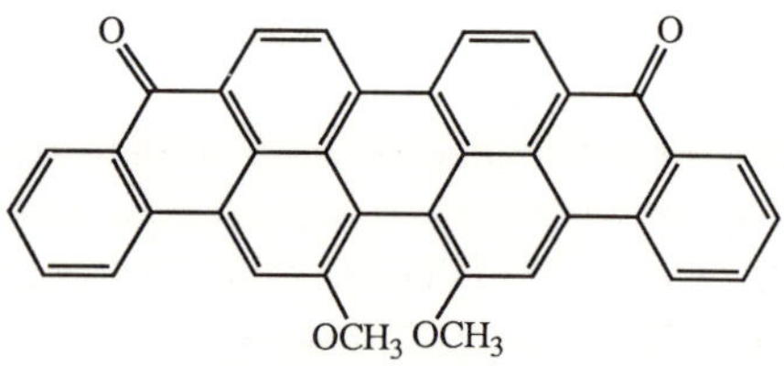

(52) (C.I. Vat Green 1)

the main dyebath in which dyeing proceeds. It can be seen from the reaction scheme above that the reduction liberates acid, which slowly neutralises the alkali present. When, as is usual, the dyebath is exposed to the atmosphere, the reduction of the vat dye and the maintenance of the leuco solution compete with oxidation, so that careful supervision of the condition of the dyebath is needed.

The reduction reaction is normally accompanied by a large colour change, due to the extensive change in conjugation which accompanies reduction. In the indigoid series the shift is hypsochromic, while in the anthraquinonoid and condensed aromatic systems the shift is bathochromic. The conjugation can impose a planarity on the large molecule which is not present in the quinonoid form, and this can be a significant factor in the colour change occurring as a result of re-oxidation. This part of the vat dyeing process is no longer left to the slow action of atmospheric oxygen but is achieved rapidly using potassium dichromate, peroxides, etc. More than just the regeneration of the carbonyl form is involved in oxidation, however. The adsorption of the leuco planar molecule by the fibre involves extensive interaction of the molecular orbital of the dye with the polymer molecule, as will be considered in more detail later. Thus, oxidation may give rise to an isomorph of the dye which, though oxidised, is stabilised in a planar configuration by dye–fibre bonding. As a result the shade may be unstable to hot washing processes, as these will supply the necessary energy for the breaking of the dye–fibre bonds and complete reversion to the more stable non-planar form. When this occurs it is accompanied by a shade change. Additionally, the freshly oxidised dye will be in a state approximating to molecular dispersion, in which state the dye will have an unusually high solubility. Hot wet treatments lead to the growth of vat pigment microcrystals which have very low solubility, and this process also includes a shade change due to the change in the optical properties of the pigment dispersion in the fibre. In order to produce a dyeing or a print with a stable colour it is necessary to subject the coloured fabric to a hot washing or 'soaping' process.

9.3.7 *Sulphur dyes*

Like the vat dyes, the sulphur dyes comprise water insoluble coloured compounds that are solubilised by reduction. In the reduced state they are used to dye cellulosic fabrics, and after the dyeing stage is complete they are oxidised once again to the insoluble pigment form. The dyes are relatively cheap to produce and easy to apply. They were first prepared by entirely empirical methods using unknown starting materials of animal or vegetable origin. Sulphur dyes based on starting materials of known structure were introduced towards the end of the last century, but since the methods of preparation lead to complex mixtures of reproducible but uncertain composition, the structures and the chemistry of the sulphur dyes have not been evaluated as yet and their classification is based on their method of manufacture rather than their chemical structure. Three main methods are used in their preparation:

1. *Sulphur or polysulphide baking*, in which the starting materials are heated to temperatures in the range 160–320°C with sulphur or sodium polysulphide.

2. *Polysulphide melting or refluxing*, which involves heating the starting materials with aqueous sodium polysulphide under reflux or under pressure in a closed vessel.
3. *Solvent melting*, whereby the water in the melt process is replaced wholly or largely by a solvent such as 2-ethoxyethanol, butandiol or dioxitol.

The dyes prepared by the baking method using sulphur are usually based on aromatic amines, mono- and di-amines, and their alkyl or acyl derivatives. Other compounds, for example, may be added to produce shade-modifying byproducts. The colour produced, generally yellow, orange or brown, is thought to be due to the presence of the thiazole (**53**), thiazone (**54**) or thianthrene (**55**) in the numerous coloured products formed. The dyes formed

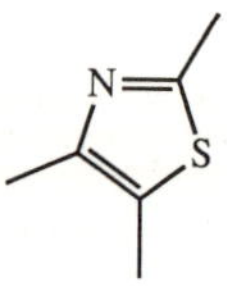

(**53**)

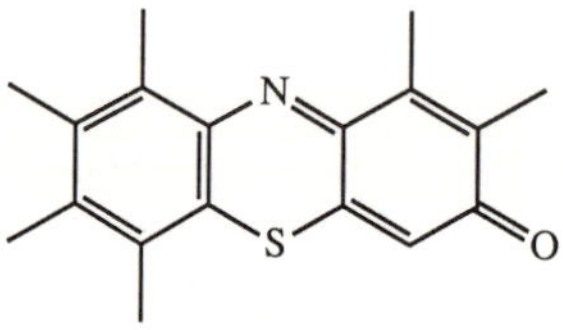

(**54**)

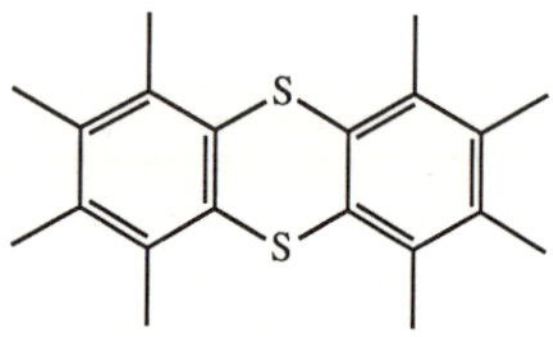

(**55**)

are polymeric in nature with disulphide bonds linking the individual coloured polymer units. In the presence of aqueous sodium sulphide these are reduced to ionised mercaptan (—S^-) groups which solubilise the dye during the dye application. The baking with polysulphide method produces a slightly different range of compounds. It can be used with nitro and polynitro intermediates more safely than straight sulphur. The shades produced are usually brown or olive.

The melting on reflux process can be used to provide a much wider range of shades than the baking processes. Starting with 2,4-nitrophenol, a black dye is produced, C.I. 53185, which it has been suggested (Rys and Zollinger, 1972) is probably the most important single dye in production "amounting to 10% by weight of world production". Violet and bluish-green as well as black dyes can be made by starting with indophenols (**56**), and indamines (**57**) can be used as a basis for bordeaux and red–browns.

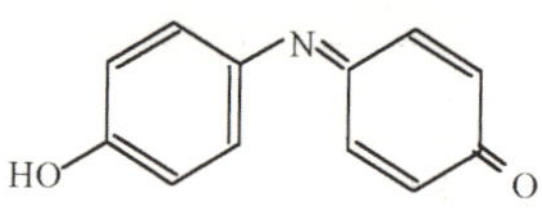

(**56**)

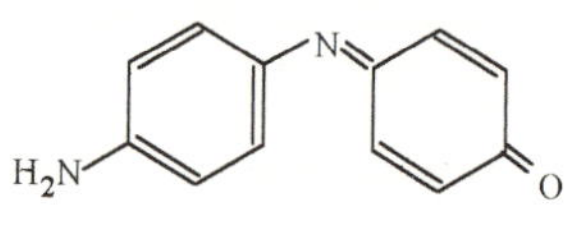

(**57**)

9.3.8 *Azoic dyes*

The azoic dyes are actually azo pigments synthesised inside cellulosic fibres in the course of the dyeing process rather than ready-formed colouring matters. The concept of colouring cotton by this method was developed in 1880 when T. and R. Holliday impregnated the fibres with an alkaline solution of β-naphthol and then with a diazo solution obtained from aniline and other primary aromatic amines, so that an azo pigment was formed within the interstices of the fibres. It was applied to the dyeing and the printing of the cotton a few years later (Kunert, 1912; Kunert and Acker, 1914). Although β-naphthol was extensively used in the early stages it has many disadvantages due to its very low affinity for cellulose, especially under the alkaline conditions needed to produce the water-soluble sodium salt. In addition, the azoic pigments formed from β-naphthol do not have good fastness to light. Despite this, the process was attractive because of the intense shades that could be produced, sufficiently attractive in fact to sustain interest during the twenty years it took before a naphthol with affinity, 3-hydroxy-2-naphthoic acid anilide, was developed. Prior to this development, fabrics impregnated with naphthol had to be dried before treatment with the diazonium salt solution in order to obtain an acceptable result. 3-Hydroxy-2-naphthoic acid anilide was marketed under the name Naphtol AS, and its use made it possible to dye cotton at 30°C and treat with the diazonium solution directly. A series of new coupling components with behaviour similar to that of Naphtol AS was rapidly developed. It should be noted that despite the use of the term Naphtol only a minority of the compounds developed contained a hydroxyl group. Initially the shades produced were orange, scarlet and red, but as new naphthols and also new diazotisable amines were developed an almost complete range of shades became possible. The original Naphtol AS remained, however, the dominant coupling component in cotton dyeing by this method, and an indication of the range of shades which could be produced using it is given in Table 9.1. The method adopted for the production of azoic prints involved preimpregenation and drying the fabric and then printing it with a thickened paste containing the diazotized amine. This left unchanged coupling component in the unprinted areas, and this had to be washed out.

Table 9.1 The variety of shades produced with Naphtol AS using different diazo components

Diazo component	Shade
o-Nitroaniline	Orange
2,5-Dichloroaniline	Scarlet
p-Nitroaniline	Red
2-Methoxy-4-nitroaniline	Bluish red
Dianisidine	Reddish blue

The advantageous affinity which had been built into many of the new Naphtols developed for dyeing thus became a liability, although Naphtol AS itself and selected other coupling components could be used. The 3-hydroxy-2-naphthoic arylamides could not be used for the production of yellow shades and a new series of compounds, the acylacetic arylamides, was developed.

The production of an azoic pigment in the interstices of the fibre leaves a certain amount of surface pigmentation which must be removed to give a dyeing which does not have poor rubbing-fastness properties. In addition, the trapped pigment is in a very highly dispersed state which is not physically stable. Consequently azoic dyeings (and prints) need to be washed at the boil to remove surface pigmentation and also to enable the pigment particles to grow to a size with minimum solubility. In this state their lightfastness is maximised and their shade is stable to further washing treatments.

The extent to which azoic dyes are used has declined considerably due to two factors. Firstly, the appearance of reactive dyes for cellulose has made a full range of fast bright shades covering the whole colour gamut available to dyers, making the contribution of azoic dyes much less important. Secondly, the textile printing designs for which azoic dyes proved highly suitable for many years are now produced using resin bonded pigments or reactive dyes.

9.3.9 *Pigments*

The distinction between dyes and pigments is not as clear cut as might be supposed. Dictionary definitions are of little help. Chambers English Dictionary (1988) defines a dyestuff as 'a material used in dyeing' while a pigment is 'any substance used for colouring or that which gives colour to animal and vegetable tissues'. In the textile as well as the coatings field, pigments are regarded as insoluble colorants while a dye is a soluble colorant at least during some essential stage of application. Thus vat and azoic dyes, while colouring textiles by pigmentation, are applied from solution like other dyes, including the sparingly-soluble disperse dyes. Pigments are used in two ways to colour textiles. For the most part they are used together with binding resins to produce a surface coloration, primarily in printing applications, but also to some extent in so called 'pigment dyeing' processes. A second more limited application is in the mass coloration of man-made textile fibres during the spinning stage of manufacture. The main uses of pigments are in the production of paints, inks and mass-coloured plastic fabrications, and in these fields both inorganic and organic pigments are to be found. Their use in textile applications makes demands of a special kind on the stability of the pigments to hot alkaline washing, bleaching, etc., and consequently, apart from the black (carbon black) and white (titanium dioxide) pigments extensively used in textile printing, the pigments employed in textile coloration are synthetic organic materials.

The chromogens used in the production of organic pigments are the same as those used in making dyes. Early synthetic pigments were frequently the insoluble calcium, barium and magnesium salts of soluble acid dyes or the phosphotungstomolybdic acid salts of soluble basic dyes, both known generically as lakes. These are still in use in printing of ephemera but are of little value for textile applications due to their frequently poor light fastness and solubility in alkali. The majority of textile pigments are based on the azo chromogen which provides yellow, orange, red and brown materials. Yellows are generally derived from acetoarylamides, typical examples being (**58**) and (**59**).

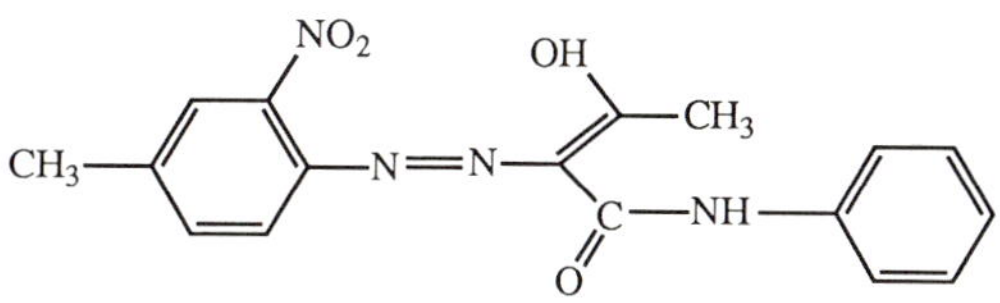

(**58**) (C.I. Pigment Yellow 1)

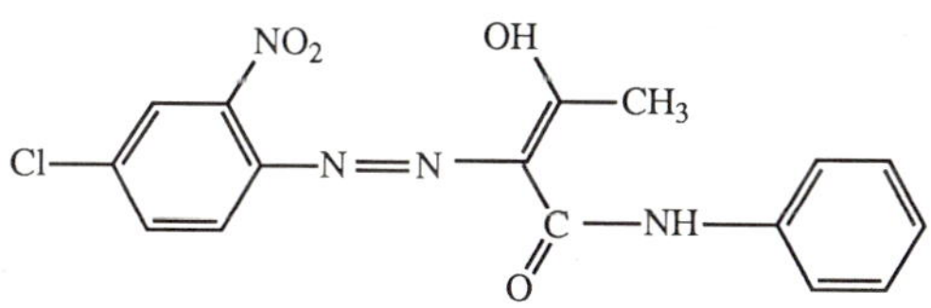

(**59**) (C.I. Pigment Yellow 3)

Yellows based on tetrazotised benzidine were in use for some years but have been generally superseded on health and safety grounds. Red pigments for textile use are normally based on the naphthoic arylamides used in relation to the azoic dyes. Blue pigments are completely dominated by copper phthalocyanine and its derivatives.

The way in which pigments produce their coloured effect differs from that of dyes, and this has very important repercussions with regard to their manufacture. Dyes generally behave as if they were in solution, selectively absorbing incident radiation passing through an essentially transparent medium. Pigments, on the other hand, produce colour due to selective absorption of light reflected at the pigment particle surface and light penetrating the particle which is subject to multiple internal refraction before emerging. The refractive index of the pigment particle relative to that of the medium in which it is embedded is of great importance, because this will affect the degree of light-scattering and transparency of the coloration. The relative contributions of these factors to the final colour depends on the particle size and the production of a useful pigment involves finishing the raw pigment to optimise colour strength and covering power. Some pigments immediately following their synthesis have very little colour strength and exhibit a very dull shade, both of which change dramatically when the particle size has been optimised. At the same time the surfaces of the pigment particles may also be modified by coating with dispersants and resins to enhance the dispersibility and the dispersion stability of the pigment in the medium in which it is to be used. Some pigment products contain as little as 50% of the colouring matter but have nonetheless enhanced colour strength. Covering power or the ability to hide the substrate is of great importance in many printing applications, since many of the fabrics printed may be strongly coloured.

Suitable pigment dispersions can be 'exhausted' on to textile by controlled destabilisation to deposit the pigment on the fabric surface. This was used extensively at one time to apply vat dyes to yarn packages by the Abbot–Cox process, now little used. However, ecological pressures on dyeing operations may lead to some revival of this approach, by using regular pigments and binders, because pigments potentially offer the advantages of high exhaustion efficiency and low effluent colour levels. Pigment dyeing is normally carried out by impregnating cotton fabrics with an aqueous dispersion of pigment and binder, drying and curing to polymerise the binder. No post-washing is needed and fixation is 100%. Achieving drying without migration of the pigment requires careful control. The binders used are generally acrylic- or polyurethane-based and of the same types as are used in pigment printing. Since pigment printing is the major textile use for these colouring matters, pigment binders are discussed in greater detail in the following section.

9.4 Textile printing

In effect, textile printing involves localised dyeing processes in which a dye or pigment solution or dispersion is applied to the fabric in a design, and dyeing proceeds from the printed 'dyebaths'. Thus, the general principles and colouring matters involved are the same as those already discussed. However, the production of the printed design introduces new considerations as compared with general exhaustion processes, in particular the questions of thickeners, binders and procedural modifications made necessary by printing conditions.

Textile printing is carried out using gravure rollers or more usually with rotary or flat screens. In the first method the ink is delivered to an engraved roller and excess is removed to leave ink only in the engraved indentations using a doctor blade. The roller is contacted with the fabric so that the ink is drawn from the engraved areas to produce the printed design. In screen printing a polyester or metal mesh is partially sealed to leave the design as an open area. The ink is delivered to the screen and then forced through the mesh using a metal or synthetic resin squeegee on to the fabric. A thickener is necessary to ensure that the printed design is precisely located with clean defined edges. The flow properties of the ink have to be such that the ink can flow in and out of the engraving in the case of gravure printing or through the screen in a controllable manner in the case of screen printing. In the latter method the shear forces on the ink as it passes through the screen are very high and the high fluidity produced has to be lost through thixotropic behaviour rapidly enough to retain definition. Several other characteristics are essential in a successful thickener, not the least of which is ready release of the dye. The characteristics of print-paste rheology are too extensive for the present chapter and are covered elsewhere (Miles, 1995).

Numerous polysaccharides derived from starch-containing plants, seaweeds yielding alginates, plant gums, and cellulose derivatives, particularly cellulose ethers, are used as thickeners in textile printing. The alginates are of particular importance in the printing of reactive dyes, since they do not react with the dyes. This is due to the D-mannuronic acid (**60**) and L-glucuronic acid (**61**) monomer units, which not only lack the primary carbinol groups which react readily with the dyes, but also supply ionised carboxylic acid groups, which prevent the physical interaction with the anionic reactive dye molecules which is a necessary precursor to reaction.

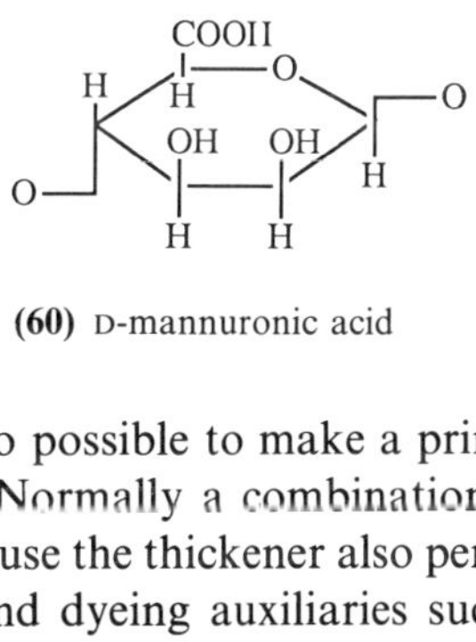

(**60**) D-mannuronic acid

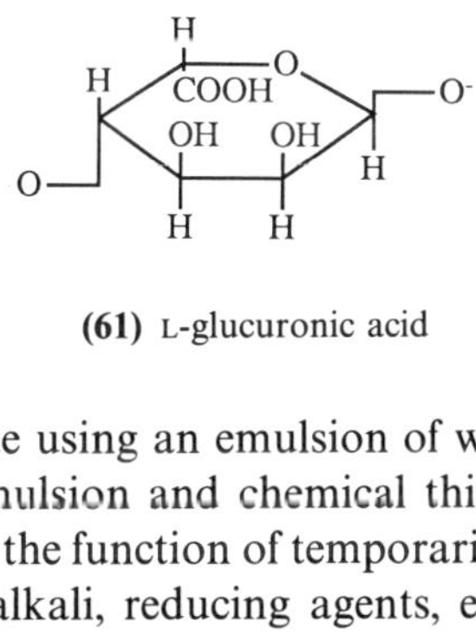

(**61**) L-glucuronic acid

It is also possible to make a print paste using an emulsion of white spirit in water. Normally a combination of emulsion and chemical thickening is used, because the thickener also performs the function of temporarily holding the dye and dyeing auxiliaries such as alkali, reducing agents, etc., *in situ* pending dye adsorption and fixation by steaming.

A combination of safety factors and ecological legislation has greatly eroded the popularity of white-spirit emulsions, which have become increasingly replaced by synthetic thickeners based on poly(acrylic acid) derivatives. Copolymers of methacrylic acid and ethyl acrylate dispersed in water under neutral or weakly acid conditions produce no viscous effect. Under these conditions the polymer chains are coiled randomly, but when the pH is raised the carboxylic acid groups along the polymer chains ionise, causing strong charge repulsion effects which straighten out the polymer chains and develop a structure in solution in which the distance between the chains is maximised. The resistance of this structure to deformation creates the viscosity required. The development of viscosity with this simple system has a very sharp end-point and its sensitivity to pH is too great for practical use. By using a bifunctional monomer crosslinking is developed, which prevents the polymer from going into solution and a more satisfactory product is obtained. Since the thickening effect is produced almost entirely by coulombic intersections, the effect is always sensitive to the electrolyte concentration. This can create difficulties when thickening dye solutions, especially where the dye is a polyelectrolyte, for example a reactive dye. However, because of their marked effect at low solids levels, acrylic thickeners are very suitable for pigment printing, which in any case is the basis for 60% of textile printing world wide.

Pigment printing involves the application of a thickened dispersion of pigment containing a polyacrylic or polyurethane binder, which condenses during a baking treatment after printing and drying. Its great attraction, apart from simplicity, is the fact that prints do not need to be washed off and dried, fixation being 100%. Acrylic thickeners are available which combine to form part of the binding system. The binders, like the synthetic thickeners, are generally acrylic copolymers produced by emulsion polymerisation. According to this technique the unsaturated monomers are dispersed in water using appropriate surfactants and then caused to polymerise using free-radical initiators, e.g. potassium persulphate/sodium bisulphite. The polymer chains grow to a high molecular weight, which can be controlled by the addition of monofunctional blocking compounds. The polymer is produced in the form of an emulsion containing 40–45% solids, the surfactant controlling the dispersion size to 120–300 nm. When the emulsion is dried, flocculation takes place followed by coalescence to form a film. The film formation is a vital part of the development of a pigment print. The binders used are capable of forming a film at room temperature. It is necessary for the film, when cured, to have a strong adhesion to the textile, be resistant to abrasion and possess a high degree of flexibility. Copolymers of butyl acrylate and acrylonitrile have many of the required properties, although the resistance to solvents (e.g. dry cleaning) is reduced if too much butyl acrylate is used. Emulsion polymers based on butadiene are also used, as well as the reaction products of mono- and di-carboxylic acids with polyols. Emulsion polymers of polyurethanes have also been used. Binders and thickeners now available enable printers to produce very fast soft prints with a soft handle, and the simplicity and economy of application has led to pigment printing dominating textile printing world wide.

Of less importance than the printing of continuous-web fabrics but nevertheless of considerable importance is the printing of precut garment panels. These have to be printed by dry processing, but as approximately 40% of panels to be printed are predyed simple pigment printing cannot be used as this would not cover the ground shade adequately. Two processes are used instead. Firstly, an adaptation of the well-established process of discharge printing was developed by Green Print International under the name DisTron, which is effectively a dry process not requiring steaming or washing off. Like conventional discharge printing, it involves the incorporation into a pigment printing ink of a reducing agent which, during the drying and curing process, substantially destroys the dyes in the ground shade. Reducing agents for discharge printing are related to sodium dithionite, as used in the dyeing of vat dyes, but processing high stability until the temperature is raised to 100–125°C, when they become active. The most generally used product is the zinc salt of formaldehyde sulphoxylic acid. This is prepared by dissolving SO_2 gas in aqueous formaldehyde. The resulting acid solution can then be reduced using zinc and acetic acid to form the zinc salt:

$$HCHO + H_2O + SO_2 \rightarrow HOCH_2SO_3H$$

$$2HOCH_2SO_3H + Zn/CH_3COOH \rightarrow (CH_2OH \cdot SO_2)_2Zn$$

This compound can be activated by dry heat to give a discharge print not requiring post-washing on cotton fabrics dyed with selected reactive dyes, usually but not exclusively of the vinyl sulphonyl type. Another agent that works very well, providing that steam heating is used to develop the discharge, is thiourea dioxide (C.I. Reducing Agent 11):

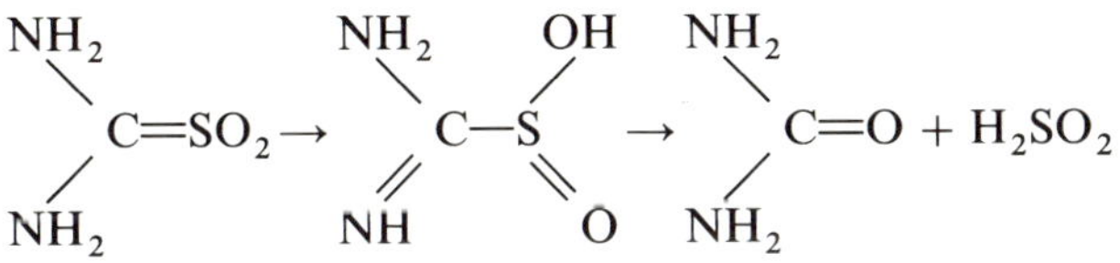

Both of these agents are also used in printing with vat dyes, since their stability at room temperature makes it possible to prepare a print paste stable to the atmosphere. The second method for printing dyed fabrics is the widespread but rather less satisfactory use of plastisol inks. These are dispersions of pigments in a base containing finely dispersed vinyl chloride copolymers in nonvolatile plasticisers such as di-octyl phthalate, tricresyl phosphate, etc. The inks contain as much as 60–70% of the plasticiser. The highly viscous inks lie on the fabric surface hiding the underlying fabric. They do not dry but on heating to 140–160°C they coagulate to give a dry print.

References

Berthollet, C.L. (1791) *Eléments de l'Art de la Teinture*, Part 1, Paris.
Bird, C.L. (1954) *J. Soc. Dyers and Colorists*, **70**, 68.
Bird, C.L. and Firth, J. (1960) *J. Textile Inst.* **51**, T1342.
Bird, C.L. and Harris, P. (1957) *J. Soc. Dyers and Colorists*, **73**, 199.
Bird, C.L. and Rhyner, P. (1961) *J. Soc. Dyers and Colorists*, **77**, 12.
Bird, C.L. *et al.* (1954) *Disc. Far. Soc.*, **16**, 85.
Bohn, R. (1901) German Patent Specification 138 119.
Boltzmann, L. (1894) *Annal. Phys. Leipzig*, **53**, 959.
Chadwick, C.S. and Neale, S.M. (1958) *J. Polym. Sci.*, **28**, 355.
Chambers English Dictionary (1988) Cambridge, UK.
Chantrey, G. and Rattee, I.D. (1969) *J. Soc. Dyers and Colorists*, **85**, 618.
Coates, J.A., Ellard, V. and Rattee, I.D. (1980) *J. Soc. Dyers and Colorists*, **96**, 14.
Colbert, J.B. (1671) *Instruction Générale pour la Teinture des Laines et Manufactures*, Paris.

Crank, J. (1956) *The Mathematics of Diffusion*, Clarendon Press, Oxford.
Craven, B.R. and Datyner, A. (1967) *J. Soc. Dyers and Colorists*, **83**, 41.
Cross, C.F. and Bevan, E.J. (1895) *Researches in Cellulose*, p. 34.
de Plasse, N. (1972) *Teintex*, **37**, 675.
Dufay de Cisternay, G.F. (1737) *Mémoires de l'Academie des Sciences, Paris*, 253.
Friedrich, K. and Stammbach, K. (1968) *J. Chromatography*, **34**, 351.
Griffiths, J. (1976) *Colour and Constitution of Organic Molecules*, Academic Press, London.
Hague, D.N. *et al.* (1971) *Nature*, **229**, 190.
Hillson, P.H. and McKay, R.B. (1965) *Trans. Far. Soc.*, **61**, 374.
Horrabin, S. (1963) *J. Chem. Soc.*, 4130.
Jost, W. (1952) *Diffusion in Solids, Liquids and Gases*, Academic Press, New York.
Kunert, F. (1912) *Rev. Gen. Mat. Color.*, **16**, 255.
Kunert, F. and Acker, E. (1914) *J. Soc. Dyers and Colorists*, **30**, 128.
Miles, L.W.C. (1995) In *Textile Printing* (ed. L.W.C. Miles), Dyers Company Publications Trust, Bradford.
Murayama, T., Dumbleton, T. and Williams, M.L. (1968) *J. Polym. Sci.*, **6A2**, 787.
Murthy, K.S. and Rattee, I.D. (1969) *J. Soc. Dyers and Colorists*, **85**, 368.
Rattee, I.D. (1953) *J. Soc. Dyers and Colorists*, **69**, 288.
Rattee, I.D. (1978) In *The Chemistry of Synthetic Dyes*, Vol. VIII (ed. K. Venkataraman), Academic Press, New York.
Robinson, C. and Moiliett, J.L. (1934) *Proc. Roy. Soc.*, A, **148**, 630.
Rosetti, G. (1548) *The Plictho*, translated by S.M. Edelstein and H.G. Borghetty, M.I.T. Press, Cambridge, Mass.
Rys, P. and Zollinger, H. (1966) *Helv. Chim. Acta.*, **49**, 760.
Rys, P. and Zollinger, H. (1972) *Fundamentals of the Chemistry and Application of Dyes*, Wiley-Interscience, London.
Schetty, G. (1955) *J. Soc. Dyers and Colorists*, **71**, 705.
Sumner, H. and Taylor, B. (1967) *J. Soc. Dyers and Colorists*, **83**, 445.
Sumner, H. and Weston, C.D. (1963) *Amer. Dyes Rep.*, **52**, 442.
Valko, E.I. (1972) *Review of Progress in Coloration*, **3**, 50.
Vicherstaff, T. (1957) *J. Soc. Dyers and Colorists*, **73**, 237.
Witt, O.N. (1876) *Berichte*, **9**, 522.
Zeneca plc (formerly ICI Ltd) (1954) British Patents 789,121 and 797,946.

General bibliography

The following references provide a more extensive account of the subject.

The Chemistry of Synthetic Dyes, Vols I–VIII, ed. K. Venkataraman, Academic Press, New York, 1978.
Review of Progress in Coloration, Vol. 1 (1971) to Vol. 23 (1993), Society of Dyers and Colorists, Bradford.
The Principles and Practice of Textile Printing, E. Knecht and J. Fothergill, 4th edn, Griffin, London, 1952.
Fundamentals of the Chemistry and Application of Dyes, P. Rys and H. Zollinger, Wiley-Interscience, London, 1972.
Colour and Constitution of Organic Molecules, J. Griffiths, Academic Press, New York, 1976.
The Physical Chemistry of Dye Adsorption, I.D. Rattee and M.M. Breuer, Academic Press, New York, 1974.
Dyes and Pigments, Vol. I (1970) to date, ed. A.T. Peters, Elsevier Applied Science, London.

10 The environmental impact of the textiles industry

D.L. CONNELL

10.1 Introduction

Current concern over environmental issues is reaching fever pitch to the extent that it affects most of the working population to some degree, even in financial and banking circles. In particular, the level of emphasis given to the environmental agenda in school curricula in many countries of the Western World is predicted to produce a new generation of consumers and employees with a considerably heightened level of awareness over environmental matters. This will inevitably fuel consumer demand and influence national and international political agenda, producing a general demand on industry and commerce to deliver ever-improving environmental performance (Elkington, 1990; Good Housekeeping, 1990).

As a major world industry, textiles will increasingly be forced to respond to this rising concern. A thorough knowledge of the basics on environmental matters will therefore become as much a prerequisite for technicians qualifying in the industry as the technology of coloration, construction, manufacture or design that they have chosen to study. No textile technologist qualifying today will be able to avoid becoming involved with environmental concerns during a career in the industry. Whilst this chapter will seek to identify some of the major issues likely to be of concern to the textile technologist, full justice cannot be done to such a broad subject. It is therefore essential that reference is made to specific works on the subject, and in particular that current developments are monitored through the trade journals. This is important in a subject area which is experiencing such a rapid rate of change.

10.2 The environment

In any logical study of a subject area, a clear overall definition of the scope of the subject is a prerequisite. Unfortunately, whereas the dictionary defines the term 'environment' as "the circumstances surrounding an organism or group of organisms", it has been and is being used in so many differing ways, consciously or otherwise, that one definition suitable as a basis for study proves difficult to find. An evaluation of the various contexts in which the term has relevance is useful in gaining some understanding of the scope of the subject. The term 'environment' can usefully be considered as relating to various areas, approximate in scope and even overlapping considerably, but nevertheless giving rise to particular sets of issues dictated by the nature of the area involved. These may usefully be described as: immediate, local, regional and global.

10.2.1 *The immediate environment*

This use of the term may best be described as relating to all situations where there is a direct interaction between an individual and his or her immediate surroundings. In this respect, it could also be called the 'personal' environment. Issues arising through this interpretation of the term stem from such concerns as hygiene and health, safety, comfort, etc. Concern is therefore created through a perceived threat to the individual.

The most immediate personal environment is that normally generated by an individual's clothing, and it is in this area, where the textile technologist is most directly involved, that the perception of threat is clearest. Whilst issues such as comfort and fit are the obvious concerns of the textile technologist, increasingly the health or hygiene issues will need to be taken into account.

The prominence of this type of issue in the consumer's mind can be seen in the publicity given to the issue of residual amounts of manganese compounds present in 'stonewashed' denim articles which had been subjected to a bleaching process using potassium permanganate. This was given particular prominence in the USA and Germany. Similarly, the amount of formaldehyde released by articles treated with crease-resist resins has become of concern and even subject to legislative regulation in Japan. There is no doubt that other such issues will become important in the future.

Beyond the individual, immediate environmental concerns relate to the workplace or the home. In both cases health issues are central. Any textile technologist becoming responsible for production will have to deal with the work-related aspects directly, as is already the case—for example: noise-related hearing impairment; allergies such as dermatitis and asthma; and the abandonment of solvents and dyestuffs suspected of being carcinogenic.

One market-related illustration of direct relevance to textiles involves the purchasing pattern for floor-covering in the Scandinavian countries in the early 1990s. Originally this area was a good market for traditional tufted or woven carpet. Studies were then produced to imply that textile floor-coverings could harbour dust mites. The droppings of these minute creatures were associated with asthma and other respiratory conditions. This, together with the potential for release of chemicals into the room atmosphere created a strong consumer reaction and the market for textile floor-coverings in that area collapsed.

10.2.2 *The local environment*

Whereas the immediate environment involves issues arising from effects on or concerns of an individual, the local environment involves issues effecting groups of people concerned with their locality. Whilst there may still be health risk issues involved, these are likely to be bound up in the most common concern: quality of life.

Increasingly in the Western World people are becoming much less tolerant of the effects of industrial activity in their vicinity, and may well take action to correct what they perceive as irresponsible behaviour. This does not have to be permanently damaging, or even externally harmful. For example, in the UK, regulation of the amount of colour being discharged into the river system is becoming stricter, even though it may not be permanently damaging as such, but is nevertheless very visible pollution. Other local issues may relate to unpleasant noise, odour, smoke or steam emission, to level of traffic to and from a factory site, or even an unsightly or badly maintained factory site itself.

10.2.3 *The regional environment*

Regional issues arise where the activities of a group, factory, industry, etc., produce effects in another location, possibly at some distance from the original action. Some examples of this type are fairly well known, such as the effects of acid rain on the forests of Norway and Sweden as a consequence of fossil fuel burning in other Western European countries, or the effects of radiation fallout in Western Europe from the Chernobyl disaster.

A more directly textile-related example of this type of issue relates to the Aral Sea on the border of Kazakhstan and Uzbekhistan. The river system feeding this lake was partially diverted and used to irrigate cotton fields. As a consequence, this lake, the fourth largest in the world at one time and home to a prosperous fishing industry, shrank to a fraction of its size and increased in salinity, to the point where all fishing ceased as a result of the virtual disappearance of the fish population.

10.2.4 *The global environment*

Although generally less perceptible to the individual, changes in the global environment affect every human being ultimately, and arise out of the activities of the human race as a whole, or large portions of it. A number of truly global environmental issues exist, of which ozone layer depletion and the greenhouse effect are two of the best known.

Although to the casual observer most of the environmental movements are motivated by threats to the global environment, many of the issues which concern them may not be global except by extrapolation. Environmentalists argue that local or regional environmental issues may have a globally cumulative effect, and therefore are more important than they appear at first sight. This makes a rational response all the more difficult as industry tries to respond positively to the environmental challenge. Support for this attitude may be found in James Lovelock's 'Gaia Hypothesis'.

10.2.5 *The Gaia Hypothesis*

The geologist James Hutton in the late eighteenth century proposed to the Royal Society of Edinburgh that the world was, or could be considered as, a superorganism, and that the flow of nutrients and water through the environment paralleled the recently discovered circulation of blood in the human body. This idea was largely discarded as science grew in the nineteenth and twentieth centuries, but was revived by James Lovelock in 1972.

That the world may be considered to be alive seems absurd when the biosphere, or living part, effectively forms a thin layer over a large sphere of inert rock. Proponents of the theory point out that a giant redwood tree largely consists of an inner core of dead woody matter, and only the thin outerbark is actually alive: 1% living matter superimposed on 99% inert material. Lovelock has continued to develop his hypothesis and provide data and theoretical studies in its support (Lovelock, 1986a,b).

The implications of the Gaia Hypothesis are considerable, in terms of both public opinion and the environmental effects of the activities of the human race. Both are important for the textile technologist concerned with environmental matters, since public opinion may affect consumer behaviour in complex ways, and hence alter purchasing patterns from an industry of global proportions.

The Gaia Hypothesis suggests that the planetary conditions suitable for life, and in particular the highly unstable combination of gases in the atmosphere, are maintained by a complex series of mechanisms involving all the living species on the planet. Any imbalance in climate or atmospheric composition results in an enhancement of the appropriate mechanism (or species activity) to compensate for the change. This is not a conscious act by the planet but a natural consequence of the unconscious behaviour of the multitude of living organisms which make up the biosphere.

One example of how humankind may be affecting this mechanism lies in the destruction of the large rain-forest belts throughout the world. It is considered by environmentalists that these act to recycle carbon dioxide to oxygen in the atmosphere and also act to control weather patterns on a global scale. In addition, they harbour a wide variety of plant and animal life whose value to and interaction with the global environment can only be guessed at (Allen, 1980; Woodward, 1989).

This view, which is widely propagated, leads to public concern that may be evidenced in complex ways. Strong lobbying against the use of valuable hardwood timber produced by such rain-forest areas is a very direct effect. More subtle effects may occur; for example, the rumour that certain fast foods were produced from land made available by rain-forest clearance has required the companies in question to mount a strong campaign to allay fears in a volatile mass market.

10.2.6 *Environmental complexity*

All the foregoing serves to illustrate the vagueness and complexity of the environmental debate. There are few clear boundaries or well-defined issues, though many vested interests on both sides would assert differently. A rational and commonsense approach is required to deal with the many technical problems which are produced from this arena, with a clear understanding of the true facts of the situation as a primary requirement.

Such a framework is now being developed and applied in industry, with some degree of success, by the pioneers of this approach. Before considering this in detail, it is first necessary to address some of the major technical issues in question.

10.3 Environmental impacts of the textiles industry

Although industrial activity has many effects on the environment, these may be divided into two main groups: use of resources, and the production of waste. In both groups, a number of aspects are important to the environmentalist. A clear appreciation of these issues is necessary before embarking on an environmental programme.

10.3.1 *Resources*

There are two main types of resource, renewable and nonrenewable, though the category into which some resources fit may be a matter of dispute. Renewable resources are those which may be replaced as they are used up, and in large part are the product of living things, such as cotton, wool and silk for example. Nonrenewable resources are those with a finite supply on this planet, and are largely mineral in origin, such as coal, copper and diamonds. In some cases, nonrenewable resources as so defined might be regarded as renewable because they exist in such great abundance on the planet, like water for example. Even here, water can be regarded as renewable in an area with abundant rainfall, but nonrenewable in a more arid region where it is always at a premium.

10.3.1.1 *Renewable resources.* Renewable resources cannot be regarded as being in infinite supply. Being produced by living organisms, the amount of any resource available will be dictated by the lifeform capable of producing it, and the timescale and the yield from the resource base. For example, to produce equal amounts of wool and cotton fibre in the same time will, it has been estimated, need some 33 times more land area for wool than for cotton. In the case of the more exotic fibres such as cashmere and angora, where yields are even lower than for wool, the ratio is even more extreme. Attempts to enhance the production rate to give higher yields do not automatically do so for very long, and may even suppress the long-term viability of the existing production rate.

Two concepts are important in this respect: sustainable development, and biological diversity. Sustainable development is the discipline of production of the required resource, cotton for example, at a rate and on land types which secure the long-term maintenance of the crop yield without any adverse environmental side-effects. Since this rate may not be that realisable by the best modern technique, sustainable development is much less widely practised than is advisable for the long term.

Biological diversity has only recently been recognised as being of vital importance in the sustainable development of world crops. Earlier, it had been thought that the important factor in agriculture was maximisation of crop yield by development of super-strains of plants and animals giving high yields, and the widespread use of these in preference to locally used strains. This leaves the crop wide open to predators and disease and has, in a number of cases, led directly to a drop in yields until new strains could be produced. This can only be done if sufficient differing strains of the species have been preserved to provide a suitable gene pool.

10.3.1.2 *Nonrenewable resources.* Although the exact extent of the amount of any mineral on the planet may not be known at present, reasonable estimates are available, from which by extrapolation it is possible to project an approximate date by which they will be depleted at current rates of exploitation. Whilst only the most pessimistic of such estimates predict that this is in imminent danger of happening within the next ten years, the vast majority agree that a number are destined to run out or become severely depleted within the next 20–50 years, that is, within a generation or two from now (Myers, 1986).

Whilst the textile industry is not so directly dependent on mineral resources as, for example, the engineering industry, minerals are needed to produce the many complex machines used extensively in textiles. In addition, certain metals are essential components in some dyestuff types, and in some dyeing techniques. However, the textile industry is very dependent indeed on fossil fuels, in common with all other world industries.

The use of fossil fuels, especially oil and gas, is a major world concern, since much of our industrialised society depends upon these resources. Whilst predictions vary, it is generally accepted that the oil supplies available are unlikely to last very far into the twenty-first century (Myers, 1986), and will certainly cost more in real terms to extract as the easily depleted sources are used up. Fossil fuels impact widely in the textile industry, through heat energy, electrical energy and motive power for transport, and therefore any environmental audit must carefully examine energy and fuel usage.

10.3.2 *Production of waste*

All living things produce waste in some form in order to continue living, and this waste is released into the environment, where it is generally dealt with by an appropriate biological mechanism. If the waste is produced in too great a quantity, this mechanism can break down from pollution overload. Should the production of waste nevertheless continue beyond this point, serious disruption of the population of living creatures in the area affected by the waste will occur, and in the extreme, most of the higher life forms will die. In nature, there is often a self-regulating mechanism, since the lifeform producing the waste is also killed and the source of the pollution is removed. Gradually the pollution load will then dissipate and the affected area will be restored.

The pollution produced by humankind is usually carefully channelled away from the population that produces it and is deposited elsewhere. The amount of care given to the disposal of this waste will dictate what happens in the area where the disposal occurs. The nature of the waste is also relevant. Basically, waste can take three physical forms: solid, liquid and gaseous. It may also fall into one of four categories:

(a) Nontoxic, biodegradable
(b) Toxic, biodegradable
(c) Nontoxic, nonbiodegradable
(d) Toxic, nonbiodegradable

No biodegradation mechanisms exist for gases, and therefore waste released in this form must first be taken into solution in rainfall and deposited on to the land or into the sea before any biological mechanism can come into play. The physical form of the waste largely dictates the means by which it is released into the environment, whilst the category of the waste will dictate its environmental effects. In general, these become more severe as toxicity increases and biodegradability decreases.

10.3.2.1 *Biodegradability.* Waste materials that are biodegradable are broken down into simpler molecules by the digestive action of micro-organisms that are capable of using them as food. These simpler molecules are then excreted as waste, which may form a food source for another organism, be converted into yet simpler molecules which are excreted, and so on. For biodegradation to occur, moisture and a suitable population of micro-organisms must be present. Usually, the micro-organisms also need an adequate supply of oxygen to survive and digest the waste. This is aerobic digestion or biodegradation. Anaerobic digestion occurs in the absence of air or oxygen, and is often associated with the production of toxic and unpleasant gases such as hydrogen sulphide. If there is an oversupply of the material to be aerobically degraded, the population of micro-organisms grows rapidly and so does the oxygen consumption until all the available oxygen is used up. At this point, anaerobic digestion takes over, and all the oxygen-breathing life in the area of oxygen depletion dies. In rivers, lakes and seas this process is referred to as eutrophication.

Whilst biodegradation of waste products results in the production of simpler molecules, this in itself is not necessarily desirable if the simpler molecule has undesirable properties. It may be that the new molecule is not readily degraded further for example, and a 'biodegradable' molecule has been traded for a nonbiodegradable one. Alternatively, the products of biodegradation may be toxic to other higher life forms, again an undesirable result.

In considering biodegradation, it is therefore important to distinguish between primary and ultimate biodegradability. Primary biodegradability is the degree to which the initial molecule is removed from the environment, whereas ultimate biodegradability is the readiness with which, and degree to which, the material is finally converted to the simplest molecules such as water and carbon dioxide. Clearly, the ease with which a material undergoes ultimate biodegradability, and the amount of oxygen used up by the micro-organisms in this process, are the most important factors in evaluating the environmental impact of a material.

Some materials are more rapidly removed from the environment than others, the rate also being dependent on a wide variety of factors for any given material. It is therefore important to have some quantified measure of biodegradability in order to compare materials and to set legislative standards. Measures of biodegradability are widely quoted for many textile chemicals, but great care must be taken in the use and interpretation of the information before conclusions are drawn.

These tests all measure the biodegradability of the product in solution in water. The measure is usually of primary degradation, being based on the disappearance of the target molecule, or some property associated with it, in a controlled experiment using a culture of suitable bacteria. These tests have their origins in the work done to provide indications of the environmental persistence of detergent materials in rivers, the original intent being to avoid the generation of unsightly foam banks in waters receiving treated sewage outfall. These tests do not provide information on the rate or degree of ultimate biodegradation. An example of this is found in the comparison between alkylphenol-based nonionic detergents and those based on linear fatty alcohols. Whereas both pass the primary biodegradability tests, alkylphenols are much more resistant to ultimate degradation than linear alcohol ethoxylates (Kravetz, 1983).

The situation becomes much more complicated when the materials evaluated by these tests differ greatly in their physical properties, by being emulsions in water rather than solutions, for example. The results from standard OECD tests may be highly unreliable. More specialised test regimes are being developed for oily materials, such as the CEC Method L-33-T-82 (Co-ordinating European Council for the Development of Performance Tests for Lubricants and Engine Fuels, London). For this reason, biodegradability data quoted without reference to the test protocol used should be viewed with caution, especially when compliance with legal limits is being considered. It should furthermore be borne in mind that biodegradation produces a large population of bacteria which must be withdrawn from the treatment plant as a slurry or wet filter cake, and disposed of in landfill or by limited spraying on agricultural land (provided that the toxic metal content meets statutory limits).

Knowledge of biodegradation of solid materials is at a much more elementary stage even than that of materials in solution. Biodegradation in the solid state occurs in soil or some mixture of solid wastes with a bacteria source and a supply of air and moisture. This process is composting, as used over many generations to turn domestic and animal waste into usable manure. However, though such a process may occur quite readily in the controlled environment of the compost heap, it has been shown that it does not follow that degradation will occur in municipal refuse tips. Virtually unchanged samples of vegetables, meat and newspaper have been found by excavating old municipal rubbish facilities, even though the material has been buried for over ten years.

From the foregoing, it can be seen that biodegradability is no panacea for the environmental problems facing the textile chemist. The picture is complex and highly technical. Where simplistic claims of 'environmental friendliness' based on biodegradability are made, the claimant is wide open to attack from sophisticated and well-informed environmental pressure groups.

10.3.2.2 *Toxic materials.* Not many of the chemicals used in textile production and processing are highly toxic, though some materials may have long-term toxic effects. For example, some dyestuffs based on benzidine were withdrawn on the basis of their suspected long-term toxic effects on users in the dyestuff industry. Toxicity in environmental terms is a very different consideration. Substances which may be harmless to humans can be very toxic to other species. Surfactants, widely used in industry, are often toxic to the more sensitive varieties of fish, such as rainbow trout, for example. Properly treated in an effluent plant these materials present a relatively low threat in the environment, but if a substantial accidental spillage directly into a watercourse were to occur, substantial loss of aquatic life would be likely.

This damage would nevertheless be localised in nature, since the material would be diluted in the water flow and biodegraded by freshwater organisms. More serious effects are possible when the material is not biodegradable. In such cases, toxic effects from accidental releases into the environment are prolonged and can generate substantial adverse publicity. Further, the toxic material may be deposited in the sediment along the stream or river bed, being released periodically by disturbances such as storms or dredging.

Concern is now being focused on a more insidious problem than the foregoing. Low-level release of non-degradable toxic materials may occur for a great many years with no obvious effects. This material is, however, gradually concentrated through the food chain. It is first ingested by detritus eaters who live on sea and river bottom deposits. These are in turn eaten by their immediate predators, who are in turn eaten by larger predators, then small fish and so on. At each stage, if the toxin cannot be easily metabolised or excreted, the concentration of toxin in the body of the creature in the food chain is substantially higher than that in its predecessor until, in the higher carnivores, the concentration reaches the point at which serious toxic effects are evident.

These materials are often referred to as micropollutants, since they may not be discharged at levels of more than parts per million or parts per billion in effluent, but are suspected of being accumulated in the food chain (bioaccumulation). The risk of bioaccumulation is at its most severe in landlocked seas such as the Mediterranean and the North Sea. Political action has been taken to deal with this pollution issue on an international scale. In the case of the Mediterranean, this began with the Barcelona Convention in 1976. In the subsequent protocol in 1980, two lists of offending substances were drawn up. A 'black list' of highly toxic substances specified materials, such as mercury, cadmium and DDT, which were totally banned from release, and a 'grey list' specified materials where release was strictly controlled. Similar arrangements exist for the North Sea.

Micropollutants present substantial difficulties for the textile industry. The chemicals used are purchased either for their primary chemical composition, such as formic acid, sodium carbonate, etc., or for the effects they produce. There is therefore no requirement or predisposition to disclose the complete chemical analysis of products used in textile wet processing. When manufacturers are pressed, basic chemical analysis data may be presented, but would not necessarily deal with trace components. However, if trace amounts of nonbiodegradable toxic substances are present in the products, these may be sufficient to render the effluent 'over limit' for this material.

As an illustration of the problem, sodium hydroxide (caustic soda), which is widely used in the textile industry, has been produced by the mercury cell process for many years. This results in trace amounts of mercury in the final product, as used by the unsuspecting textile processor. This trace contaminant is passed through the system and into the factory effluent. Only at this stage may it be picked up in analysis of the effluent by the monitoring authority.

A lengthy process of backtracking is then required to pinpoint the source of the contaminant from the wide range of dyestuffs and auxiliaries that may have been used. This is no easy matter if the production routes of the chemicals are not known to the textile chemist seeking the answer to the effluent problem, as is usually the case. In the case of sodium hydroxide, the level of mercury in the product is very low, and can vary between manufacturers. Fortunately, in this case, there exists an alternative, in material made using a new, mercury-free diaphragm cell process, and the problem has been identified with manufacturers who are working hard to reduce mercury levels with immediate effect, prior to switching eventually to the new process.

A more complex problem is presented by the introduction of various toxic, nonbiodegradable organic materials into the textile process, and hence into the effluent, via the textile goods being processed. These are materials such as pentachlorophenol and a variety of insecticides. In a situation of this nature, the source of the micropollutant must first be identified, and proof obtained that the source is none of the chemicals used by the wet processor.

Although pentachlorophenol is applied to cotton as a rotproofing agent, and various insecticides, such as permethrin (actually one of the more biodegradable of such materials) are applied to wool as moth-proofing agents (Shaw, 1990), this is only carried out in a small number of the processing plants with these materials in their effluent. In the remainder of the cases, the pollutant is introduced into the effluent by the dry goods being processed. The pollutant may have been introduced on to these goods at an earlier stage in processing, perhaps as an antifungal agent for cotton goods to prevent rotting during storage in hot, humid climates, or as a preservative agent in a processing oil or sizing agent, for instance.

To complicate matters further, the source of the pollutant may lie at an even greater remove from the wet processor, being applied to the natural fibre during its growing stage to protect the plant or animal from attack by insect pests. The textile technologist has then to become detective to try first to pinpoint the source of pollutant in the factory, and then to ascertain the point of introduction on to the goods, to find means to obviate the problem. The easy answer of issuing an outright ban on shipment of goods bearing this material may not be practical, if the use of the undesirable chemical is widespread, and at the very least requires a method to be developed for detection of the offending chemical on shipments of goods being received by the factory.

This may well prove to be a recurring problem for textiles, as raw materials are sourced from new Third World countries with little previous experience. Since the polluting materials are generally cheaper than more environmentally acceptable alternatives, there is strong commercial pressure to use the former.

Further complications exist when chemical reactions occur during the textile processing route, thus producing molecules in the effluent which were not introduced into the system directly or indirectly on the goods to be processed. Processes using highly reactive chemical species do exactly this. For example, chlorine is used in the bleaching of cotton and other fibres to modify the coloured impurities in the grey goods chemically and produce a whiter fibre. Alternatively, it may be used with wool as part of a process aimed at imparting shrink-resistance (Shaw, 1990). The exact chemical reactions involved are generally not known, are definitely complex, and undoubtedly follow various pathways. Recent work has shown that these processes produce new, chlorine-containing molecules in the effluent stream. This mixture of materials of unknown composition has an unknown environmental impact.

The regulation of such materials in effluents is therefore based on the 'precautionary principle'; where there is evidence to suggest that some material might possibly have an adverse environmental effect but no studies have been done, the emission of the material into the environment is regulated as if the material is proven to be undesirable. For regulation to be possible, some accepted test method is required.

In this particular case, measurements are made in terms of AOX (adsorbable organic halogen) and EOX (extractable organic halogen), with AOX being the more common measurement. The test methods are based on detection of halogen in the combustion gases from samples isolated out of the substrate being examined following a specified isolation procedure. This gives an estimate of the amount of organohalogen compounds being carried into the environment from the process. In no way, with this technique, is any attempt made to evaluate the environmental impact of the organohalogen compounds.

10.3.2.3 *Nontoxic nonbiodegradable materials.* Some chemicals, for example silicones, used by the industry are not broken down by the usual biodegradation systems. If tested for biodegradability in the usual procedures, even where these have been properly modified to be suitable for the form of the material to be examined, the chemicals will be shown to be poorly degradable. A simplistic approach based on the precept that all chemicals used in wet processing should be readily biodegradable would result in an immediate ban on use of the product. This would not necessarily be the most desirable course of action in true environmental terms. Firstly, no adverse effect on the environment may exist. If the chemical has no known or suspected effect on flora and fauna in streams and rivers, its release into the aqueous environment could be seen as neutral. Secondly, if the material is nonbiodegradable, it will probably not participate in eutrophication, and may well be preferable to an alternative, rather more degradable, material that does. Thirdly, a nonbiological route for degradation may exist which is not shown up in controlled laboratory test environments aimed at measuring biodegradation. Such evidence may be beginning to emerge in the case of silicones.

10.3.3 *Air pollution*

Although associated with images of smoking factory chimneys and vehicle exhausts, the most serious air pollution is invisible and therefore not so immediately evident to the general public. Some of the issues created by air pollution have been given considerable prominence in the media in recent years, and therefore issues exist and will continue to arise which the textile technologist must take into account. Although biodegradation mechanisms do not exist to remove airborne pollution, this interacts with the biosphere very directly, in a manner dependent on the type of pollution, giving rise to specific issues.

10.3.3.1 *Acid rain.* In any combustion process, byproducts are formed from impurities in the components. Those giving rise to acid rain are the oxides of sulphur (SO_x) and nitrogen (NO_x). They are produced from the oxidation of sulphur- and nitrogen-containing impurities in the fuel used for combustion and, in the case of NO_x, from oxidation of a small amount of the nitrogen in the air. These oxides are emitted into the atmosphere along with the products of combustion, and eventually interact with water droplets in the atmosphere to produce dilute solutions of the oxyacids of nitrogen and sulphur. This often falls as rain. The low pH value of the rain then alters the natural pH balance of the soil upon which it falls. At first, the alkalinity of the soil will counteract this effect, but this will eventually be exhausted. The alteration of pH, and the action of the weak acidic solution on plants, results in damage to vegetation, especially trees. In addition, the low pH of the water table enables toxic minerals to be released into the groundwater, affecting water supplies and poisoning aquatic life (Pearce, 1987).

All combustion processes will produce these gases to some extent, though the amount produced can be controlled by regulating the quality of fuel burnt and the combustion conditions. They can be removed from exhaust gases by post-combustion cleaning or scrubbing, but this introduces extra cost, may decrease the efficiency of some combustion processes, and generates solid or liquid waste which is not biodegradable and must be disposed of. It is therefore not a viable proposition to regulate this form of pollution in all cases.

10.3.3.2 *Ozone layer depletion.* Chlorofluorocarbons (CFCs) were first developed by DuPont in the 1930s. They are chemically very inert substances which do not dissolve in water, do not burn and are not taken up by any of the usual mechanisms in the biosphere. This led to their being used widely as refrigerants, solvents and propellants for aerosols. However, in 1974, a paper by F. Sherwood Rowland suggested that these apparently harmless gases may pose a very severe threat indeed (Stevenson, 1990).

The biosphere on earth is protected by a layer of ozone in the high atmosphere. This ozone is created by the action of sunlight on oxygen in the

Earth's atmosphere, and results in the removal of much of the ultraviolet light in sunlight before it reaches the earth's surface. Ultraviolet light is inimical to many forms of life, and causes sunburn and skin cancer in humans. Sherwood Rowland showed that, in the conditions prevailing in the upper atmosphere, CFC molecules would be attacked and liberate chlorine atoms. These chlorine atoms would interfere with the formation of ozone drastically, in an approximate ratio of 100 000 ozone molecules destroyed per chlorine atom generated, before the chlorine atom was removed from the process by other mechanisms.

This resulted eventually in the Montreal Protocol in 1987, which aims to regulate and eventually ban CFCs altogether. In the meantime, severe depletion of the ozone layer, especially at the Earth's poles, continues to occur (Gribbin, 1988a).

10.3.3.3 *The Greeenhouse Effect.* The temperature of the Earth's surface is expectedly cold when compared with that of its neighbouring planets. Calculations suggest that the natural temperature if life were absent would be 290°C, instead of the actual mean of 13°C (Myers, 1985). Regulation of planetary temperature is a complex combination of many influences which either reflect some heat energy from the sun back into space, or absorb it through the mechanism of photosynthesis. Some gases in the atmosphere act against this mechanism by absorbing and retaining the heat. The most notable of these is carbon dioxide, produced by all living things. The level of carbon dioxide in the atmosphere is regulated by a number of natural mechanisms such that it forms only a tiny proportion of the Earth's atmosphere, and therefore does not retain too much of the heat (despite the 25% increase in heat output from the sun over the 3.5 billion years that life has existed on the planet).

The situation has been changing since the Industrial Revolution occurred. Humankind now produces vast quantities of carbon dioxide at an ever-increasing rate through the burning of fossil fuels. This is carbon dioxide that has been locked up in the mineral content of the earth from earlier times. Simultaneously, there has been a progressive net destruction of natural forests. These actions result in a gradual increase in the amount of carbon dioxide in the atmosphere, resulting in, it is claimed, a tendency for mean global temperatures to increase through the absorption of more heat by the extra carbon dioxide, the global warming effect. It is not known exactly what the end result will be if this effect proves to be real. Various scenarios have been put forward, all of which involve considerable disruption to humanity and loss of life (Gribbin, 1988b).

Other gases may also contribute to heat retention in the atmosphere. Methane is produced in the digestive tracts of ruminant cattle, generated by rotting waste matter (such as in landfill sites), and accidentally released during the extraction of natural gas deposits in the earth. Methane is thirty times more effective at retaining heat than carbon dioxide. CFCs are even more powerful, being some 17 000 times more effective than carbon dioxide. Carbon dioxide is, however, considered to be the main threat due to the amount being released into the atmosphere each year.

10.4 Approaching the problem

The foregoing has illustrated the complexity of the environmental issue; the vagueness of the meaning of the term itself, the emotional, sometimes illogical, view which may be taken, and the various, sometimes conflicting, issues that are already known. One thing is certain. This is a major issue which will affect humankind through the coming decades, and one which the textile industry must address. Like all business issues, it is best addressed in a logical and ordered manner, with all the various actions defined and an action plan put into place to overcome them. This programme must also be flexible, like all other business strategic planning, to allow for future legislative and consumer pressures.

In the absence of properly structured assessment of environmental issues, there is an inevitable focus on fashionable issues, resulting in false or unsound actions or claims. These serve to mislead public opinion and eventually serve to discredit efforts by industry and commerce to produce a sound response to the environmental issue. Tools are now emerging which enable a rational approach to be taken to these complex issues, so that commercial organisations may act positively and constructively in their activities. These are the environmental audit and life-cycle analysis.

10.4.1 *Environmental audits*

Auditing, formerly a purely financial term, has been widened in definition with the burgeoning quality movement in world industry. The environmental application has further widened its definition. An environmental audit is a detailed evaluation of all aspects of an organisation's activities, in which all the effects, risks and likely outcomes from the organisation's activities are identified and evaluated in terms of their effect on the environment. Degrees of risk may have to be evaluated and effects quantified.

The object of an environmental audit is to determine the environmental performance of the organisation and act as a prelude to the development of an environmental action plan. This action plan will be aimed at improving the performance of the organisation in environmental terms. This may well lead to policy decisions being taken, such as using recycled paper or paper produced from a renewable source of timber, and will certainly result in some form of waste- and energy-reduction programme.

Various programmes exist for environmental auditing, some produced by commercial or industrial organisations, for example the US and UK Chemical Industry Association's 'Responsible Care' programme. Others come from national programmes, such as the British Standards Institute's BS 7750 and the American Textile Manufacturers Institute E3 programme. Whilst they may differ in style, emphasis and complexity, all have a common structure with three main parts.

(a) *Policy and goal setting.* Initially the organisation involved must decide its overall approach to environmental issues in a manner understandable by all its members. This will take the form of a simple policy statement endorsed by the head of the organisation. This may also specify some initial goals which the organisation aims to achieve.
(b) *Auditing.* In this part of the process, the basis of assessment is set, in terms of areas to cover and standards judged acceptable. The actual performance against these standards is then determined, and areas are identified where action is needed to remedy problems or improvements could be made. The audit process should also identify what solutions are available.
(c) *Strategy and implementation.* An action programme is drawn up to make changes to those activities or aspects of the organisation that are judged not to meet the standards set in the first phase. Interim goals or rates of improvement will be set, and investment, product redesign, research or other actions will be required.

This whole process is cyclical in nature, with the strategy and implementation phase followed by a further policy and goal-setting phase, conducted in the light of the knowledge gained through the first cycle. Standards may well be better defined as a result of the clearer understanding gained, and this will result in a more effective audit phase, and so on.

This type of approach is applied to all the aspects of the operation of an organisation, but is focused on the organisation. Another approach, for manufacturing industry in particular, looks at the products made by the organisation, not only whilst in the manufacturing stage within the company, but throughout the whole life of the product. This is termed life-cycle analysis.

10.4.2 *Life-cycle analysis*

Whereas an environmental audit is concerned with an organisation, a life-cycle analysis looks at a product. The process is basically a form of audit in which all aspects of the product are considered. The first step is to consider the raw materials involved in the manufacture of the product and the environmental impact involved in its production. Following this, the production process by which the product is made is then considered, and again the environmental impacts are assessed. The study will then go on to consider the environmental impacts involved in the sale and use of the product, and, finally, the environmental impact involved in its eventual disposal.

This 'cradle to grave' approach, as it is also referred to, is valuable particularly in identifying the aspects of a product's life cycle with the strongest environmental impact. It is then logical to seek to lessen the environmental impact of a product by addressing this particular aspect first. For example, it has been determined that the major energy and water use in the life of a domestic washing machine occurs largely during its use, not manufacture. The outcome of this study is a new generation of machines that aims to reduce the water and energy consumption during the use period of the product's life cycle. The application of this type of analysis has also resulted in the production of certification schemes in various countries to provide consumer guidance in the form of 'eco labels'. The first types of product to be covered by such a scheme in the EU were washing machines and dishwashers.

In the textile industry, the issues are very complex indeed, and a true life-cycle analysis of a product can be a formidable, if not impossible, task. A simplified example of the problem is provided by considering the relative merits of 'greenness' of articles manufactured from cotton rather than polyester. The simplistic view taken by the average consumer may be that because cotton is the naturally produced fibre, it is therefore better in some way than the artificial version. This is supported by the perception that in the personal environment (when clothing is being worn) cotton is the more comfortable and acceptable fibre. The truth is, of course, much more complex (Van Winkle *et al.*, 1978).

During the use phase, for example, cotton requires more energy when being laundered, often being washed at a hotter temperature, usually requiring ironing, and taking more energy to dry after washing than a comparable synthetic fibre garment. This is due to cotton's ability to absorb moisture. After use, however, cotton is readily degraded by composting in soil, and thus presents no serious long-term waste-disposal threat, unlike the degradation-resistant synthetic fibre garment.

An examination of the production phase reveals more problems for cotton, however. Scouring, bleaching and other wet preparation procedures are required for cotton, while the intrinsically cleaner and whiter synthetic fibre demands relatively little preparation. During dyeing, cotton coloration is relatively difficult technically, with poorer dye yields resulting in wastage of expensive (in both money and energy terms) dyestuffs, and the use of additional dye fixing agents to ensure proper colour-fastness. This contrasts unfavourably with the relative ease with which many synthetic fibres can be dyed to good fastness standards and dye yield. Some fibres such as polyester, however, require dyeing at higher temperatures, thus taking more energy in the process.

Raw-material production presents another set of problems. Cotton may, at first glance, seem to have the more favourable position, since synthetic fibres are in general produced from non-renewable sources. Cotton requires land on which it has to be grown, and there are arguments that allowance should be made for the diversion of this land from food production. In addition, the use of fertilisers, insecticides and fungicides in cotton growing represents another hidden environmental penalty.

In order to obtain a proper picture of the relative environmental merits of the two types of fibre, a very complicated evaluation must take place, comparing identical textile articles in the same market area. All aspects of environmental impact must be looked at, and quantitative assessment of the impacts produced. This process inevitably involves some degree of assumption, and the placing of values on previously nonquantified actions or situations. As a consequence, the results can themselves induce debate. Such evaluations have been carried out by various workers. Van Winkle *et al.* (1978) have itemised early contributions to the debate, which still continues (Greenwood, 1991; Kummer, 1991).

The question of how far back to take a life-cycle analysis is difficult. Purists argue that the process should go back as far as possible, but this can be a very time-consuming and laborious process, as the cotton example ilustrates. It has already been argued in many quarters that the textile industry would be better advised to take a more local view based on the environmental audit approach, to determine what may properly be done to improve the impact of its activities.

10.4.3 *The action programme*

Whatever approach is taken to address the environmental issues raised in audits or life-cycle studies, at some point action must take over from analysis. This can be the point at which mistakes are made if some thought is not given to the action programme. As far as waste and pollution are concerned, a general consensus has been reached about the types of action to take and their order of priority where relevant.

(a) Remove or omit the operation, part or item. Starting from the question 'What happens if you don't do it?', this approach is often the best choice environmentally, but may not always be practicable. It has, however, led to a reduction in the amount of packaging materials used in some textile areas. Whilst some operations such as sizing weak yarns cannot be avoided, others such as prebleaching before dyeing may feasibly be omitted, but may require an alteration of standards or style.

(b) Replace an offending process or product with an environmentally more acceptable one. When seeking to use this approach, it is important to ensure that the issues are properly and fully addressed, and that the alternative is provably better overall. In textiles, there have already been some moves in this direction, such as the switch from chlorine to hydrogen peroxide in the bleaching of cotton, for example (Weck, 1991).

(c) Reduce or minimise the amount of or intensity of some component or process. For example, application of this approach has resulted in the reduction of the amount of polythene used in packaging of textiles by moving to thinner gauge polythene sheeting, with no adverse effects. Significant benefits are also to be gained by using this approach in the development and adoption of low-temperature dyeing techniques.

(d) Recycle materials after their use rather than discard them. In the textile industry, there has been a long history of recycling of textile goods into new textile materials, with extensive technology developed for this purpose (Spangenberger, 1991). New approaches are being developed today (Hoenig, 1993).

(e) Re-use waste or reject material for other purposes or as a lower-grade feedstock in the same process. Again, this is an area with a long history within the textile industry (Hoenig, 1993). Recent developments are centred on the possibilities presented by the conversion of textile waste into nonwovens (Watzl, 1992). Other successful approaches are centred on the use of certain wastes as fuel for energy production, as either heat or electricity or both.

(f) Treat the undesired item or waste. This is the so-called 'end of pipe' approach, and is traditionally the primary reaction of industry to demand to improve effluent to air or water. The weakness of this approach is that 100% removal is very difficult, and therefore the factory is trapped into a spiral of rising cost to meet ever-rising discharge consent standards.

(g) Dispose of the waste to landfill. Inevitably this is a course of last resort. Land for this use is rapidly disappearing, and the restrictions on the management of such sites are constantly escalating as regional government seeks to protect the community from the long-term consequences of this activity. In addition, landfill sites are a source of methane, a greenhouse gas, and thus landfilling of waste can generate future pollution.

10.4.4 *Action areas*

These will be highlighted by the environmental audit of the company's activities. A carefully conducted audit will aim to quantify the size of the impact on the environment of all activities, and thus should then make it possible to draw up a prioritised list of actions. It is then necessary to consider how to approach each problem. A multistage programme may be necessary to achieve best environmental performance in an area. The following should serve to illustrate some of the issues which may be involved.

10.4.4.1 *Sizing—an example.* Sizes are used to facilitate the conversion of yarn into fabric, and must then be removed at some stage before the fabric or garments are finally sold. Sizing is considered an essential step in fabric production, to enable yarn to be woven at the higher speeds made possible by modern loom machinery. These are necessary in order for the industry to deliver the large quantities of high-quality, low-priced fabric demanded by modern society. However, this process does present significant environmental difficulties. Stiebert (1975) claims that up to 80% of a textile finishing plant effluent may, in extreme cases, be due to the use of sizing agents. This is further exacerbated by the use of cheap, highly effective sizing materials such as poly(vinyl alcohol) (PVA), which are not very biodegradable. There is thus a direct conflict between the needs of modern industry to become ever more efficient and the demands by environmentalists for cleaner industrial processing. Positive approaches are being taken to address this problem, and these can be divided up into the action types specified in section 10.4.3.

(a) Remove. Although it is not possible in most cases to weave without size, some progress has been made within a limited range of qualities. In the case of linen yarns, where high yarn strengths are available, a switch to specially developed dressings made from biodegradable materials with good water removability has transformed some sectors of the industry, especially in Northern Ireland.
(b) Replace. Becker and Grunert (1980) sought to remove the problem from the effluent by using a size finish which remains on the fabric. This, of course, is technically difficult, because the finish must not interfere with any of the subsequent processes applied to the cloth, and must give a handle and performance acceptable to the final customer.
(c) Reduce. A novel approach to the problem of aqueous size removal has been reported (AATCC, 1973). PVA size is broken down to harmless gases by passing the woven fabric through a low-temperature plasma. In this process, it is reported that approximately 60% of the applied size is removed, leaving only 40% to be removed by conventional wet processing. Unfortunately, this process is therefore additional to those normally used and presents an on-cost to the goods, as well as using more energy in the plasma step. A careful comparative analysis would be needed to ascertain whether there was a net gain in environmental terms, which serves to illustrate the difficulty with which researchers are faced when attempting to develop alternative technology for environmental reasons.
(d) Recycle. Since the polymers used in sizing are effectively unchanged by the process, it is theoretically possible to remove them from the liquor and re-use them in a further sizing operation. A wide variety of methods has been proposed by various workers in this field (Becker and Grunert, 1980). Removal of sizes in aqueous solution is complicated by the high dilution of the retrieved size compared with the concentration required for effective sizing. Various methods have been proposed for overcoming this, using carefully designed removal procedures and/or liquor-

concentration techniques. In all cases, there is an inevitable degradation of the size material by contamination and molecular-weight degradation. Unlimited recycling is thus not possible, though recycling will greatly reduce the amount of load on the environment. A variant of this technique is to use solvent, either solely or in conjunction with water. In the former case special solvent-soluble size molecules must be designed before the process can be used. In both cases recycling is eminently possible, but there is an additional environmental risk introduced: the possibility of organic solvent being discharged into the atmosphere, a risk which must be evaluated and controlled.

An additional drawback of recycling is that it is only really feasible in vertically integrated operations or where special contract agreements are drawn up. It is of little interest to the commission finisher who will have no ready outlet for the recovered size, and little or no control over the type of size used on the goods processed.

(e) Re-use. There are as yet no direct uses of size removal liquors. However, biologically degradable sizes will produce sludge that may then be used as a liquid fertiliser on agricultural land.

(f) Treat. Whilst such materials as PVA are considered to have no ready biodegradability, claims have been made (*Textile World*, 1974) that bacteria capable of degrading PVA can be developed in a purpose-made effluent treatment plant. This is a slow process, but removal rates of over 90% have been claimed. The main disadvantage of this approach is that it is only possible in large vertically integrated plants where sufficient effluent volume of a consistent nature is available, and a large capital plant may be constructed.

Efforts are under way to produce size molecules which are more readily biodegradable than the current materials (Peppmoeller *et al.*, 1992; Habereder *et al.*, 1984). Based on either synthetic molecules, or modified natural products, they are claimed to have similar properties to the traditional materials. They must nevertheless be treated in effluent plant for removal from the environment.

(g) Dispose. Generally, no landfilling of size waste is undertaken, since the liquors are usually too dilute. Where biological treatment is undertaken, sewage sludge is produced, and this may well have to be landfilled if no other disposal route is available.

10.5 Conclusion—the way forward

As the above example of sizing shows, considerable creativity can be liberated when a clearly defined environmental issue is given due prominence in the industry. As the environmental approach of the textile industry becomes more sophisticated, more such issues will emerge. It must be stressed, however, that the best environmental solution may not be achievable without co-operation across the industry. In some cases, environmental problems can be created by a lack of communication between suppliers and customers in the supply chain. It is likely that this will result in the same type of partnership arrangements as are currently being created in the pursuit of total quality management and just-in-time programmes. It is certain that the accelerating pace of internationalisation of both the textile industry and environmentalism will serve as powerful pressures to change the technical face of textile processing in the coming decades.

References

AATCC (1973) *Text. Chem. Col.*, **5**(11) 239.
Allen, R. (1980) *How to Save the World*, IUCN–UNEP–WWF.
Becker, E. and Grunert, H. (1980) *Textile Praxis Internat.*, **35**(5), 593.
Elkington, J. (1990) *Business*, August, 109.
Good Housekeeping (1990) September, 170.
Greenwood, P.F. (1991) *Textiles*, **20**(3), 22.
Gribbin, J. (1988a) *New Scientist*, 5 May, 1.
Gribbin, J. (1988b) *New Scientist*, 22 October, 1.
Habereder, P. and Bayerlein, F. (1984) *Melliand Textilberichte* (Eng. Edn), **13**(12), 749.
Hoenig, W. (1993) *Melliand Textilberichte*, **74**(3), 202.
Kravetz, L. (1983) *Text. Chemist and Colourist*, **15**(4), 57.
Kummer, C. (1991) *Textilveredlung*, **20**(3), 22.
Lovelock, J. (1986a) *A New Look at Life on Earth*, Oxford University Press, Oxford.
Lovelock, J. (1986b) *New Scientist*, 18 December, 25.
Myers, N. (1985) *The Gaia Atlas of Plant Management*, Pan Books, London, pp. 12, 110 and 112.
Pearce, F. (1987) *New Scientist*, 5 November, 1.
Peppmoeller, R., Gossens, B. and Winik, K. (1982) European Patent Application 0511589.
Shaw, T. (1990) *Textile Asia*, **21**(10), 30.
Spangenberger, M. (1993) *Melliand Textilberichte*, **74**(8), 751.
Stevenson, R. (1990) *Chemistry in Britain*, **26**(8), 732.
Stiebert, A. (1975) *Melliand Textilberichte*, **56**, 925.
Textile World (1974) **124**(2), 43.
Van Winkle, T.A., Edelanu, J., Prosser, E.A. and Walker, C.A. (1978) *American Scientist*, **66**, 280.
Watzl, A. (1992) *Melliand Textilberichte*, **73**(7), 561.
Weck (1991) *Textil Praxis Int.*, **42**(2) 144.
Woodward, I. (1989) *New Scientist*, 6 May, 1.

Index